To: Tony Hilvers

Thank you for your support and
encouragement of my work.

Rajan
                            July 1998

QUICK RESPONSE MANUFACTURING

....

A Companywide Approach to Reducing Lead Times

RAJAN SURI

Productivity Press • Portland, Oregon

Additional copies of this book are available from the publisher. Discounts are available for multiple copies through the Sales Department (800-394-6868). Address all other inquiries to:

Productivity Press
P.O. Box 13390
Portland, OR 97213-0390
United States of America
Telephone: 503-235-0600
Telefax: 503-235-0909
E-mail: service@ppress.com

Cover design by Chris Hanis
Text design by Susan Swanson
Art composition by Smith & Fredrick Graphics and
    William H. Brunson, Typography Services
Page composition by William H. Brunson, Typography Services
Printed and bound by Edwards Brothers in the United States of America

Library of Congress Cataloging-in-Publication Data

Suri, Rajan.
    Quick response manufacturing : a companywide approach to reducing
    lead times / Rajan Suri.
        p.   cm.
    Includes bibliographical references and index.
    ISBN 1-56327-201-6 (hardcover)
    1. Production management.   2. Production scheduling.
    3. Manufacturing cells.   4. New products.   I. Title.
    TS155.S89   1998
    658.5—dc21   98-18321
                                                               CIP

To my parents,

Man Mohan and Suniti;

my wife, Lisa;

and my children,

Devika, Francesca, Malini, and Nathaniel

Contents

Publisher's Message

FOR MANUFACTURING FIRMS, competing by *implementing speed* means finding ways to reduce lead times. While there has been much publicity about implementing speed in manufacturing, there remain many misconceptions about how to accomplish it. In fact, many key policies in use at manufacturing companies today are working against lead time reduction—and most managers don't even realize it. This is where Quick Response Manufacturing (QRM) comes to the rescue. QRM is an expansion of the time-based competition (TBC) strategy of reducing lead times. Not only will QRM set your company on the right course of reducing lead times for both existing and new products, it will provide managers with a whole new approach to implementing speed. From QRM's single principle of minimizing lead time emerge the tools, principles, and methodology for changing the whole organizational structure. As author Rajan Suri says, "lead time reduction cannot be done as a tactic. To significantly impact lead times firms must change the traditional ways of operating and redesign organizational structures. . . . This means QRM has to be an organizational strategy led by top management."

From the shop floor to the office to your suppliers, *Quick Response Manufacturing: A Companywide Approach to Reducing Lead Times* reveals the weaknesses of traditional manufacturing methods (or mindsets) and provides the QRM principles that must replace them. QRM is also designed to work in environments where products are highly customized and with highly differing specifications, or where there is a large number of possible products with variable demand for each one. As a result, you'll be jettisoning cost-reduction strategies, reorganizing production, implementing cellular manufacturing, changing the management mind-set, and learning how to use lead time reduction to make all your business decisions. Rajan Suri also provides the tools that adjust your accounting practices, purchasing policies, and performance measurements to support all these changes. And all the while you are learning about QRM, the author will be discussing the manufacturing system dynamics that explain why the QRM approach is so effective. Particularly, you'll deal with the Response Time Spiral—a pernicious stumbling block to reducing lead time that the QRM approach eliminates.

QRM also dramatically alters your capacity and lot sizing policies, along with a new material planning and control approach called *Paired-cell Overlapping Loops of Cards with Authorization* (POLCA). POLCA uses aspects of MRP but does not use JIT/kanban principles. In fact, there is a very different focus in lean manufacturing (or JIT) and QRM, which is discussed throughout the book (in particular, Chapters 8 and 9). Broadly stated, while lean manufacturing focuses on eliminating non-value-added waste to improve quality, reduce cost, and reduce lead time, QRM focuses on reducing lead times to improve quality, reduce cost, and eliminate non-value-added waste. Even though QRM relentlessly pursues lead time reduction, it does not require scrapping all the lean manufacturing programs you've already invested in. Instead, QRM builds upon the ideas of kaizen, TQM, TPM, etc., making it an innovative next step in your company's constant quest for competitive strategies.

To help the reader understand and apply these principles the author provides plenty of tools, examples, and how-to's. There are QRM main lessons and on-going quiz questions that examine and challenge traditional manufacturing beliefs. You'll learn why seven companies are using QRM—John Deere, Beloit Corporation, and Ingersoll, to name a few. There are numerous other case studies, examples, and hypothetical situations that show you why QRM succeeds in implementing speed where other methodologies fail. You'll be given formulas to measure lead times; a new performance measure, called the QRM *number,* that is based on rewarding the reduction of lead times; and two quick fixes to adjust your existing accounting systems to enable your QRM efforts. Finally, you'll be given the tools and principles to change your mind-set to support QRM, and fifteen specific steps to implement a successful QRM program.

With over a decade of helping companies develop and implement QRM, Rajan Suri brings considerable knowledge and expertise to this book. He is professor of Industrial Engineering at the University of Wisconsin-Madison and director of the Manufacturing Systems Engineering Program that also has ties to the School of Business. Rajan Suri also serves as director of the Center for Quick Response Manufacturing, a consortium of over 40 firms created in 1994 to work with the University on understanding and implementing QRM strategies. He has served as editor in chief of the *Journal of Manufacturing Systems*, and is currently associate editor of the *International Journal of Flexible Manufacturing Systems* and area editor of the *Journal of Discrete Event Dynamic Systems*. He has consulted for leading firms such as 3M, Alcoa, Allen Bradley, ABB, AT&T, Ford, Hewlett Packard, McDonnell-Douglas, IBM, and Pratt & Whitney, and has also engaged in consulting in Europe and the Far East.

We are very pleased that Rajan Suri chose Productivity Press to introduce this new comprehensive approach to implementing speed. We believe that Quick Response Manufacturing is a way for companies to break from the traditional mind-set of making decisions based on economies of scale and cost-reduction to one that emphasizes economies of *speed* and *time* reduction—an approach that also improves quality, reduces cost, and eliminates non-value-added waste. We also wish to thank all those who participated in shaping this manuscript and bringing it to bound book: Diane Asay, editor in chief; Gary Peurasaari for development editing; Susan Swanson, for text design and production management; Chris Hanis for cover design; Sheryl Rose for copyediting; Jane Loftus for proofreading; William H. Brunson of Typography Services for page and figure composition; and Lee Smith of Smith & Fredrick Graphics for additional figure composition.

Steven Ott
President and Publisher

Foreword

TREK BICYCLE makes the best bicycles in the world—we just have a very clumsy process of doing it.

After working 13 years for the company in sales, I took over as the president of the company in June 1997. From a sales perspective I always knew that we were behind the curve in manufacturing but I had no idea what to do about it, or where to start. We were suffering from high manufacturing costs, long lead times, and a complete failure to build to schedule. These problems were compounded by the fact that we are in a seasonal business that has annual model changes for 100 percent of our bicycle products. As an organization, we rationalized the majority of our problems on our suppliers.

Not long after I became President, the head of our manufacturing engineering department, Jeff Amundson, suggested that I talk to Rajan Suri, who is the head of the Quick Response Manufacturing Center at the University of Wisconsin. For years, Jeff had been trying to get Trek management interested in Quick Response Manufacturing (QRM). Rajan was kind enough to meet with me and he quickly convinced me of the benefits of QRM. According to Rajan, implementing QRM would lead to shorter lead times, which would lead to better forecasting, lower inventory levels, higher margins, less working capital, more competitive products, and lower overall costs throughout the business. Not only would QRM address the majority of our manufacturing issues, but by reducing lead times it would place us in a position to do a better job with customer service, product development, and back office operations. This sounded good to me. My only question was, What is the next step? Rajan said he wanted to take one section of our business and apply the QRM principles to prove that it works. This would help in convincing the people within the organization that QRM was the way to go.

I listened to this approach, and then drew a triangle on the chalkboard in my office. I told Rajan the triangle represented an iceberg and all he wanted to address was the tip of the iceberg. I told him what we really needed was for him to address the entire iceberg. At the time I was very fearful that our business was under attack and that if we did not get our house in order, we would be in serious trouble. Rajan agreed

to consult with us for a few days a month to help us with QRM. This
meeting marked the beginning of a new era at Trek Bicycle.

Being firmly convinced that the future of our organization rests with
Quick Response Manufacturing, I set an overall corporate goal of reducing our finished goods bicycle lead times from 128 days to 30 days in
one year. The best definition of leadership I have ever heard is that it "is
the ability to make the dream a reality at the grass roots level." So we
took the QRM message to our management group and then to the entire
company. I would start every meeting with the fact that our current
lead times are 128 days and because of it we had high inventory levels,
low fill rates, and low margins. The problem was the long lead times
forced us to guess what our customers needs were going to be 128 days
ahead of time. Try guessing what a bicycle dealer is going to need for
June in February. It's like asking the manager of the Dairy Queen to put
his order in for the July 4th weekend in March.

The message was simple—we needed to reduce our lead times from
128 days to 30 days and by accomplishing this objective we would realize the following four benefits:

1. Higher gross margins
2. Less inventory
3. Higher sales due to higher fill rates
4. Lower manufacturing costs

The last part of my message was that this was the direction the company was going and if you wanted to be a part of it, GREAT, if not you
should work somewhere else. I have given this talk so many times I can
do it in my sleep, and sometimes I do.

As part of the strategy for adopting QRM, I had Rajan put on seminars for our entire management team, along with all managers and
group leaders in manufacturing. I thought we were in for a real battle.
The majority of our manufacturing group had been with the company
for a long time living with long lead times. Certainly QRM would be a
real tough sell. WRONG! In my 13 years at Trek I've never seen a concept so readily accepted in such a short period of time. Its simplicity
and logic makes it an easy sell.

After the training we tasked each of our manufacturing teams with
putting together lead time reduction plans using the QRM methods

developed by Rajan. A funny thing happened—not only did the teams turn in plans that met the corporate lead-time reduction goal I had spelled out, but also the majority of the teams beat the goal by a substantial margin. Better than that, we have developed a QRM culture within our entire organization, and the army is running over the generals with talk and action plans of reducing lead times far below the corporate target.

In the six months since we adopted QRM as a corporate strategy, Trek Bicycle has made significant progress. On the manufacturing side we have reduced our overall lead times by 50 percent, and we will reduce our lead times by another 25 percent by the end of the year. In the office we are reducing the amount of time it takes to develop new products, process credit applications, and to perform other tasks such as the time taken to process orders and credits.

The best part of QRM is it is a simple concept that applies to every single aspect of your business—a concept we are using to take Trek to the next level. I'm sure that after reading Rajan's book, *Quick Response Manufacturing: A Companywide Approach to Reducing Lead Times*, he'll convince you, as he did me, that QRM is the way to go. And if you do adopt QRM I hope you enjoy the same kind of success we have had.

John Burke, President
Trek Bicycle Corp.

Preface

IN 1989 I WAS WORKING ON LEAD TIME REDUCTION projects at several manufacturing companies in Finland. Dr. Suzanne de Treville of the Helsinki
University of Technology, who had invited me to Finland to work on
these projects, showed me an article by George Stalk that had appeared
a few months earlier in the *Harvard Business Review*. This article documented an emerging strategy of competing on speed, called *time-based
competition*, or TBC. The article had a profound effect on me. Although
I had been working in the area of lead time reduction for several years,
the article changed my view of this topic. Suddenly, it put a lot of my
work in perspective. Reducing lead time was no longer a shopfloor
strategy relegated to the manufacturing process experts–no, clearly it
was a broad and powerful corporate strategy, with tremendous benefits
for the whole company. As more articles appeared on the subject of
competing on speed, I read them all eagerly. Converted to this new way
of thinking, I thought that the combination of this new knowledge and
my past experience in lead time reduction would enable me to convince
top managers of the merits of embarking on implementing speed, as
well as assisting them in achieving it. With much enthusiasm, I referred
my industry contacts to these publications, gave them an overview of
the benefits, and waited for them to prove the theories with success in
reducing their lead times.

But soon I realized that things weren't going as well as the theories
predicted. By now, more and more articles–and even whole books–were
being published on TBC, lead time reduction, and competing on speed,
and yet something was missing. Many of the companies I observed were
still having fundamental problems implementing speed. True, some
companies had successes in pilot areas, but by and large, in the rest of
the company it was still business as usual. And other companies weren't
even getting to first base. Why this gap between the widely published
theories and practical implementations?

I thought back over the work that Suzanne de Treville and I had
done in Finland. A key observation we made early in our projects was
that top managers in manufacturing companies were not aware of some
fundamental properties of manufacturing systems. There is the property,
for example, that trying to utilize machines and people to 100 percent

capacity in an effort to maximize asset utilization might actually work against a speed strategy; or the nonlinear relationship between lot sizing policies and product lead times. Although academic research literature has reported many of these properties, they were relatively unknown to managers. In trying to communicate these "manufacturing system dynamics" to managers, we also realized that a short explanation never sufficed. We had to engage in a process of education, explaining the basis for many of the relationships and then working up to their effect on lead time.

Over the next five years, I continued to work with consulting colleagues along with managers from several companies on implementing lead time reduction—not just on the shop floor but in many other areas, including materials procurement and office operations such as order processing and quote generation. In conducting these projects, we encountered many unexpected obstacles. It became clear to me that not only were managers missing the insight on manufacturing dynamics, but more than that, policies in *all* areas of manufacturing companies were working against lead time reduction. These policies were so deep in the marrow of management ideals that I had difficulty convincing senior managers that such was the case. In fact, quite the opposite. As their lead times grew, managers believed they needed to push harder in imposing the very policies that were the cause of long lead times.

At about the same time, along with a few Midwest companies and some of my colleagues at the University of Wisconsin-Madison, we had launched the Center for Quick Response Manufacturing. One of the missions of the center was to help companies implement quick response throughout their operations. Over a few years, we engaged in dozens of lead time reduction projects with companies of all sizes, in diverse industries and in various segments of their organizations. Our experiences—in conducting the analyses for these projects, making recommendations, and trying to convince managers of the merits of implementing them—reinforced my views on the depth and extent of managerial policies that were obstacles to lead time reduction.

As I reflected on all these experiences, a pattern began to emerge. There were a small number of policies in key areas of the company that, again and again, seemed to be at the root of the misconceptions about speed. These same policies then impeded our attempts to implement quick response in the organization. I continued to gather data from managers all over the country on the policies they used in their attempts to implement speed. The data confirmed my hypothesis: They clearly showed that key policies in use at manufacturing companies were working counter to lead time reduction. In 1995 I documented ten "bad" policies in a short article, "Slaying the Beast," in the American Production and Inventory Control Society's publication *The Performance Advantage*. Along with each bad policy I gave a short explanation that provided insight into why the policy worked against lead time reduction, and then proposed the policy that needed to take its place if the company truly wanted to implement Quick Response Manufacturing (QRM). The response to this short and simply worded article was overwhelming. In the three months following publication, hundreds of managers from all over the world contacted me, asking if I had more information on this new QRM approach. Reprint requests for the article soared into the thousands. The most touching response was from Gary Amundson, at that time a senior manager in an equipment manufacturing company:

Dear Rajan,
... It was nearly twenty years ago that I began my career in manufacturing ... I was very excited eight years ago when I was given the opportunity to move into materials management. ... I believed materials management was the hub of the manufacturing wheel and success was totally dependent on us. There were many long and difficult hours with some successes, but the process was by brute force. Meeting after meeting and continual expediting of items that were hot (which seemed to be just about everything) made up the daily routine.

The stress was overwhelming and I was burned out. I was disappointed that I was unable to solve the complex world of materials management. I was contemplating a career change. Late one evening, after another tough day, I found your article, "Slaying the Beast," in a manufacturing magazine. ...

Words cannot express my appreciation, Rajan, but thank you for waking me. I am back on the right track. Your ideas have revitalized me. It has been a real pleasure finding the correct way through Quick Response Manufacturing.

As a result of my short article in *The Performance Advantage*, and subsequent articles that I wrote on the subject, demand for more details arose. Managers wanted to know how they could learn more about the principles of QRM so they could fully understand the basis for the new policies. Since at that time the details of QRM were not documented and really only existed in my own thoughts, I instead offered to spend time with the managers, explaining the principles with case studies and examples. This activity soon developed into a full-scale program of seminars on the principles of QRM. Once again, I was surprised at the impact these seminars had. There were, after all, so many seminars available on new manufacturing strategies. What was special about this approach?

Whit Kellam, human resources manager at Thompson Equipment Co., explained his enthusiasm after attending a seminar in June 1996: "After the seminar, we made more improvements in our entire order cycle time than we had made in the last 50 years. The information enabled us to explain QRM concepts better, receive management and employee buy-in, and thus implement QRM practices faster with better results. The seminar has paid off in huge dividends for us, both at the employee and stockholder level."

Reflecting on these developments, it appeared that while there were many publications on speed, there was still a need for a comprehensive theory on implementing speed in a manufacturing company, covering all areas of the enterprise. In order to provide such a comprehensive and useful theory, I would have to meet several goals. First, such a theory would have to overcome the many misconceptions organizations have about how to implement speed. Then it would need to show how their existing policies were obstacles to lead time reduction. But simply doing away with existing policies would not be helpful. The theory would have to provide new and practical policies to take their place. Even with all this, managers were still missing the detailed how-to's for implementing the new concepts. The books on speed and TBC stated general principles, but were missing stepwise implementation details. So, the comprehensive theory would need to be supplemented with many practical details covering issues of how to start, specific tools, and step-by-step methodologies. And finally, the theory would need to put all these principles together in a consistent framework in terms of management's goals, performance measurement, and organization structure.

Quick Response Manufacturing: A Companywide Approach to Reducing Lead Times attempts to achieve the above goals through a comprehensive approach that I call QRM. Part One of the book sets the stage for the rest of the QRM approach. With so many new manufacturing strategies being introduced these days, it is necessary to differentiate QRM from other programs such as JIT (lean manufacturing) and reengineering. You will see that QRM is particularly well suited to low-volume and even one-of-a-kind manufacturing environments, a direction that many of us feel will be the focus for the company of the future. Chapter 1 explains QRM in relation to other manufacturing strategies, and also documents the degree to which current management beliefs are working against lead time reduction. In addition, the chapter will give you a quick overview of the main QRM principles. In the next two chapters, I create the environment for learning about QRM. Chapter 2 documents the benefits of embarking on a QRM strategy—including many nonobvious benefits—which will reinforce your desire to implement QRM. Chapter 3 provides the background on existing organizational policies and structures. You will be introduced to the first set of "response time spirals," a dysfunctional phenomenon that we will encounter in various forms in later chapters. You will need this background throughout the book as we confront attitudes of "we believe that this traditional policy is vital if our company is to make money."

In Part Two I delineate the first set of QRM principles, beginning with addressing the manufacturing and materials portions of the organization, since these form the heart of the manufacturing enterprise. Chapter 4 sets down the principles for reorganizing the shop floor. An important aspect of these principles is the creation of the *cellular organization.* Much attention has been focused on cellular manufacturing in the last decade; however, I have found that practical methodologies for implementing cells are lacking. Equally lacking are insights on how to deal with many seeming obstacles during implementation, especially for low-volume or custom manufacturing environments. Therefore Chapter 5 provides a step-by-step approach to creating cells and addressing many implementation concerns. Chapter 6 then shows how to engage in creative rethinking to enable cell formation. With numerous practical examples, I show how you can combine engineering and management principles to engage in such rethinking. You will also encounter novel concepts such as time-slicing and time-sliced virtual cells as ways to overcome obstacles to cell formation.

To support the QRM organization on the shop floor, you also need to change your policies for capacity and materials planning, as well as material control. But without some basic understanding of manufacturing system dynamics, you would be hard pressed to justify these changes. Chapter 7 will give you a simple, yet theoretically rigorous understanding of these dynamics. You will understand the need for a whole new systems approach to capacity planning and lot sizing, and you will also be armed with the requisite knowledge to explain these concepts to your colleagues. Chapter 8 will explain how a QRM program affects your material requirement planning (MRP) system. Then Chapter 9 addresses the issue of material control and replenishment. You will see that despite its popularity through the success of JIT methods, a pull (kanban) system is not appropriate in the QRM environment. Instead, in this chapter I propose a new system called *Paired-cell Overlapping Loops of Cards with Authorization* (POLCA), which combines the benefits of both push (MRP) and pull systems, while also overcoming their main drawbacks. The final portion of material management extends outside your company to its suppliers and its customers. Chapter 10 will put in place the QRM principles for the external movement of material. Since organizations have ingrained purchasing and sales policies that have been in place for decades, I use detailed examples to reveal the pitfalls and drawbacks of these "old" policies. Only then do I state the new QRM principles you must use in their place.

Part Three moves into the office operations of your company. My experiences with over a decade of lead time reduction efforts show that manufacturing firms tend to underestimate the opportunity for lead time reduction in the office side of the company, instead pressuring the shop floor for faster and faster response. In Chapter 11 I document this phenomenon and give you insight into its causes. (One exception to this neglect of office operations has been the recent popularity of reengineering; we will see, however, that QRM differs from reengineering in many ways, and goes beyond it in others.) Chapter 11 provides you with the core principles for achieving quick response in the office. Again, a key construct for attaining speed in the office is the use of cells, or what I call the quick response office cell (Q-ROC). In addition to providing the Q-ROC principles, the chapter gives you a step-by-step methodology for implementing a Q-ROC. Similar to my approach for the shop floor, I then give in Chapter 12 the tools to assist you in finding the office opportunities, and discuss implementation concerns and how to overcome them. After this I bring together the ideas in these two chapters with a detailed case study from Ingersoll Cutting Tool Company. And in Chapter 13 I extend the system dynamics principles to the whole organization.

Thus far, I have focused on quick response to demand for existing products. In a QRM company, you can customize these products in terms of options, or they may have custom-engineered features, but they are still inside the envelope of currently manufactured products. On the other hand, studies on competitiveness have shown a high correlation between successful companies and the speed with which they can develop and introduce new products. So in Part Four I describe how QRM principles can be extended to the issue of new product introduction (NPI). Although there is a substantial body of literature on NPI— such as design for manufacturing and assembly (DFMA) techniques and concurrent engineering approaches—we will see that QRM brings its own flavor to the issue of NPI. The insights gained in the preceding chapters show that these existing methods need to be modified to work with QRM. In addition, I give some entirely new principles for NPI, principles that have their roots in our knowledge of QRM.

Finally, for your company to be successful in implementing QRM, it has to tackle the issue of performance measurement. This is tied to the fact that you must change organization structure as well as challenge accounting systems. In fact you will see that organization structure, performance measurement, and cost systems are three facets of the same issue: the core operating philosophy of a company. And this issue is affected most by the mind-set of the company's management. Chapters 15 and 16 in Part Five will address all these aspects in the context of QRM. I will introduce a new performance metric, the QRM number, and argue that it must become the primary metric in a company that is serious about quick response. I will also show that you do not need to replace your entire accounting system. Rather, two simple fixes will take you a long way toward having your accounting system support QRM policies.

The book concludes by providing you with a set of concrete steps for embarking on your QRM program. Whereas the how-to's and step-by-step methodologies of previous chapters focused on projects in specific problem areas, Chapter 17 is a roadmap for top management. Although each company will need to engage in its own learning process to bring QRM ideas into the organization, I can provide some guidelines based on experience with implementing QRM in diverse industries. I have seen that adhering to a few core principles can make the difference between a resounding success and a disappointing failure. The steps laid down in this chapter will give top management clear direction on how to maximize its chance for success in QRM implementation.

From my experiences with over a decade of helping companies implement QRM, and from the feedback I have received from seminars to employees at all levels of organizations, one lesson is repeated again and again: The biggest obstacle to implementing QRM is neither financial resources nor technology, it is mind-set. The mind-set of everyone in the company—from shopfloor workers and supervisors to office workers and top managers—needs to be refashioned for QRM. It is my hope that this book, with its educational approach, statement of precise principles, detailed examples, and industry case studies, will help in creating the mind-set to support QRM in your company. I wish you success as you embark on this exciting QRM journey.

Acknowledgments

A WORK OF THIS MAGNITUDE would never have been possible without the input and support of many, many people over many years. For more than two decades, many individuals have generously shared their knowledge with me and provided me with access to their organizations. In particular, the contributions of a few key people stand out as I reflect on my own journey toward deriving the QRM principles.

The first thanks must go to my early supporters, who—even a decade ago when my ideas were unproven—were willing to risk their reputations and experiment with those ideas. In chronological order, they are: Wali Haider, Fred Choobineh, Kathy Stecke, Dave Callahan, Stan Barwikowski, Mike Mills, Ken Anderson, Gary Kapusta, Chris Kuhner, Rich Harper, Mike O'Loughlin, Sam Bansal, Jeanette Nymon, Susan Finger, Ashoka Mody, Mike Jones, Sunil Varma, Jim Riihl, Jim Schneider, Mike Wayman, Mike Diehl, Bob Dovich, Merle Clewett, and Vickie Vincenz. I am particularly indebted to Suzanne de Treville, who started me on this journey when she introduced me to TBC, and many of the ideas here have their roots in our joint seminars and consulting projects.

I also owe a great deal to my colleagues who worked on consulting assignments with me and whose insights helped me in developing my thoughts about QRM. They are Greg Diehl, Peter Wallace, Mike Tomsicek, Larry Ho, James Cheng, Peter Golden, Christos Cassandras, Donna Rae, Uday Karmarkar, Karla Bourland, Masami Shimizu, Mark Spearman, and Wally Hopp. In addition, colleagues and staff at the Center for Quick Response Manufacturing have contributed to pushing the state of the art in QRM and supporting this work in many ways: Frank Rath, Urban Wemmerlöv, Raj Veeramani, Rajit Gadh, Joan Bailey, Susan Landes, Thomas Dewar, Ying-Tat Leung, Bor-Ruey Fu, Ramki Desiraju, Chandu Rao, Bash'shar El-Jawhari, Shyam Bhaskar, and Ananth Krishnamurthy. Several colleagues at the University of Wisconsin-Madison have been a source of good advice and support over the years: John Bollinger, Greg Moses, Jerry Sanders, Steve Robinson, Dave Gustafson, Arne Thesen, Mike Smith, Larry Casper, Marvin DeVries, and Harry Steudel. My father-in-law, Craig Perkins, provided valuable feedback on the first version of my manuscript. Many organizations have supported the work at the Center for Quick

Response Manufacturing, and they are listed in the Appendix section at the end of the book.

The staff at Productivity Press did a wonderful job of bringing this work to fruition. I would like to thank Diane Asay for her acquisition and support of this project, and Gary Peurasaari for his superb editing and reorganization of my manuscript. I would like to also thank Sheryl Rose, copyeditor, and Susan Swanson, production editor, for their vigilant attention to all the details.

Finally, I would like to thank my wife Lisa, and my children Devika, Francesca, Malini, and Nathaniel, who showed a great deal of understanding in putting up with this writing project for two years, and gave me their support to help in getting it done.

Part One

A New Way of Thinking Stems from One Principle

1

QRM: Not Just Another Buzzword

QUICK RESPONSE MANUFACTURING (QRM) finds its roots in a strategy used by Japanese enterprises in the 1980s. In the late 1980s this strategy was documented by several U.S. authors and became known as *time-based competition* or TBC.[1] The basis of TBC is the use of *speed* to gain competitive advantage: A company that uses a TBC strategy delivers products or services faster than its competitors. Although you can apply a TBC strategy to any business, including banking, insurance, hospitals, and food service, the focus in this book is on its application in a manufacturing firm. I call this specific application of TBC strategy *Quick Response Manufacturing* or *QRM*. By focusing on manufacturing companies, QRM sharpens the principles of TBC as well as adds a number of new dimensions.

You're probably thinking, "QRM: yet another buzzword? We've done JIT and TQM, I've read about Reengineering, and I've been to seminars about Agility. Do I really need to learn about another continuous improvement program?" Or even, "Not another three-letter acronym. In manufacturing, these fads come and go like tides on the ocean." But it is imperative for you—a manager working in *any* area of a manufacturing company—to understand QRM.

The first thing you need to know about QRM is that it is here to stay. The reason is simple: Modern society and technology have produced impatient consumers. They're always looking for newer products, more features, better functionality, and products customized to their needs. This is what they expect. As a result, manufacturers must respond quickly to their dealers or distributors. In turn, manufacturers expect their suppliers (subassembly and component manufacturers) to have quick response. These suppliers, now as customers, expect quick turn-around from *their* suppliers—and so on, all up and down the supply chain. In dealing with hundreds of companies in dozens of industries I hear the same story: *We're under pressure from our customers to cut our lead times.*

Essentially, QRM relentlessly pursues the reduction of lead time in all aspects of your operations. However, to gain more insight, it is useful to address the definition of QRM in two contexts: externally (as perceived by your customers) and internally (in terms of its implica-tions for organizational policies). Externally, QRM means responding to your customers' needs by rapidly designing and manufacturing products customized to those needs. As you read this book you will see that in so doing, QRM goes beyond the established goals and even the capabilities of JIT. Equally important is what QRM means inter-nally to your organization. Whereas JIT (or lean manufacturing) focuses on the relentless pursuit (continuous improvement) of elimi-nating non-value-added waste to improve quality, reduce cost (and reduce lead time), QRM focuses on the relentless pursuit of reducing lead times throughout your operation to improve quality, reduce cost, and eliminate non-value-added waste. However, there is much more to QRM than these short definitions might imply. An analogy serves to drive home the point.

The Toyota Production System (TPS), on which just-in-time (JIT) is based, has as its core principle the elimination of waste throughout the manufacturing system. From this one principle stem the manifold sup-porting structures needed to implement JIT, such as continuous improvement, Total Productive Maintenance (TPM), SMED or quick changeover, zero defects, etc. Similarly, for QRM you will find that from the single principle of minimizing lead time come implications for orga-nizational structure, manufacturing systems, purchasing policies, office operation structures, capacity planning and lot sizing policies, and much more. Remarkably, the policies that QRM recommends end up being, in many cases, quite different from those in place in most manu-facturing organizations today.

What is unique about QRM is that it espouses a *relentless emphasis on lead time reduction* that has a long-term impact on every aspect of your company. Although QRM uses the viewpoint first proposed in TBC philosophy we now can capitalize on a decade of observing manufacturing companies that have applied TBC to go beyond the original TBC strategy. QRM has refined TBC by:

- Focusing only on manufacturing.
- Taking advantage of basic principles of system dynamics to provide insight into how to best reorganize an enterprise to achieve quick response.
- Clarifying the misunderstandings and misconceptions managers have about how to apply time-based strategies.
- Providing specific QRM principles on how to rethink manufacturing process and equipment decisions.
- Developing a whole new material planning and control approach.
- Developing a novel performance measure.
- Understanding what it takes to implement QRM to ensure lasting success.

So you say, "I know the importance of quick response. You don't have to sell me on that. But I've already implemented tons of programs in my firm. We've learned everything there is to know about JIT. We've been through five kaizen workshops. I've personally read books on time-based competition and time-based manufacturing." Be that as it may, let me present you with a simple fact. During 1995–96, I interviewed more than 400 U.S. executives and managers in dozens of industries, and even though all of them were from firms that were trying to cut their lead times, more than *70 percent of the policies in use by these managers and their companies were major obstacles to lead time reduction.* In fact, some of these policies were the very cause of long lead times. Worse yet, in most cases these managers had no understanding or even awareness that these policies were the source of the problem. *If you don't know that you have a problem, you can't begin to fix it.* If more than two-thirds of the policies in use at an average U.S. firm are preventing it from cutting its lead times, what's the chance that your company too suffers from this malady? Yet companies keep trying program after program without lasting success.

HOW QRM DIFFERS FROM OTHER CONTINUOUS IMPROVEMENT AND QUALITY PROGRAMS

QRM should not be viewed as a radically new initiative that requires you to scrap all the other programs you've invested in, like JIT, total quality management (TQM), and kaizen. On the contrary, the QRM program builds upon many of these ideas and should be seen not as a new step but rather as a large step further in your company's constant quest for competitive strategies. I will discuss these tools, methodologies, and techniques in more detail throughout this book, but a brief explanation of the differences between QRM and these other programs is presented here. Table 1-1 provides some additional points.

1. *JIT or flow and theory of constraints.* There are several limitations to JIT. First, most firms that adopt JIT or flow techniques need somewhat repetitive manufacturing and somewhat stable demand. In contrast, companies having a wide variety of products with demand that varies considerably can apply QRM methods—even one-of-a-kind manufacturers. I have assisted in implementing QRM programs at many companies where each order is a custom job. (In such cases, rates and constraints change significantly from day to day, and theory of constraints approaches also are not readily applied.) Second, JIT has become synonymous with the use of a kanban system, which has several drawbacks for QRM. An alternative approach is needed. Third, Western implementations of JIT tend to focus on the factory floor and a company's suppliers. QRM goes beyond the shop floor to examine all company operations, including office operations, up-front efforts such as quoting, and overall company policies.

2. *Business Process Reengineering (BPR).* Reengineering also looks at office operations. But the unifying perspective of using lead time as a yardstick, in both shopfloor and office operations, results in specific principles that are more concrete in their application than the general principles of reengineering.

3. *TQM, kaizen, and TEI (Total Employee Involvement).* The unifying perspective of using time as a yardstick also helps contrast QRM with these programs. Quality and kaizen targets can become arbitrary, and TEI programs can become ineffective if employee involvement is not aimed at the right target. None of these

Table 1-1. QRM and Other Manufacturing Management Approaches

Approach	Comparison with QRM
JIT (Just-in-Time) Flow TOC (Theory of Constraints)	Best applied with stable demand, higher volume products (need to have the line rate or "rhythm" and need to identify the constraint). QRM can be applied to one-of-a-kind custom products too (with rates and constraints that change from day to day). Western implementations of JIT focus on factory floor and suppliers. QRM expands to the whole organization. JIT has become synonymous with kanban. If used with QRM, the kanban method needs significant modification. However, if JIT/Flow has been implemented, it paves the way for QRM.
TQM (Total Quality Management) Kaizen TEI (Total Employee Involvement)	Quality and kaizen targets can become arbitrary, and employees can lose motivation for ever-improving targets. TEI programs can lose momentum or not show results. None of these builds on understanding manufacturing dynamics. Without this, quality improvement or employee involvement produce only limited gains in responsiveness. By understanding the dynamics, QRM motivates specific quality improvement efforts and focuses employee involvement to achieve goals. If in place, TQM, kaizen, and TEI can thus support a QRM program.
BPR (Business Process Reeingineering)	Both QRM and BPR share principles from earlier literature on time-based competition (TBC). However, QRM focuses these principles, and expands on them, for manufacturing firms. Principles of BPR are not clear, nor are there well-stated implementation steps. Many BPR efforts have failed due to lack of understanding of these points. BPR does not use insights about system dynamics. In contrast, QRM has focused principles and clear steps for implementation, and is guided by our understanding of manufacturing dynamics
SCM (Supply Chain Management)	Seeks to coordinate production/inventory throughout the supply chain. The focus is on optimizing *across* facilities, rather than how to improve *within* a facility, like QRM. However, SCM complements the QRM approach to suppliers and customers.
Agile Manufacturing	An evolving concept: Examples of agile behavior have been given, but core principles of how to implement it are still being developed. Agility may take us beyond QRM, but many managers still do not support the core principles of QRM. After a company has mastered QRM it can target agility. By then, the principles may be better understood.

approaches, and also none of the books on TBC and reengineering, contain sufficient explanations of the *dynamics* of product delivery systems. Both office and shopfloor operations are governed by some simple rules. Once understood, these rules are invaluable in redesigning the delivery system for QRM. By capitalizing on this insight, quality, kaizen and TEI programs can derive specific, meaningful goals from QRM principles.

4. *Supply chain management (SCM)*. This is another current hot issue that seeks to coordinate production and inventory throughout a supply chain. The focus in SCM is optimizing decisions *across* facilities, but little is said about how to improve *within* a facility. QRM can complement SCM methods by making each facility more responsive. In addition, you will see that QRM methods can change the fundamental character of the supply chain, rather than simply optimizing the chain under current conditions.

5. *Agile manufacturing*. This is another recently introduced approach that is still evolving. Leading proponents of agility can give examples of agile behavior, but they are still developing the core principles of how to implement it. In contrast, QRM consists of precise, logically derived, and detailed principles, as well as a methodology for implementing it. Besides, the progress an enterprise makes by successfully adopting QRM will provide a good foundation if it eventually wants to pursue the agility concept.

IMPLEMENTING QRM: WHAT MANAGERS BELIEVE

For manufacturing firms, competing with speed means reducing lead times—both the time to bring new products to market and the time to manufacture an existing product from its raw materials. However, the beauty of QRM is that the very act of looking for ways to speed up existing procedures results in manifold benefits. With QRM techniques you can achieve substantial reductions in lead time—more than 75 percent in new product introduction time and 90 percent in time to fill orders for existing products. Furthermore, successful QRM programs result in quality improvement and cost reduction as well. *Implementing QRM simultaneously achieves low cost, high quality, and rapid delivery.*

You would expect that the QRM program would be a priority for any manufacturing firm. But making QRM a priority and making it work are two entirely different matters. Although there has been much publicity about competing on speed, and many books and articles have been written on the subject,[2] our experience with hundreds of managers shows that there are many misconceptions about how to implement QRM, and that these misconceptions prevent successful results.

How can I say this with authority? In the last 15 years as I worked on lead time reduction programs, and encountered difficulties with some of them, I began to find the same issues surfacing again and again. We would be in the throes of an apparently successful project, lead times would be down by 20 percent and we would be gearing up to see 50 percent or better numbers, but then, wham! Our program would come to a grinding halt with improvement stalled at the 20 percent figure. What caused the program to run into so many obstacles? Who was responsible?

The search for answers led me to develop a simple true and false quiz on implementing QRM. I would like you to complete the quiz on the next page after reading the following instructions. For each of the assertions in the quiz, ask yourself: "Do the key managers in my company consider this statement to be true or false?" Let's set some ground rules to make sure you are being completely ruthless in your evaluation. Take the first statement in the quiz as an example:

---

1. Everyone will have to work faster, harder, and longer hours, in order to get jobs done in less time.

❏ True   ❏ False

As you look at this, you surely think, "We all know that to be false. We need to work smarter, not harder." But then ask yourself, "Do we frequently use overtime? Does it take a lot of expediting to get jobs out on time? Do we often work weekends?" If the answer to any of these is yes, then it is clear that key managers in your company believe item number 1 is true! Use this same probing mind-set as you approach each of the remaining items. Now mark your answers in the boxes and evaluate the results.

Quiz on Implementing QRM

For each statement below, ask yourself: Would the key managers in my company consider this statement to be true or false? Mark your responses in the boxes, then compare them with the answers provided.

1. Everyone will have to work faster, harder, and longer hours in order to get jobs done in less time.

    ❏ True   ❏ False

2. To get jobs out fast, we must keep our machines and people busy all the time.

    ❏ True   ❏ False

3. To reduce our lead times, we have to improve our efficiencies.

    ❏ True   ❏ False

4. We must place great importance on "on-time" delivery performance by each of our departments and suppliers.

    ❏ True   ❏ False

5. Installing a material requirements planning (MRP) system will help in reducing lead times.

    ❏ True   ❏ False

6. Since long lead time items need to be ordered in large quantities, we should negotiate quantity discounts with suppliers.

    ❏ True   ❏ False

7. We should encourage customers to buy our products in large quantities by offering price breaks and quantity discounts.

    ❏ True   ❏ False

8. We can implement QRM by forming teams in each department.

    ❏ True   ❏ False

9. The reason for implementing QRM is so that we can charge our customers more for rush jobs.

    ❏ True   ❏ False

10. Implementing QRM will require large investments in technology.

    ❏ True   ❏ False

For successful implementation of QRM it is necessary that your company's key decision-makers understand that *every single one of those assertions is false*! This may be obvious to you in some cases, such as item number 1, where you know you have to find ways to work smarter. But what could be wrong with improving efficiencies? Or, for that matter, with on-time delivery? And what about teams? Aren't they all the rage these days, in everything from shopfloor work to office operations? How could all those assertions possibly be false? In fact, each item for which you answered "true" will, sooner or later, become an obstacle to the success of your QRM program. Not only do some of these beliefs scuttle a QRM program, but worse yet, in some cases the belief in one of these assertions *increases* your lead times. As your lead times get longer, the same senior manager (or managers) who truly believed in one of these ten assertions push even harder on those beliefs, thinking they are not enforcing them enough, eventually resulting in even longer lead times. The process becomes a vicious circle.

How well did your firm score? Give your company a score of 0 for each true and 1 for each false. Count up the number of times you checked the false box, and that is your score. This score is on a scale of 0 to 10, where 0 denotes a company that will have to undergo a gargantuan change to succeed at QRM, and 10 denotes a company that is a "real veteran" of QRM.

In reality, most companies will score somewhere in between. Don't despair if your score is low. I have given this quiz to hundreds of employees around the United States, and the typical score for a North American company is between 3 and 4. Interestingly, this average remains true across industry segments, from equipment manufacturers to parts suppliers, from electronics assembly firms to plastic injection molders, from high-tech semiconductor fabricators to low-tech "metal bashing" firms. The average score also seems to be independent of company size, with firms ranging from fifty employees to several thousand.

PERCEPTIONS OF IMPLEMENTING QRM IN NORTH AMERICAN COMPANIES

Figure 1-1 summarizes the scores of 425 employees at all levels in a number of industries. As you can see, the bulk of the respondents (over

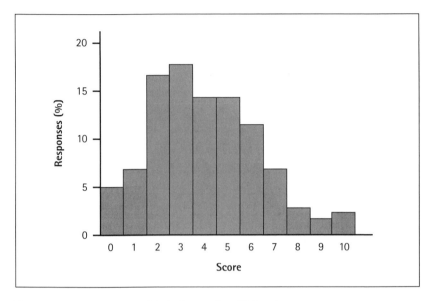

Figure 1-1. Distribution of Quiz Scores for All Employees

60 percent) scored between 2 and 5, with an average score of 3.9 for the whole group. Are there differences in this average across industries? As far as I can tell, no significant differences exist either by industry or by size of company. As an example, Table 1-2 shows scores from U.S. plants of three equipment manufacturers with differing size and ownership characteristics: Still, the average scores for these companies are similar.

Finally, there is the average score for a cross section of second-tier suppliers. Now the responses are from managers and employees of several small companies who make parts for one of the equipment manufacturers. The suppliers represent a number of discrete parts industries.

Table 1-2. Quiz Scores for U.S. Plants of Equipment Manufacturers

Company Description	No. of People Sampled	Average Score
Large American-owned company	28	3.3
Large non-American-owned company	16	3.7
Small American-owned company	53	3.2

Again, there is no significant difference—the average score for a sample of 16 respondents is 3.1.

Perceptions of Senior Management

The previous data involved employees at all levels, without differentiation. Now let us focus on managers. In order to do that, I need to point out an interesting bias effect that I have seen repeatedly in the responses to the quiz. Figure 1-2 summarizes the quiz scores of 67 presidents or CEOs of high-tech electronics companies from the United States, Canada, and Mexico. The average score is 4.8 and the graph is clearly tilted toward the high scores. Figure 1-3 shows the scores for a group of 30 middle managers who work for the same set of companies.

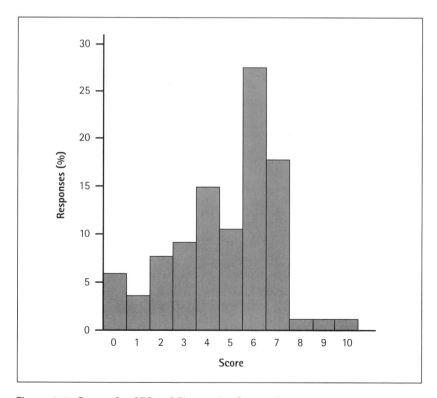

Figure 1-2. Scores for CEOs of Electronics Companies

The contrast between the two graphs is noticeable. Figure 1-3 is clearly tilted toward the low scores, with an average of 3.5.

What these graphs tell us is that these top managers have an overly optimistic view of where their organization is with respect to QRM-related philosophies while the employees feel differently about how their bosses view the same issues. This trend is not confined to electronics companies. When I gave the QRM quiz to 49 presidents and CEOs of manufacturing companies in diverse industries around southern Wisconsin, their average score was 5.0–a significant bias in statistical terms.

This difference in perception is not just between upper management and the rest of the company. Typically, a wide range of perceptions exists among employees in a company. For instance, Figure 1-4 shows

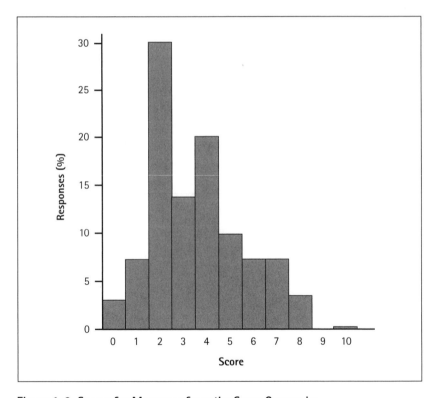

Figure 1-3. Scores for Managers from the Same Companies

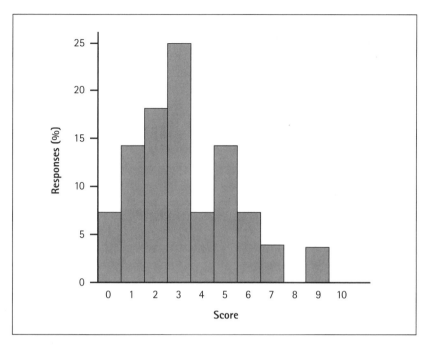

Figure 1-4. Scores for a Sample of Employees at an Equipment Manufacturer

the scores for a cross section of 28 employees from a single company, an equipment manufacturer. The average score is 3.3. Figure 1-5 similarly shows scores from a company that makes packaging. In this case there are 19 respondents with an average score of 3.7. Both these figures tell the same story. Even employees in the same company can have widely different perceptions of management policies. The message is clear. *The view from the bottom up is not the same as the view from the top down.*

Perceptions of Specific Policies

How do individuals feel about specific issues in the quiz? Table 1-3 analyzes the responses to each of the ten questions.[3] To emphasize beliefs that are not consistent with QRM, we focus on the frequency of "wrong" answers, that is the proportion of responses that said "true." It is the perception of the respondents that in some policy areas more than

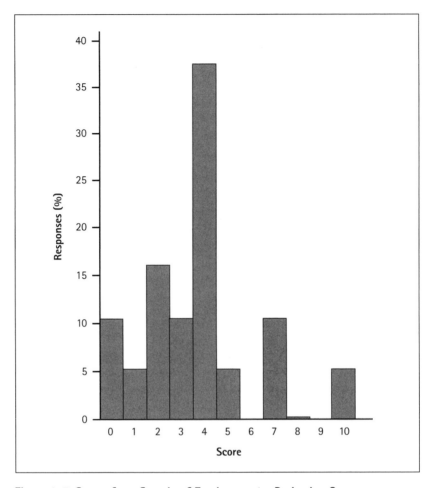

Figure 1-5. Scores for a Sample of Employees at a Packaging Company

90 percent of managers are supporting policies that are inconsistent with implementing QRM.

Implications of the Surveys

The preceding graphs and tables highlighted two issues. One is the difference in perception within an organization; the other is the specific nature of the beliefs. The difference in perception has been demon-

Table 1-3. Analysis of Responses to Individual Quiz Questions

Quiz Item Number and Context	Percentage of "Wrong" Answers[†]
1. Work harder	67
2. Maximum resource utilization	73
3. Improve efficiency	91
4. "On-time" emphasis	96
5. Use of MRP	66
6. Supplier quantity discounts	67
7. Customer quantity discounts	55
8. Teams in each department	68
9. Charge for rush jobs	20
10. Use of technology	30

[†]Percentage of answers that were marked true.

strated in two directions: up and down, and across. For the difference in perception between senior managers and the rest of the organization, you might ask: Which piece of data is more significant, what managers say about their policies or what employees perceive the managers' policies to be? For the proper functioning of an organization, not only should managers be operating on the correct principles, but employees should perceive that these principles are in operation. Thus the responses given by the rest of the organization truly measure the health of an enterprise with respect to QRM principles.

Whether between managers and employees, or between employees in different parts of the organization, you can tackle differences in perception in a relatively short time with better communication of policies by management and a few cross-organizational workshop sessions. After you eliminate the "perception gap" the organization as a whole can work toward instituting its policies. However, the peril is greater where the managers' own beliefs are typically in error, for example, in quiz items 3 and 4. Not only are the wrong principles in operation but the managers may not know that these principles are wrong. This situation can take much longer to rectify. In the following chapters I will provide discussions, arguments, and examples that should clearly reveal the basis of each principle and help convince senior managers of the validity of the "false" answer. More important than the correct response to each quiz item is an in-depth understanding of why it is the correct response.

QRM PRINCIPLES

We live in an impatient society, and that, in fact, is one of the drivers of the need for QRM. Many busy executives want a quick overview before deciding if they want the details, or if, indeed, they think QRM is just another buzzword. This section will give you a quick stroll through the ten QRM principles that counter the ten traditional beliefs presented in the quiz. Each of these beliefs and principles will be discussed throughout the book, providing you a better understanding of QRM, along with key principles that will ensure successful implementation.

Traditional belief number 1: Everyone will have to work faster, harder and longer hours, in order to get jobs done in less time.	*QRM principle number 1:* Find whole new ways of completing a job, with the focus on lead time minimization.

Quick Overview: Our organizations are not designed to manage time. Organizational structures, accounting systems, and reward systems are based on managing scale and cost. A legacy of scale/cost-based management systems, and the greatest enemy of QRM efforts, is the functional organization with specialized departments. Another legacy is the *response time spiral*, an increasing spiral of lead times that results from scale/cost-based management systems. Taking time out of the system requires completely rethinking how you organize production, materials supply, and white-collar work.

Traditional belief number 2: To get jobs out fast we must keep our machines and people busy all the time	*QRM principle number 2:* Plan to operate at 80 percent or even 70 percent capacity on critical resources.

Quick Overview: The first reaction of most managers is, "We can't afford to do that. We will be wasting our resources and our costs will go up!" QRM will eliminate the complex series of dysfunctional interactions—

long lead times, growing queues, jobs spending a lot of time waiting for resources—that result from the present 100 percent utilization policy. QRM will show how idle capacity actually serves as a strategic investment that will pay for itself many times over in increased sales, higher quality, and lower costs.

Traditional belief number 3:	*QRM principle number 3:*
To reduce our lead times, we have to improve our efficiencies.	Measure the reduction of lead times and make this the main performance measure.

Quick Overview: This traditional belief goes hand in hand with the desire for maximizing resource utilization. The problem is not the concept of efficiency, but that most measures of efficiency work counter to lead time reduction. The QRM principle may seem a rather bold step. But as a case study in Chapter 8 will show, in one company, lead times for a line of spare parts dropped from 36 days to 6 days using reduction of lead time as the main performance measure. To accomplish this, though, it is important for everyone in a manufacturing firm, especially senior managers, to understand the dynamics of factory operations. They need to study the interactions between capacity utilization, efficiency measures, and lot sizing policies, and their effects on lead time. You will learn that lot sizes appropriate for QRM bear little relation to the values calculated by the economic order quantity (EOQ) formula, which fails to consider many costs of large lots and ignores the value of responsiveness. Nor can good lot sizes for QRM be predicted by an MRP system, since it assumes fixed queue times regardless of workload.

Traditional belief number 4:	*QRM principle number 4:*
We must place great importance on "on-time" delivery performance by each of our departments and suppliers.	Stick to measuring and rewarding reduction of lead times.

Quick Overview: Almost every book on modern manufacturing discusses on-time delivery and says that it is a cornerstone of JIT. What I have observed, though, is that while on-time performance is desirable as an outcome, emphasizing it as a performance measure is dysfunctional. Human nature being what it is, instead of trying to reduce lead times, internal departments and external suppliers alike tend to pad their quoted lead times so that their on-time deliveries look good. As a result, the Response Time Spiral takes over the organization. With QRM, organizational changes, along with a novel performance measure, promote shorter lead times. These shorter lead times, in turn, kill the response time spiral, and delivery problems disappear—resulting in on-time performance.

Traditional belief number 5: Installing a material requirements planning (MRP) system will help in reducing lead times.	*QRM principle number 5:* Use MRP to plan and coordinate materials. Restructure the manufacturing organization into simpler product-oriented cells. Complement this with a new material control method that combines the best of push and pull strategies.

Quick Overview: MRP systems serve an important function of assisting with materials supply but don't expect them to solve lead time problems because the underlying model in MRP is flawed. In the redesigned organization, MRP is used for a higher level of planning and providing authorization but not for micromanaging work centers. Teams should run their own cells, and they should be provided with simple tools to manage their capacity and continually improve their responsiveness. A novel material control strategy called POLCA combines the best of push and pull methods to limit congestion while at the same time providing a high degree of flexibility, enabling even custom-engineered products to be made.

Traditional belief number 6: Since long lead time items need to be ordered in large quantities, we should negotiate quantity discounts with suppliers.	*QRM principle number 6:* Motivate suppliers to implement QRM, resulting in small lots at lower cost, better quality, and short lead times.

Quick Overview: The more you purchase items in large batches, the longer the suppliers take to make them, motivating you to put in orders for even larger batches. This creates another dysfunctional spiral that is made worse by traditional purchasing policies and incentives.

Traditional belief number 7:	*QRM principle number 7:*
We should encourage customers to buy our products in large quantities by offering price breaks and quantity discounts.	Educate customers about your QRM program, and negotiate a schedule of moving to smaller lot sizes at reasonable prices.

Quick Overview: This is the reverse of number 6. Now you are the supplier. The customer's behavior of ordering larger batches will degrade your delivery performance. With QRM you form strategic partnerships with your customers and show how QRM will allow them to receive smaller batches at lower cost.

Traditional belief number 8:	*QRM principle number 8:*
We can implement QRM by forming teams in each department.	Cut through functional boundaries by forming a quick response office cell, which is a "closed-loop," collocated, multi-functional, cross-trained team responsible for a family of products. Empower it to make necessary decisions.

Quick Overview: Some of the team implementations that follow traditional belief 8 are the result of the quality (TQM) movement. True, a team with all its members in one function department may result in local quality improvements. For the purpose of QRM, however, such a team will do little to cut overall lead time for office operations. Instead, the team for QRM must be the office cell with characteristics specified above. (Cells and other QRM changes are not restricted to the shop floor.) Such quick response office cells are the only way to get significant reduction of lead times for jobs such as estimating and quoting, order processing, and engineering. *Closed-loop* means that all the required steps can be done within the team, which means you will have to cut across functional boundaries and change reporting structures.

Traditional belief number 9:
The reason for implementing QRM is so that we can charge our customers more for rush jobs.

QRM principle number 9:
The reason for embarking on the QRM journey is that it leads to a truly lean and mean company with a more secure future.

Quick Overview: Although customers may pay more for speedy delivery, and this may be a good short-term result of better response, it should not be the main reason for engaging in QRM. Searching for ways of squeezing time out uncovers quality problems and wasted efforts. Fixing these results in higher quality, lower WIP, less waste, lower operating costs, and greater sales. JIT and related methods have put a lot of emphasis on elimination of waste, but those approaches ignore certain types of waste caused by long lead times. With its broader definition of waste, QRM can create an even leaner enterprise that will remain a formidable competitor for years to come.

Traditional belief number 10:
Implementing QRM will require large investments in technology.

QRM principle number 10:
The biggest obstacle to QRM is not technology, but "mindset." Combat this through training. Next, engage in low-cost or no-cost lead time reductions. Leave big-ticket technological solutions for a later stage.

Quick Overview: New technologies, such as rapid prototyping and CAD/CAM, offer great opportunities for time reduction. These are important, but several steps must precede them, like education. In particular, you must realign the mind-set of all employees, from the shop floor to the boardroom, from desk workers to senior managers, to QRM principles. To bring about the mind-set change, organizations will need to thoroughly rethink existing performance measures. Performance measurement is intimately tied in with the cost accounting system, which is an obstacle to implementing an effective QRM program. QRM goes beyond activity-based costing (ABC) to address this issue.

IMPLEMENTING QRM—THE PREREQUISITES FOR SUCCESS

Although the idea of competing on speed has been with us for almost a decade there are many misconceptions about how to implement this strategy. Much has been written about competing on speed, but not enough information is provided on many supporting topics critical to its successful implementation. QRM tackles these topics by laying down the following prerequisites for a successful implementation:

- *There must be a companywide understanding of the basics of QRM: what it means, why it is necessary, how it works.* You need to provide such an understanding to everyone, not just to manufacturing workers and managers. To implement QRM a company needs active involvement from senior executives, staff, and workers in all functional areas, including marketing, sales, accounting and finance, purchasing, materials, design, engineering, and manufacturing.
- *Workers and managers need to understand some basic system dynamics of manufacturing systems.* Specifically, they need to know how capacity planning, resource utilization, and lot sizing policies interact with each other and how they affect lead time. Without this, there will be no buy-in to the key techniques and policies of QRM.
- *The QRM program has to be implemented in both shopfloor and office operations.* Office operations constitute a significant portion of the total lead time for products, yet they are often overlooked as an opportunity for lead time reduction.
- *QRM policies should be incorporated in all areas.* From purchasing to shipping, from equipment purchase to employee hiring, from accounting to performance appraisal you have to rethink how you operate.
- *Both shopfloor and office employees need to thoroughly understand the concept of work cells.* Even though the concept of cellular manufacturing has been with us for two decades, and many companies are implementing cells to reduce lead times, I continue to find lack of organizational will to implement cells, as well as incorrect implementations or outright failures. I attribute all of these problems with implementing cells to a misunderstanding of a few

basic principles. Educating all employees on the principles of work
cells has turned failures around to resounding successes.
- *Obstacles to implementation should be anticipated as much as possible, so everyone is prepared to combat them.* This takes us back to
  the danger of traditional beliefs rearing its ugly head.
- *Even though you should create companywide QRM education and
  awareness, top management should not attempt to reorganize the
  whole company for QRM right away.* Instead, QRM implementation
  should begin by focusing on a market segment where there is an
  opportunity via a quick response strategy, and a small part of the
  company should be reorganized using QRM principles to serve this
  market. In this way, by trying QRM in one or two areas, management can minimize its risk and investment while it proves to itself
  and the rest of the company that this approach really works. After
  absorbing the lessons from this experience, you can reorganize
  additional parts of the company for QRM.
- *Concrete steps for implementing QRM should be identified at the
  start of the initiative.* By building on lessons learned from implementing QRM at dozens of companies, we are able to provide a
  roadmap for successful implementation. It is important for management to review the entire map early, so that they buy in to the
  whole plan.

IMPROVEMENT FOR THE ENTIRE ORGANIZATION COMES FROM A SINGLE THEME

As mentioned at the beginning of this chapter, a notable aspect of the
QRM approach is that all the principles stem from a single concept—
reduce lead times. Many other manufacturing management approaches
appear as a collection of disjointed ideas—a recipe made up of many
different ingredients. Managers and employees have to remember a list
of assertions: the "seven Ks," or the "five Ss," or just a "laundry list" as
some put it. In contrast, the entire set of principles in this book is
derived from one theme, yet these principles are powerful enough to
span the entire organization, from the shop floor to the office, from
order entry to accounting, from purchasing to sales. Such an approach
is more palatable to managers than a disparate collection of ideas,
because it enables them to stick with a consistent message to the organization. Although I will develop details of how the QRM approach

applies to different parts of the organization I will never stray from the path of the single theme of reducing lead times. In this context it is important to remember that QRM espouses not just the reduction of external lead times (as perceived by the customer) but also internal lead times (elapsed times for jobs to flow through all parts of the company).[4]

Having read this far, it should be clear to you that *lead time reduction cannot be done as a tactic*. To significantly affect lead times firms must change their traditional ways of operating and redesign organizational structures. Such changes require total commitment from top management. *This means QRM has to be an organizational strategy led by top management*. I have begun with a 10-point quiz. In the following chapters I continue with this common thread through different sections of the organization, ending with a methodology for implementing QRM.

The next chapter will give examples of unexpected benefits of quick response in production, in new product introduction, and even in office operations such as estimating and quoting. I'll also examine some crucial issues of manufacturing cost and its interpretation. Early understanding of these issues by senior management can make or break a QRM program.

Main QRM Lessons

- Seventy percent of policies in use by managers today work counter to lead time reduction.

- QRM is not JIT, kaizen, flow, TQM, TBC, or reengineering. It can build on these concepts if you are using them, but there are significant differences and new principles. It is essential to move beyond these methods to QRM techniques.

- QRM is not just for manufacturing managers. Top executives, managers in all areas of the company, and employees need to understand QRM and how it is implemented.

- North American companies, regardless of industry and company size, typically score less than 4 out of 10 on a quiz that measures their management's understanding of QRM.

....

2

....

Benefits of QRM

To GAUGE THE FORCES AT WORK in the marketplace today, consider what the CEO of a Massachusetts-based high tech electronics company told me in the spring of 1996:

> We are reevaluating how we can compete. We have tried to compete on price, but there are too many big players in the market. They have both the capital to buy the high-capacity machines and the volume to support those machines and keep costs down. Worse, when there's a downturn in the market, everybody cuts price and we don't have the staying power of the big players. As for quality, that's a given these days. Customers expect it from everyone and it's no longer a differentiator.

What this senior executive is saying, and many companies have experienced, is that competing on price can be a very slippery slope. Further, it's all downhill for everyone! Second, he notes that the quality movement has taken a firm foothold in industry and is no longer a significant competitive edge, particularly since customers have become sophisticated in their expectations with regard to quality. A month later, this CEO told me:

> In the last few weeks we have identified that a key strategy to differentiate us would be speed in delivery. We are going to deliver custom-engineered products—designed, manufactured, and shipped—faster than anyone else in this business.

Another example is provided by the phenomenal growth of Dell Computer Corporation. Started by Michael Dell in 1984 with $1,000 in

savings, Dell grew to sales of $14 million in 1989, and to an amazing
$2 billion in 1993. How did Dell manage this?

> Most people have assumed that Dell sold on price alone. It did in
> the beginning, but no longer.... By the mid-1980s Michael Dell real-
> ized that while selling on price might be a good way of breaking into
> a market, it was no way to build a future. Someone else could cut
> prices a percentage point lower.[1]

In searching for the strategy that would eventually propel his company into the stratosphere, Dell found the Achilles heel of the computer sales process. Retailers thought that customers would pay a
substantial markup in return for being able to go to a store and touch
and feel a machine. But a lot of buyers were old-time users coming
back for their second or third machine, quite knowledgeable about
what they wanted. More critical was Dell's amazing responsiveness
compared to the retailers.

> ... if a customer wanted a customized version, he often had to wait
> for the retailer to send in an order and for the factory to get around
> to filling it and shipping it. With Dell, it was different. The retailers
> were stunned when they saw how quickly the Texas upstart could
> deliver customized products—in substantially less time than it would
> take them to place the order and wait for the manufacturer to ship it.
> ... Dell could respond to the small customer orders—the onesies and
> twosies that we wouldn't attempt to deal with. Suddenly, everyone
> realized that what he had wasn't just a product, it was a process.

These two examples—the CEO of the electronics company and the
Dell strategy—are critical to our study of QRM because they differentiate
QRM from JIT, lean manufacturing, synchronous manufacturing, flow
manufacturing, and other recent manufacturing techniques. Most such
efforts focus on minimizing the inventory in the system with a relatively stable and predictable demand. However, QRM allows you to
focus more and more on individual, customized production, while still
maintaining low inventory and fast response. Also, these other initiatives have typically been confined to the shop floor, with only minimal
changes in the rest of the company. Yes, some areas such as materials
planning were affected, but for the rest of the organization, such as cost
estimating, order processing, or engineering, it was "business as usual."

QRM affects the entire organization. No department can conduct business as usual once a QRM program is fully in place.

BENEFITS OF QUICK RESPONSE IN PRODUCT INTRODUCTION

Clearly if your firm's lead time in bringing new products to market is shorter than that of your competition, you can capture market share while the competition plays catch up with your product (see Figure 2-1). You may even be able to "skim" the market by charging high prices during the period when there is no competing product. All of this can result in excellent profits for your firm.

However, what is less frequently appreciated is a different benefit. With a shorter time to market, your firm and its competition can hit the market at the same time with similar products, yet your firm's product will contain newer technology.

Consider the example in Figure 2-2. There are two firms, one with a four-year time to market, and the other with a two-year time to market. Both firms wish to introduce a new product in 1998, aimed at a similar market niche. However, the second firm is able to start development two years after the first. Most design and manufacturing decisions are cast in concrete in the first 5 percent to 20 percent of the total time to market. As a result, while the first firm locks into 1994 technology, the second firm uses 1996 technology. This includes decisions on materials, controls, manufacturing processes, and manufacturing equipment. As examples of the advantages to the product, the second firm's product designs could incorporate microprocessor and control technologies that had not been developed in 1994, as well as plastics, composites, and

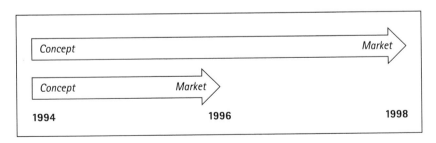

Figure 2-1. Beating the Competition to Market

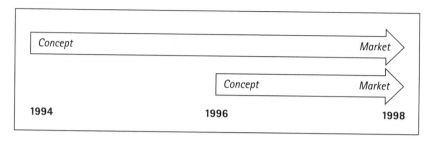

Figure 2-2. Hitting the Market at the Same Time, But with Newer Technology

other new materials that also were not available then. It could plan its production based on new injection molding, laser machining, or other processes that were not well developed in 1994. These and other decisions will result in a product that will be perceived by consumers as being more up-to-date and having superior functionality than the first firm's product. It is also likely to have higher quality and lower cost. All this can result in a very successful product offering.[2]

Another benefit often underestimated in firms that are able to cut their product introduction lead times is that they are able to do this while at the same time using fewer resources for the development and introduction processes. In some cases, firms have cut their lead times in half *and* used half the number of people[3]—a fourfold improvement in productivity. Since most of the resources used in product introduction are not accounted for as direct costs, but rather as overhead, this results in a substantial reduction in a firm's overhead. This is a critical observation as far as the implementation of a QRM program goes, and we'll return to this point later in the chapter.

BENEFITS OF QUICK RESPONSE IN EXISTING PRODUCTION

The advantages of a short lead time to produce and deliver existing products does not refer only to manufacturing—it includes all the steps from receipt of order to delivery to the customer. Typically this may include activities such as design engineering, process planning, production planning and scheduling, procurement, fabrication, assembly, packaging and shipping.

Again, there are some obvious benefits of quick response. Responding fast to your customers clearly promotes customer satisfaction. There is nothing like a happy customer. If you can fulfill your customers' needs every time they contact you, there will be little reason for them to go elsewhere. In fact, your performance reputation may assist in taking orders away from your competitors if they are sluggish in their deliveries.

Moreover, if your customer has an urgent need for your product, you might even be able to charge a price premium and further add to your profits. Consider a company that supplies manufacturing equipment and aftermarket parts. Suppose one of its customers' production machines breaks down, and the customer's entire production is halted as a result. This customer might be happy to pay a few thousand dollars more for the privilege of getting the needed spare part delivered in 48 hours instead of the standard two weeks. Now you are in the enviable position of having a happy customer who is also paying more.

Although these are good reasons for considering quick response or having your company embark on a QRM program, these short-term reasons are not the key reasons for implementing QRM. Less obvious, but more fundamental, is the fact that implementing QRM improves integration of your whole enterprise and searches for ways of squeezing time out of the whole process, thereby uncovering sources of inefficiency, quality problems, and wasted efforts. As these are eliminated, costs decrease and quality improves. However, this explanation is only the tip of the iceberg. As firms like Beloit Corporation and Ingersoll Cutting Tool Company have discovered, the more an organization gets into QRM, the more opportunities it finds. For instance, an initial study at Ingersoll Cutting Tool seemed to indicate that order processing time could be reduced by just 30 percent, but with deeper exploration, it was found that the reduction could be as much as 70 percent, and finally, the team that was charged with implementation devised ways of achieving a 90 percent reduction in lead time.[4] Similarly, the CEO of Northern Telecom stated, "We found that all the things that were vital to our long-term competitiveness had one thing in common: *time.* Everything we wanted to do to improve operations had something to do with squeezing time out of our processes."[5]

The QRM Detective

The *QRM detective* is about the process of seeking out and finding opportunities for improvement. All it requires is asking one question: "Why is this order waiting here?" After this, the QRM detective just follows up with common-sense questions to find the root cause. Let's say our QRM detective is pointing at a batch of twenty castings that are sitting in the middle of the shop floor with no obvious destination machine. He or she would turn to the supervisor of the machining operation and ask:

QRM Detective: "Why is this order waiting here?"

Supervisor: "There was a blowhole in several of the castings."

QRM Detective: "Why not continue with the rest of the castings?"

Supervisor: "The defect wasn't discovered until the third machining operation. We were concerned about putting more time into this job and have it be wasted. We have contacted the supplier—a foundry on the West Coast—and asked for a whole new batch of castings. So production of the whole job has been halted pending the arrival of the new castings."

QRM Detective: "Has this happened before with this foundry?"

Supervisor: "Yes, several times. I've complained to our purchasing department, but this foundry always underbids the others, and purchasing wants to keep giving them a chance."

At this point the message is clear: Your purchasing department needs to make it a priority either to work with this foundry to improve its quality or else find another foundry that has a better track record. In addition, it is time to reevaluate purchasing policies, so that buyers are not driven by cost alone.

Now let's say our QRM detective spies several batches of parts sitting in front of a CNC lathe. Again, he or she starts by asking the machine operator the key question:

QRM Detective: "Why are these orders waiting here?"

Machine Operator: "The machine is down."

QRM Detective: "Is it down often?"

Operator: "Yes, about once a month, for several days at a time."

QRM Detective: "Why does it take so long to fix?"

Operator: "Because a few years ago we reduced all our inventories as part of a JIT program. This included all the spare parts. Often we just need a $20 belt or a hose, but we have to wait for it to be shipped by the manufacturer."

Again the message is clear. The firm needs to reevaluate its policies for storage of critical spare parts, as well as its preventive maintenance policies, and possibly even operator training in case the operator's procedures are contributing to the failures.

But this detective work is not limited to the shop floor. The QRM detective is free to roam anywhere in the company, including office operations. Now we find our QRM detective in the design engineering department, where he or she discovers a large stack of folders on a desk, with partially completed drawings in them.

QRM Detective: "Why are these folders of orders waiting here?"

Engineer: "They are all waiting for information from our inside sales department."

QRM Detective: "Didn't these folders come *from* the inside sales department?"

Engineer: "Yes, they did!"

QRM Detective: "Then why do you still need their input?"

Engineer: "In each case, some details about the customer requirements are missing. I can't complete the design until those details are supplied."

Here too, the message is clear after only three questions. You need to improve this process so that you deliver the right information the first time. Perhaps a set of forms needs to be designed, or the inside sales staff needs to be trained. The problem may even go back to the field sales reps. In any case, the opportunity for improvement has been uncovered.

Everyone in the company should be encouraged to become a QRM detective. By asking these kinds of questions a company will continually be uncovering improvement opportunities. You can then tackle these opportunities using well-established TQM or continuous improvement problem-solving techniques such as pareto charts, fishbone diagrams, and affinity diagrams.[6]

As you tackle these opportunities and improvements are realized, two things happen in the organization. First, quality improves significantly. By quality I don't mean simply the number of parts rejected at a given step, or how close you can observe specified tolerances in a machining process. That is a narrow view of quality. By quality I mean *doing things right the first time*, that is, a definition based on the concept of total quality. Second, operating costs start to go down. To understand this, you need to be aware of a significant statistic. In the average American company, 25 percent of total operating cost is spent on fixing problems that should never have occurred in the first place.[7] Consider the magnitude of this statistic. For a company that has an annual operating budget of $100 million, this means that $25 million is spent each year fixing problems that should not have occurred at all. Suppose the company makes a profit of $10 million a year. If it can reduce the wasted efforts by 50 percent, it will cut $12.5 million out of its annual expenses and more than double its profit.

These wasted efforts take on many forms. Some are obvious, others are more insidious—they've been around for so long you don't even recognize them as non-value-adding; you just think of them as standard operating practice. Let's consider some instances of these wastes and related practices in the specific context of QRM, and perhaps you may recognize a few that are close to home.

WASTE DUE TO LONG LEAD TIMES AND LATE DELIVERIES

A cornerstone of JIT is the elimination of waste. However, in the JIT literature descriptions of the types of waste to be eliminated ignore many of the wasted efforts that are the result of delays in delivery, which is important in understanding QRM. To show the difference between the

JIT and QRM approaches I'll first give some classic and obvious examples of resources devoted to fixing problems that should not have occurred in the first place:

- People whose job it is to do rework.
- Supervisors of the people doing rework.
- Time spent by managers and executives on managing the rework department and its employees.
- Scrap material as a result of poor quality.

You have probably thought of all these situations as wasteful, and they are typically given as examples in the JIT literature too. But now let's look at some less obvious examples of waste:

- Time spent by planners and schedulers to reschedule jobs due to rework that caused delays.
- Time spent by managers to resolve conflicts between competing delayed jobs that need access to common resources.
- Expeditors: people whose entire job consists of keeping orders moving in spite of unexpected delays.
- The cost of overtime pay to expedite delayed jobs, or for herculean efforts to complete jobs in order to prevent them from being shipped late. One Martin Marietta plant found that 90 percent of the overtime it was using could be eliminated when it adopted QRM approaches.[8]
- Time spent by salespeople to placate customers who are upset because of quality or delivery problems. This can be much worse than you think. The vice president of a Midwest company that manufactures sophisticated testing machinery recently gave me an astounding statistic. This company has a sales subsidiary in Asia. The company estimated in 1996 that *a full third* of the resources of this subsidiary were spent dealing with customers who were upset with late deliveries. Why would it take so much? Here is how he explained it to me:

When we first receive notification from the U.S. plant that we are going to be late, we need to take the customers out for lunch or dinner and gently tell them the news. But then we hear from the plant again about a further delay. This time it takes two or three meetings with the customers to calm them down. The worst, however, is when we find out that there is yet another delay. Now I have to dedicate a salesperson full time to this customer just to make sure the entire sale isn't revoked. But it doesn't end there. The next time this customer wants a machine, if they even want to talk to us, it takes our salespeople three to four times the average sales effort to close the sale, because of the customer's bad experience. And this haunts us in the general marketplace too. We find more and more customers hearing about our track record and it is taking us longer and more effort to close sales with other companies as well.

- Jobs that have been institutionalized so that you don't question them anymore, but are essentially non-value-added. Consider a large East Coast factory of an aerospace company that I visited a few years ago. The factory consisted of many departments sprawled over almost a whole square mile. One department in the factory had 50 workers. Of these, 4 workers had the job title "Parts Chaser." I was tempted to ask whether parts in this factory had wings and these poor workers had to run around with butterfly nets attempting to capture them. The truth was that in such a large factory parts did get misplaced. Often one department would insist that it had completed a batch of parts, but the next department would insist equally strongly that it had never received them. The job of the parts chasers in each department was to find where incoming "lost" batches of parts had landed. Note that fully 8 percent of the salaries in this department were going into this particular job of fixing problems that should never have occurred. Similar ratios could be seen for other departments. Yet the management of this company did not see anything wrong. Quite the contrary: Both management and the union had institutionalized this job to the point where parts chaser was a union job classification.

No wonder the typical company spends 25 percent of its budget on wasted efforts and yet doesn't realize it. Although we can find many such examples in diverse industries, a study of contrasts between two industries serves to highlight the magnitude of waste as well as the lack of general awareness of this waste.

A Low Tech Industry with High Awareness, and a
High Tech Industry with Low Awareness

Any mention of the clothing industry immediately conjures up visions of Third World sweatshops and other forms of labor-intensive production. And yet the predominantly low tech apparel industry, as a whole, has spent a lot of effort in quantifying waste due to long lead times.

> Because production must typically be planned so far in advance, production plans in this industry are invariably based on highly speculative forecasts; the time-lags and errors inherent in this process result in ... forced markdowns, stockouts and high inventory levels ... the total cost of inefficiencies throughout the apparel pipeline amounts to approximately 24 percent of net retail apparel sales annually, or over $25 billion."[9]

In contrast, an industry that is known for short product life cycles and is as high tech as can be, the electronics industry, seems unaware of its own waste due to long lead times. This waste occurs in two nonobvious ways. First, some chip fabrication plants can have lead times of nine months or more for a popular chip. These plants often feed contract electronics assembly companies, who make printed circuit boards that go into products made by other divisions of the same companies that own the semiconductor factories. The long lead times for chips create excess costs throughout this supply chain, hurting the whole industry, including the very companies that make the chips!

Second, the long lead times reduce the rate of learning within each semiconductor plant, which slows their improvements in yield. In semiconductor manufacture, the key to profitability is getting a good yield as early as possible in the life cycle of a product. In their attempts to maximize traditional measures of efficiency and utilization of their expensive equipment, semiconductor plants have lost sight of the fact that these very efforts lead to the long lead times.[10] I'll explain this causality later in the book.

In this high tech industry management's awareness of these systemic effects seems to be low. In one meeting with senior executives of a leading semiconductor company, a consultant's suggestion that they consider running their equipment at something less than 100 percent utilization to achieve longer term gains was greeted with laughter and statements like, "We wouldn't be in business long enough to see those gains!"

Although JIT is, among other things, a powerful way to eliminate non-value-added waste it does not focus on these less obvious, time-consuming, and insidious forms of waste, waste that so often becomes standardized and part of the company's culture. To break from this mold you must look at why this form of waste is occurring.

This brings us back to the discussion about the more fundamental reason for embarking on the QRM journey: The QRM mind-set, as embodied in the QRM detective and adopted across the organization, reveals opportunities for improvement. Creating the mind-set of the QRM detective is an important first step for an organization. By beginning here, and working on the opportunities uncovered by your QRM detectives, you get higher quality and less waste of the type mentioned in the examples. In addition, less rework and shorter lead times result in lower work-in-process (WIP). As you will see in later chapters, you will also achieve substantial reductions in inventories of raw materials and finished goods. The payoff of all these factors is a significant reduction in total operating cost.

RESULTS OF QUICK RESPONSE STRATEGIES

Firms that are successful in implementing QRM become formidable competitors in their markets. They are leaner and they deliver high-quality products faster than their competitors. Imagine the following dialog between a field sales representative for a firm that makes custom sheet metal products and a buyer at a customer's company.

Buyer: "Have you had a chance to review those drawings for the electrical cabinets that we need?"

Sales Rep: "Yes, our engineers and I have looked them over. They are within the scope of our capabilities and we can definitely make them for you."

Buyer: "And what about delivery time? We need these rather soon to assemble them into our product. We have just received an order from an important customer who needs a shipment of our products urgently."

Sales Rep: "That should be no problem. I know that you usually need three to four weeks before you can get all the other components for your final assembly. We'll have the cabinets to you in two weeks so that you're not held up at all."

At this point the buyer is amazed. The other cabinetmakers that she has been speaking with have all been quoting 10–14 weeks, some even 16 weeks. She finds two weeks unbelievable and is a bit skeptical. She decides to probe the rep a bit further.

Buyer: "Are you sure you can still meet our expectations of quality if you rush this order through for us?"

Sales Rep: "Absolutely! I can give you several references from recent customers. They have been amazed at how good our quality has been over the last year. And you should know, we won't be cutting any corners to rush your order through. We have consistently been making high-quality cabinets with lead times of two to three weeks for some time now."

Now the buyer feels there must be a catch to this. So she asks, guardedly:

Buyer: "That sounds impressive. But does that mean we have to pay a premium to get our order through so quickly?"

Let us pause a moment to think about the rep's answer. Sometimes customers that have an urgent need are willing to pay a premium to get rapid delivery, such as the company with the broken machine discussed earlier. Should the rep mention a premium? Traditional sales training would say, "Why not? The customer is in trouble and may be willing to pay. Test the water and see how the customer reacts." But the QRM principle is a firm no. You should definitely *not* add any premium for delivering faster than the competition. Let's understand this further.

Successful companies know that the name of the game for persistent success is long-term market share, not short-term profitability. Suppose the rep does quote a price with a premium. His quote comes in at $880 per cabinet for 20 cabinets, or $17,600 total. The buyer already has a quote from another cabinet company, for $750 per cabinet for a total of

$15,000. This company has pledged to expedite the cabinets through their shop in four weeks, but it usually takes them 12–14 weeks and they do not have a good track record of delivery. So the choice facing the buyer is, either get the cabinets on time, or be willing to risk a delay in her own company's production but save $2,600. This is a difficult trade-off to evaluate and at this point it is possible that the buyer will need to think this through in detail and perhaps consult some other managers in her company.

The QRM approach, on the other hand, is to make the buyer's decision a "nobrainer." The rep should quote a price that is consistent with the marketplace, not higher. Now the decision for the buyer is simple. Not only is the price the same as that of other companies, but the quality is superb and the delivery time is what she needs. At this point the buyer need look no further. She does not need to run any complex justification past any managers. The choice is clear.

As a result of such thinking, companies that adopt such QRM strategies, both in their production and in their pricing, find themselves seizing market share at an amazing pace and increasing their profitability. Herein lies another reason for the sales rep not to quote a premium price. If the rep's price is at the same level as other companies in the market, but his company has adopted a QRM program, then his cost of production is steadily decreasing. Thus, even while maintaining the market price, his company is already making more profit relative to the competition.

It is not surprising then, that statistics show companies adopting quick response strategies doing very well relative to their competition. One study showed, for example, that such companies grow at a rate three times the average for their industry. At the same time, their profitability can be twice the industry average.[11]

Another study of 561 plants on three continents showed the impact of lead time reduction on a plant's competitiveness. Three aspects of each plant's performance were studied: its rank in the industry, its rank in the company, and its improvement in productivity. Along with this, a large number of operational characteristics were measured for the plants. Of all the characteristics in these 561 plants, only one showed strong correlation with all three performance aspects: The percentage of lead time reduction by a plant over the past five years had a strong correlation with all three aspects, industry rank, company rank and productivity gains.[12]

My claim that QRM leads to improvements that result in cost reduction is not just based on logical arguments and anecdotal evidence. Researchers have actually been able to quantify the impact of lead time reduction on cost reduction. A study of a number of companies that reduced their lead times found that on average, there was a 2:1 ratio between reductions in lead time and cost. In other words, a 50 percent reduction in lead time resulted, on average, in a 25 percent reduction in overall product cost.[13] Another study of 75 component manufacturers showed that companies implementing quick response strategies had 70 percent fewer "overhead people"—salaried and indirect labor—than companies that operated using traditional scale- and cost-based strategies.[14]

These are just aggregate performance figures over a broad range of companies. Actual performance, if a company targets a market that is ripe for a strategy of quick delivery, can be much higher.

BENEFITS OF QUICK RESPONSE IN SECURING ORDERS

When companies try to speed up the delivery process, an area of operations that is often ignored is the aspect of securing an order in the first place.[15] Looking at the big picture for a moment, you should investigate speeding up the entire cycle from first customer contact to receipt of payment, if you wish to fully adopt a QRM strategy. This cycle can be broken down into six major stages (see Figure 2-3). Five of these stages

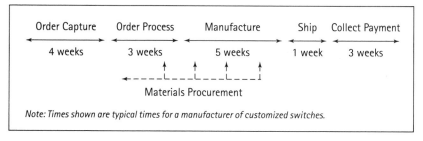

Figure 2-3. Major Stages from Customer Contact to Collection

run sequentially; one, materials procurement, runs in parallel with some of the stages. Although the details of these stages vary for different companies, a typical description of each stage is as follows:

Stage 1: Order Capture: This begins with initial customer contact and ends with receipt of a purchase order or other contract from the customer. It includes one or more cycles of cost estimating and quoting, and other negotiations to close the sale.

Stage 2: Order Process: Here the order is entered into a computer system, engineering tasks are completed if needed, components and materials that need to be procured are identified, manufacturing process routings are developed, the stages of production are scheduled, job tickets are printed, and a "shop packet" is released to the shop floor. (Note that actual release of the order to the shop floor might be contingent on arrival of certain materials.)

Stage 3: Materials Procurement: Although some standard materials might be kept on hand, items that are particular to a custom order usually need to be purchased. These items have been identified in the previous stage; this stage involves procuring them, and typically proceeds in parallel with part of the order processing and manufacturing stages.

Stage 4: Manufacture: This includes fabrication of components, and assembly and test of the finished product.

Stage 5: Ship: Here the order is packed and shipped to its destination.

Stage 6: Collect Payment: In this final stage, accounts receivable obtains payment from the customer.

This entire six-stage cycle should be the target of a QRM strategy, not just the production portion. This is especially significant since we have found that the first two stages often account for 50 percent or more of the total time in this cycle.[16] Let's take a look at the typical activities that occur in the order capture cycle stage:

- Determining the customer's requirements.
- Performing a rough design.
- Estimating the cost of making the product.
- Putting together a formal quote, which includes an appropriate price markup as well as legal terms and conditions.
- Getting the customer to accept the quote and issue a purchase order. This is most important.

You know the obvious benefit of reducing the order capture time, namely, a faster cash flow cycle for the company. There are, however, many less obvious benefits:

- *Reducing order capture time cuts down the organization's overhead.* A lot of organizational resources go into the above activities. Since these activities are performed to obtain jobs, they cannot be charged to a specific job, and are usually lumped into overhead costs. As you will see in Chapter 11, reduction of time in this stage comes from methods that also reduce the amount of effort expended in these activities.
- *Often the order capture and order processing parts of the cycle take extra time and resources because customer requirements have not been properly determined.* The order capture part of the organization is also not exempt from visits from the QRM detective. Consider a scene in the cost estimating department:

QRM Detective: "Why are these folders of quotes waiting here?"

Estimator: "They are all waiting for field sales reps to call me back. In some cases I've been playing phone tag with a rep for over a week!"

QRM Detective: "Why do you need the reps to call you?"

Estimator: "In each case, information that they put on the form is inconsistent. It doesn't make sense with earlier information on the same form. I can't do a cost estimate until the contradiction is straightened out."

As before, the message is clear after only two questions. You need to improve this process so that you deliver the right information the first time. Perhaps the forms need to be improved, or the field sales force needs to be trained. In any case, the QRM detective has uncovered an opportunity for improvement.

- *Sometimes customers can't afford to wait to get a host of competitive bids.* Consider this story. A buyer for a manufacturing firm is looking for an engineered component to go into a production machine that has broken down. She calls around for quotes. Most companies say it will take two to three weeks to look over the drawings and respond with a detailed cost estimate. In contrast, one company, we'll call Custom Metal Products Co. receives a faxed drawing of the component while speaking with the buyer over the phone and is able to give her a quote right away, during the initial phone call. The buyer has dealt with Custom Metal before, has faith in its quality and delivery, and knows from past experience that Custom Metal's prices are reasonable. Her expertise also tells her that this quote "is in the ballpark." Given the urgency of the situation, she decides she can't afford to wait another two to three weeks to get all the quotes in, and she issues a verbal purchase order to Custom Metal during this first phone call. She tells them that they will receive a fax of the purchase order the next day. Custom Metal, having also dealt with this buyer in the past, knows her word is as good as a firm order, and begins processing the order right away. This cycle could well have taken five weeks—say, three weeks to get all the quotes in, a week to review them and negotiate with different companies to try to get a better deal, and a week to issue the purchase order. You have reduced an entire five-week process to about 15 minutes, a *99.9 percent reduction* in the time for this stage. Sounds too good to be true? In Chapter 12 I give an example of a company that did just this. Such a procedure is not created overnight, though. You will learn the techniques for putting in place this type of quoting process.

- *Companies report that when they take a long time to make a quote they often find that they have lost the sale anyway.* As Bill Vogel, president of Vogel Wood Products of Madison, Wisconsin said to me, "If we take longer than a week to make a quote, we might as well not bother, because the job has already been given to someone else."

- *A rapid quoting process creates a favorable impression on the customer.* If, as a company that professes quick response in delivery, you take a long time to turn a quote around, you have left some doubt in the customer's mind about your responsiveness. On the other hand, if the customer routinely deals with companies that take two weeks to respond, and you issue your quote in a day, you have made a lasting impression on the customer about your abilities. The image that you are creating will pave the way for additional business.

BENEFITS OF QRM: FOREWARNED IS FOREARMED

Although there is an impressive potential for QRM strategies, many companies have initiated a QRM program, only to kill it in less than a year. Why? Because of inappropriate and misleading measurements. This is why you need to be clear about what needs to be observed and measured under a QRM program and how it affects the traditional measurement systems. This way you will be prepared to help others understand QRM and keep your company from swerving off the QRM path.

To make this discussion concrete, let's again consider the company that makes sheet metal cabinets and analyze the cost of production. Specifically, I will look at what is known as *cost of goods sold* or *COGS*. This is the company's best estimate of what it took to make a particular product, including accounting for all the nonproduction activities, such as order entry and engineering. For the purpose of this example, I will also include sales and general administrative costs in the *COGS* value. Put this way, if the COGS is accurate, and the company sells all of its products at exactly the COGS value, then it will exactly break even at the end of the year. Whenever the company sells a product at a price exceeding the COGS, and assuming again that the COGS is accurate, then the company is making a profit. Likewise, if it sells below the COGS, it is suffering a loss. So it is important for managers to know the COGS, and companies typically have elaborate accounting systems to track this value for each product.

In the cabinet company, let's take a line of cabinets whose COGS is typically around $1,000 (see Figure 2-4). For this line the cost of materials is $500, or 50 percent of the COGS. (This is fairly typical of manufacturing firms. The materials component of the COGS can vary between 30 percent and 75 percent, but it is close to 50 percent for a large proportion of firms.) Next we find that the cost of direct labor (i.e., shopfloor labor that works directly on manufacturing operations for this

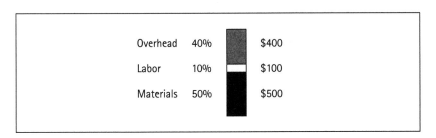

Overhead	40%	$400
Labor	10%	$100
Materials	50%	$500

Figure 2-4. Breakdown of Cost of Goods Sold (COGS) for a $1,000 Cabinet

line) is $100 or 10 percent of the COGS. (Again, this is typical. In U.S. industry, direct labor usually ranges from 7 to 20 percent of the COGS.) That leaves $400 in cost. Where does that come from? This indeed is the cost of the rest of the organization being "allocated" to this line of cabinets by the accounting system. It is usually called *overhead*.

Another name for this $400 component of the COGS is *burden* because it is allocated by "burdening" the direct labor cost with an additional amount. In the above example, if historically the company has found that for each $1 in direct labor, it spends $4 in other organizational costs (not including materials whose costs are accounted for), then it will establish a 400 percent burden on direct labor. The COGS is then derived by the formula:

$$\text{(Material cost)} + \text{(Labor cost)} + (4 \times \text{Labor cost})$$

So, in the above example, the $400 was derived not by tracking the other costs exactly, but by allocating them by multiplying $100 by 4.

Now, it is not unusual that when you implement a QRM program for a line of products on the shop floor, the direct labor content for the products actually increases by 10–30 percent (see Figure 2-5). This may sound surprising, given what I just said earlier in this chapter—I'll explain this discrepancy in a moment—but let's see what might happen in a company that was not prepared for this occurrence.

In our company, suppose the direct labor cost increases by 25 percent to $125. Since the materials cost and burden rate in the accounting system have not changed, the new COGS will be calculated according to the formula above as:

$$\text{COGS} = \$500 + \$125 + (4 \times \$125) = \$1,125$$

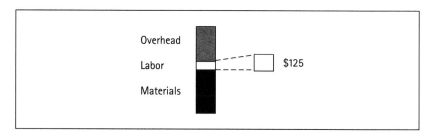

Figure 2–5. Impact on COGS After Three Months of a QRM Program

or an increase of $125. This will soon raise flags in various managers' offices, ranging from the controller to the VP of finance, to the VP of manufacturing, and even the president or CEO. "What's going on?" they will ask. "The cabinets that were made for $1,000 are now costing us $1,125—ever since we started that newfangled QRM program! We've made 200 of those cabinets in the last few months. That means we've lost around $25,000 since we initiated this program." In the vernacular of accounting, managers will begin to see "variances" all over the place, and the VP of manufacturing might even be given a warning by the president, to whom it seems the corporation is hemorrhaging badly. "Fix this problem in the next two months or you'll be out of a job," the VP is warned. What choice does the VP have? In the absence of detailed understanding of QRM by senior management, the VP can only revert back to the old way of doing the job, and so the QRM program is scuttled.

I have seen this happen at several companies. So what is going on? Is the COGS really going up? It shouldn't be, if you believe all the arguments just made about the benefits of QRM. This is what is happening to the three major cost components in the case of the cabinet line, and why.

1. *Direct labor.* There are three main reasons this goes up. First, when you reorganize production for QRM, many of the tasks normally allotted to indirect labor are now done by direct labor. Examples are material handling, quality checks, supervision, and even ordering of components. Second, because of the need for additional skill sets in the direct labor pool, wages will need to be increased. Third, a key strategy for QRM is to maintain a level of idle capacity in the resources. So, instead of planning for utilizing labor at 100 percent of available capacity, one might aim for an average of 80 percent. This means the cost of the remaining 20 percent of labor time is being spread over this line of cabinets.

2. *Overhead.* Precisely because you have given some of these indirect tasks to the shopfloor labor, however, you can expect overhead resources used by this line of cabinets to decrease. But the effect is greater than that. You eliminate or minimize many additional activities such as expediting, managers' time for problem solving and rescheduling jobs, and other examples of non-value-added activities given earlier. Also, decrease in inventory of raw material,

work-in-process, and finished goods means a decline in carrying costs. It would be conservative to say that the overhead costs incurred by this line of cabinets would decrease from $400 per cabinet to $350.

3. *Materials.* The story does not stop with the above two cost components. One might think, since there is no change in design or manufacturing processes, the materials cost will not change. However, with quality improvements, less rework and scrap, and other initiatives taken on by shopfloor workers under the QRM organization, one finds a 5–15 percent savings in total material cost. So in our example, using an average figure of 10 perent, the materials cost would be reduced to $450.

The result of these three effects is a total cost of:

$450 (Materials) + $125 (Labor) + $350 (Overhead) = $925 (Total)

or a savings of $75 per cabinet, not an increase of $125 per cabinet as reported by the accounting system.

The problem, clearly, is that the accounting system uses historical data for both materials and overhead—data that are not keeping pace with the changes created by the QRM methods of work. You usually need a year or so of experience to document some of the new data, and even then the problem is only partially fixed because some of the reductions in overhead cannot be easily traced to our particular line of cabinets (products), and are spread over all the products instead.

Admittedly, this is a simplified example. Most accounting systems are more sophisticated than this and break the COGS down in more detail. However, this problem still occurs. In fact, accountants have documented this phenomenon themselves in the specific context of lead time reduction.[17] It should also be emphasized that a newer approach to cost allocation, called activity-based costing or ABC, will not necessarily fix this problem either. It might make the effect a bit less extreme, but the effect will occur nevertheless.

One Solution—Use a Strategic Overhead Allocation Method

So what *is* the solution? It's one thing for management to give a green light to a QRM program and quite another navigating through the difficult implementation process. The solution to understanding the QRM

program involves a combination of factors: an educated, even enlightened, management that is aware of the ramifications of QRM; participation and buy-in from the accounting and finance department *before* the
program is initiated; advance notice to the board or owners of the company about this effect, and patience (or courage!) from this same body
of people until enough time has passed for the company to get a better
handle on true costs and benefits resulting from the QRM changes.

There is one additional bold approach to this COGS problem, which
you can use in conjunction with the above. I call this a *strategic overhead allocation method*. The idea is for management to estimate the
impact of the program on the materials and overhead costs for the next
year based on analysis for a line of products conducted prior to the
initiation of the QRM. The company then uses this estimate to strategically reset the costs allocated to this line of products. Now management will no longer feel the pressure from unexpected "variances" for
this product line.

Let's see how this could work for our metal cabinet example.
Management knows that the current overhead cost in a $1,000 cabinet
is $400. It reasons that this will certainly not increase, and probably
will decrease. So it decides on a conservative figure of $350 per cabinet
for the overhead after the QRM organization is in place for the cabinets.
However, this figure is now set as a fixed overhead allocation, not as a
burden that is a multiplier based on labor content. If cabinets of widely
differing complexity are being made, then management could provide
a table of new fixed overhead numbers based on cabinet style.
Similarly, management decides on a conservative 3 percent reduction in
materials cost. Now the traditional material costs can be accumu-
lated for each cabinet, but the aggregate is reduced by 3 percent before
reporting the total.

Of course this will require other adjustments in the costing system. If
overhead allocated to this line of products decreases, someone else has
to pick up the cost. This is not entirely bad, since it will put pressure on
the rest of the organization to find similar improvements. Also, if material costs are being reduced by 3 percent, where is this cost being balanced out in the system? One way to tackle either or both of these
issues is to create a *strategic pool* to collect such variances. If the estimated QRM benefits are achieved or exceeded, then at the end of the

year the improvements will be such that actual costs will have gone down and improvements will "cancel out" or even exceed the amount in this strategic pool. The burden however, has been shifted—in a manner of speaking—from labor to top management. Because now, as the year goes on and before concrete results are in, management will see this strategic pool growing, accumulating dollars that have not been accounted for, and it will take some courage to stay the course. So it is clear that you need to do the strategic allocation in concert with and not instead of the other management approaches we listed.

The good news is that there is light at the end of the tunnel. As I will show, companies experience 50–200 percent increases in productivity and 30–50 percent reduction in floor space and other overhead costs. Better still is that in the long term, as a result of many incremental productivity improvements, the direct labor content often goes down to below the original level. In our last cabinet example, where we started at $100 for direct labor per cabinet and went first to $125, you might find after two years that you have come below $90. And all this while the shop floor has accepted more responsibility.

To end with an inspirational example we look at the Beloit Corporation in Beloit, Wisconsin, a manufacturer of paper-making machinery. Beloit used QRM principles to reorganize how they made formed metal tubes that are needed as components for their machines. Where it used to take them 12 departments and 3 interplant moves to make these tubes with a lead time of four weeks or more, today two workers make the tubes in one small area with a lead time of under three days. Better still, after two years of experience with the new approach, they have doubled their production. This is a 100 percent increase in productivity combined with an 80 percent reduction in lead time. At the same time a great deal of material handling and transportation has been eliminated, and rework and quality problems have dropped from frequent to almost nonexistent.[18] If Beloit Corporation had used a strategic pool, I'm sure at the end of the year management would also have been delighted with the final values in this pool.

Main QRM Lessons

- Understanding the nonobvious benefits of quick response; specifically, many types of waste are eliminated, resulting in higher quality and lower cost.

- Identifying the manifold forms of waste due to long lead times and late deliveries. Companies typically ignore or underestimate the magnitude of this waste.

- Understanding the nonobvious benefits of quick response in securing orders, another oft-neglected area of operations.

- Discovering why many QRM programs are "killed" within a few months of inception, and how to avert infant mortality for your QRM program.

3

The Response Time Spiral—
Legacy of the Scale and
Cost Management Strategies

BEFORE IDENTIFYING THE KEY TECHNIQUES that will enable you to implement QRM, you must first understand how manufacturing strategy evolved over the last few decades and why companies are organized the way they are today. This short excursion will provide a perspective on existing manufacturing and management systems that you need to know when you redesign your company for QRM. It will also make you rethink most principles that you assume are good for business success and provide invaluable information as you tackle organizational resistance or remarks such as, "We've done it this way for 20 years and it has served us well. Why do we need to change this procedure?" As you will see in this chapter, the very strategies that made our companies successful in the twentieth century are the ones that may hold us back in the twenty-first century.

The detailed examples in this chapter will help you arm yourself for the discussions with your colleagues who may be skeptical of the need for change. To get things rolling let's look at one of these twentieth-century truths. Consider the first item on the quiz in Chapter 1:

1. Everyone will have to work faster, harder, and longer hours, in order to get jobs done in less time.

In an attempt to get products out faster many organizations continue, even today, to try to reduce the standard times for all their operations. This is the *worst* way to implement speed. Another problem is that shopfloor labor and clerical workers who hear management speak about implementing "speed" envision the same old traditional speed-up approaches using "efficiency" studies and stopwatches. But the QRM approach is quite different.

....
QRM Approach: Find whole new ways of completing a job, with the primary focus on lead time minimization.
....

Why must this be the approach? First, let's agree that the traditional stopwatch approach is not going to work. If you attempt to speed delivery using such a method you will find there is a limit to what a human or machine can do. The best you could expect by pushing people to speed up their work or by optimizing machining operations would be a 15–20 percent improvement over current speeds. On the other hand, QRM aims for 50–95 percent improvement.

Thus, we have to find a "whole new way." The key is that the new way must be designed to minimize lead time. This invariably requires major organizational restructuring. Why? Because our organizations were designed for different strategies, namely, to utilize scale economies and minimize cost. The CEO of Northern Telecom says of their journey toward QRM,[1] "The deeper we probed, the more opportunities we saw. Ultimately we didn't just change our existing processes. We looked at them in totally new ways and redesigned our entire organization." To fully comprehend the need for restructuring, you must begin by looking at the evolution of postwar U.S. manufacturing strategy.[2]

THE ERAS OF SCALE AND COST STRATEGIES

In the 1960s the world was in a high-growth phase because of the ongoing recovery from World War II. At the same time, industrial countries like France, Germany, Italy, Japan, and the United Kingdom were still rebuilding their industrial capacity which had been devastated during the war. They were using what little capacity they had left to rebuild and resupply their own countries. As a result, the United States was the only industrially advanced nation that could supply global markets. In addition, now that the Depression and war were over, consumer demand was exploding. Whether it was capital goods such as machine tools and building supplies or consumer goods such as toasters and automobiles, the world came to U.S. manufacturers to satisfy its needs.

In this golden era, the opportunities for manufacturing firms were seemingly limitless—the more a firm produced, the more it could sell. U.S. firms dominated world markets, and the name of the game in manufacturing was scale (see Figure 3-1). Put in managerial terms this meant, "How big can you build a factory and still manage it?" Many of the giant U.S. corporations such as Ford, General Motors, General Electric, and IBM grew and flourished during this era.

As European and Japanese firms entered world markets in the 1970s, their economies were still less developed than the U.S. economy, so they were able to produce goods more cheaply, albeit not necessarily better. Hence low-cost (and often, not very good quality) products crept into the lower end of world markets. In order to stave away this erosion of their market share, U.S. firms focused their competitive strategy on cost/price issues. Firms instituted companywide cost-cutting programs. Department managers were given cost-reduction targets. Controlling and reducing cost became the order of the day for everyone.

These first two steps in postwar competitive strategy, scale-based and cost-based thinking, laid the foundations for management systems that are firmly entrenched today. Organizational structures, manufacturing layouts, manufacturing systems, management methods, reporting systems, performance measures, and reward systems, are all based on managing scale and managing cost. The implications of this will become clearer as I discuss specific examples later.

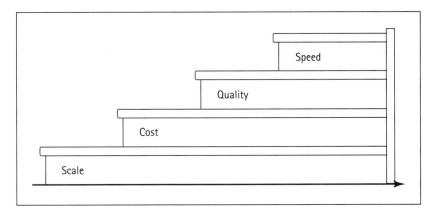

Figure 3-1. Evolution of the Postwar Manufacturing Strategy

Even though the above sequence of developments took place at U.S. firms, the management structures that resulted were not confined to U.S. organizations. After the war the United States discovered it could export not just products but management principles as well. This was no new trend; consider accounting practices as an example. In the early 1900s companies throughout the Western world were still using accounting practices that had changed little from those developed by Venetian merchants in the 1500s. Then in the mid-1920s, General Motors, under Alfred P. Sloan, perfected the system of accounting for multidivisional firms, along with the principles of "management by the numbers," and spread these principles across the world. In a similar way, in the 1950s and 1960s the United States found great demand for American-style business schools and MBA programs in Europe and Asia. These business schools provided a fertile ground for United States management methods of managing large-scale operations by the numbers, of utilizing scale economies, and of controlling costs to be sowed and to blossom in the management of companies all over the world.[3]

Two Major Paradigm Shifts

The 1980s brought the first postwar paradigm shift in manufacturing. Primarily because of the results realized by Japanese firms, the manufacturing world learned that there does not have to be a trade-off between cost and quality: If one focused on improving quality, then cost competitiveness would follow. This realization did not come without significant pain. Many U.S. firms lost substantial market share and closed down several plants before their managements fully understood the importance of this focus on quality.

More recently, another paradigm shift has occurred, leading to a focus on speed. To take the quality paradigm a step further, a company competing on speed usually sees improvements in both quality and cost—in addition to shorter lead times. This may seem too good to be true but as explained in Chapter 2, with the help of the QRM detective, you can uncover the sources of inefficiency, quality problems, and wasted efforts by constantly searching for ways to squeeze out time. But before I discuss this new paradigm you must complete our perspective on today's manufacturing systems by taking a close look at one final and important concept.

THE RESPONSE TIME SPIRAL FOR THREE DIFFERENT MANUFACTURING ENVIRONMENTS

A precursor to successful implementation of QRM is for management to fully understand the implications of the *Response Time Spiral*.[4] The Response Time Spiral is a legacy of the scale and cost management strategies. I'll begin by discussing how this spiral affects companies that only make to specific customer order, then I'll consider companies that operate in two other typical environments that others represent, in a sense, the two ends of the spectrum of manufacturing companies. At one end is a make-to-stock environment, where the products are standard and are made to stock based on sales forecasts; at the other end is an engineer-to-order environment, where each order is completely custom designed for individual customers and then manufactured. The make-to-order environment that I will be discussing lies between these two extremes.

In the make-to-order environment jobs are produced against specific customer orders. There is no significant engineering involved; however, there is no making ahead to forecast either. Perhaps there are so many different variations or options in the product that it doesn't make sense to attempt to make ahead. Or perhaps there is some amount of customization involved, but it is so minor that it can be done without significant design engineering effort. The customization can be seen more as a routine step in creating the product routings and completing the manufacturing engineering, so the company does not consider itself as engineer-to-order. In our example I will review the operations of a make-to-order company plagued with long lead times and late deliveries. Even if your company operates in one of the other environments, it is useful to review the following description because I will build on it when I discuss the other environments.

The Response Time Spiral for a Make-to-Order Company

Modern manufacturing companies are complex organizations requiring long lead times to resolve conflicts between activities needing the same resources. For example, if a customer order for 1,000 parts needs to be milled before it is sent for painting, you must be sure the milling department has sufficient capacity to perform the milling for the 1,000 pieces. This is done by specifying a lead time that allows for the order to "get into the queue" in the milling department. When its turn arrives, you need time for a machine to be set up, and finally, time for all the pieces to be milled. Typically, a given order goes through a number of departments, each with its own need for sufficient lead time, with a resulting long total lead time for the order.

These long lead times require planning ahead. For example, you usually need to plan the personnel, equipment, and materials purchases several months in advance of having actual orders in hand. In addition, as orders move through departments, each department needs some visibility into the future. A shop supervisor, for instance, would be highly frustrated if she gave permission for half her work force to go on vacation in a particular week, only to find in midweek that a major order had arrived from an upstream department and was needed by the downstream department at the end of the week. Similarly, it is necessary to have materials on hand to service customer orders. Finally, if sales are expected to increase, you need to purchase equipment well ahead of time, since most capital equipment has lead times of many months.

However, as lead times lengthen, the accuracy of such plans necessarily declines. To be convinced of this you need only look at a final assembly schedule published today specifying which products will be assembled twelve weeks from now and in what quantities. Then ask your assembly supervisor if he or she is willing to bet a month's paycheck that when they get to that week, they will be assembling exactly these products in these quantities. I'm sure the supervisor, based on years of experience at your company, would not even wager $1 on this. In fact they would probably be happy to wager a month's paycheck that the schedule will be *different* from the one you are looking at.

What are the sources of these changes? One source, of course, is the customers themselves. But even assuming customers "behave," a machine might fail in an upstream department, so the downstream department decides to work on another job that doesn't need the upstream input. When the upstream job finally arrives, it is assembled in a different week. Or material delivery may be delayed for a job, again requiring departments to work on something else. Another cause could be rework. A job assembled last week failed the final test and is back in final assembly after one of its components has been fixed. It was in the original schedule for last week, not for this week, but you have to work on it now, and this then has a ripple effect on the schedule for other jobs.

Because of numerous such problems, inventories also increase. There are two reasons for this. First, there is increasing need for safety stock at all levels: raw material stocks, component and subassembly stocks, and sometimes even finished goods stocks, even though the company professes to be only make-to-order. (For instance, a final assembly supervisor may keep extra subassemblies on hand for a commonly requested customer option.) Second, as a result of planning changes, some inventory is not used as expected. As an example, subassemblies may be completed for an order whose final assembly date is then moved back four weeks due to delay in delivery of another component needed for the final assembly. As a result, the company starts accumulating large amounts of inventory. Now comes the kicker. Orders are received from important customers that management wishes to satisfy, and these orders are for a shorter than standard lead time. In a typical scenario, the VP-sales calls the VP-manufacturing:

> *VP-Sales*: "I know we normally quote eight weeks, but we have to get the order for Universal Motors done in four weeks. The last two orders for them were quoted at eight weeks and were still shipped three weeks late, so Universal Motors is already quite upset with us. They've told me this is a critical order for them, and if we can't meet the four weeks' delivery we'll lose the whole account!"

The VP-manufacturing has little choice but to rush out to the shop floor. He lines up all the required department supervisors, who speak with their work force, and together they perform a heroic feat, working nights and getting the order out in just three weeks. But what a trail of devastation is left behind. The other jobs that were in process in all the departments had to be set aside. Setups that had taken hours to fine tune had to be torn down. The "rush job" got out with impressive speed, but the other jobs have been further delayed.

"Rush jobs," "hot jobs," "expedited orders"—these are all different names for the same phenomenon: unscheduled jobs being put into the factory at shorter than normal lead times, thus needing to be expedited.

These jobs crowd out scheduled jobs, which experience many delays, leading to even longer delivery times than originally planned. After some time, the sales organization finds itself burdened with complaints of late deliveries. Perhaps because of its reputation for being late, the company is even losing significant numbers of customers to the competition. Eventually this becomes enough of a crisis that a high-level management meeting is called.

A typical scenario might look like this. The situation is so desperate that the president himself decides to chair the meeting. As many of us working in manufacturing know, the first department to get the blame is usually manufacturing.

Scene: *A corporate conference room. Sitting around the conference table are the VPs of all the major departments, including marketing, sales, engineering, manufacturing, materials, and finance. At the head of the table is the company president. It is clear from the faces of the people around the table that the mood in the room is grim.*

President: "We have reached a new low in our delivery performance. My colleagues in sales tell me that if we don't get our act together very soon, we'll soon have no customers." [Turning to look at the VP-manufacturing] "I see from last year's reports that manufacturing's on-time delivery performance has been dismal. We are promising our customers 8-week delivery, but we usually take 12 to 14 weeks and sometimes longer. Why can't you get your department under control and run it more efficiently?"

The VP-manufacturing, however, has come prepared. He didn't get this high in the organization without having to fight this battle before. He has a number of reports and charts with him, and he begins by displaying one of them on an overhead projector (see Figure 3-2).

VP-Manufacturing: "I don't think it's fair for you to make that statement without seeing more details on how our department is actually performing. Here, in the first chart, is the utilization of all our critical work centers. As you can see, they are all being used at close to available capacity. In fact, our uptime and machine utilizations are as good as any company in town. And in a few cases, where you see

utilizations over 100 percent, these are where my work force is willing to come in on weekends when we have a crisis. In fact, I personally sacrificed a Memorial Day weekend and a Labor Day weekend
this year because I didn't think it fair to ask workers to come in if I
wasn't willing to come in myself."

As this information begins to sink in, the VP continues with another
chart (see Figure 3-3):

VP-Manufacturing: "Now let's look at how my workers are doing.
This chart shows our labor efficiencies for 1996. See the dotted line?
That's the standard for our industry. We used to be there too, a few
years ago, before I got this job. But you can see that since I became
VP, I have improved our efficiencies to where we are beating the
industry standard by almost 10 percent. It would be hard to imagine
doing much better than that."

The VP-manufacturing continues in this vein with a few more charts
and tables. As all these numbers and statistics find their mark, the pres-

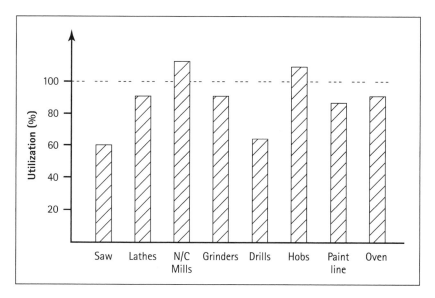

Figure 3–2. Machine Utilizations for 1996

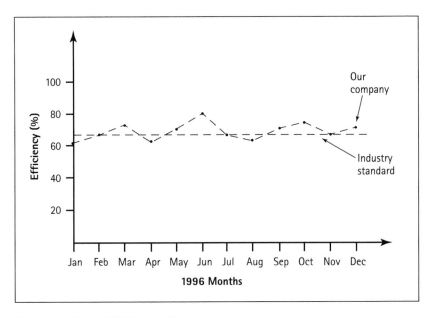

Figure 3-3. Labor Efficiencies for 1996

ident and the other top managers are stumped. It *does* seem unfair to blame manufacturing. They are doing a splendid job, even better than the competition! Then what could be the source of the problem? They throw many different hypotheses at the wall, but none sticks. The managers remain baffled in their attempt to understand the delivery problem, until finally one of the VPs, who has been in the company a long time, speaks up.

> *Senior VP*: "I think I'm beginning to understand what is happening. I've been with this company over 20 years. When we started, we were in a small building with just a dozen or so machines. We knew each one of our customers personally and I could manage both the engineering and the production to make sure things got out on time. Why, it hardly took us but three weeks to get most jobs out! But now we are a big company. Instead of a dozen machines we have a dozen *departments* in manufacturing alone! We have hundreds of machines and thousands of customers. We can't expect to run like a tiny company! We need to have better planning. It takes time for a big company to

line up its resources so it can run as efficiently as possible. Our 8-week delivery standard was set over 10 years ago when we were much smaller. Today, with our current size, 8 weeks is no longer appropriate. Just look at the charts we have seen today—with all the best efficiencies and impressive machine utilizations it still takes us over 12 weeks to get an order out. This tells me that we should be quoting 12 weeks. That way we'll create the right customer expectation to begin with, and we'll also be able to meet that target delivery date, so our customers will be happy."

Needless to say, the managers do not meet this suggestion with a lot of enthusiasm. Least happy of all are the VPs of marketing and sales. They do not want to hear that the company will be quoting *longer* lead times. But no one can find a better answer. They certainly can't expect manufacturing to surpass its already exemplary record on the use of labor and machines.

In the absence of a viable alternative, the group of executives decides that setting the right expectations and being on time is better than having customers who are increasingly more frustrated with late deliveries. So a new policy decision emerges from the meeting: Increase the quoted lead time from 8 weeks to 12 weeks. The longer lead time seems to work for a while. Jobs start coming out on time, but the organization has scarcely had enough time to bask in its newfound glory when another pattern of late deliveries starts to emerge. Now, instead of taking 12–14 weeks to deliver on an 8-week job, the company is taking 16–18 weeks to deliver on a 12-week job! *What is going on here?*

The explanation is that the longer quoted lead times require planning over an even longer time horizon. The longer out you have to predict, the more inaccurate the prediction is likely to be. If one is somewhat uncertain about adhering to a final assembly schedule 8 weeks from now, then one will be even more uncertain about adhering to a final assembly schedule 12 weeks from now. A second, even more fundamental point is that putting more time into the system provides more time for things to change. The most obvious instance of this is that, for a given customer order that is supposed to spend 12 weeks in the system, the new policy gives senior management a full 12 weeks to insert rush orders ahead of this job—previously, they had only an 8-week window to do this!

In a nutshell then, the longer lead times result in even more changes to the planned schedule, leading to more inventory, more unscheduled jobs, and later regularly scheduled jobs. It is quite possible that a few years later another top management meeting takes place and another new policy is issued: Increase the quoted lead time from 12 weeks to 16 weeks. And so the spiral unfurls. Before I discuss the solution to this problem, let's see how this spiral affects companies in the other two environments.

The Response Time Spiral for a Make-to-Stock Company

The effect of the Response Time Spiral in this environment is similar. The major difference is that the driver is sales forecasts, not the planning ahead of final assembly schedules. The reason is that in a make-to-stock company lead times imposed by customers are much shorter than those that manufacturing can achieve. Customers could be end-consumers, distributors, or retail outlets. Consider a company, we'll call it DHM Inc., that supplies dehumidifiers to retail outlets such as Kmart, Sears, and WalMart. When there is an unexpected heat wave combined with high humidity, and the dehumidifiers at the retail outlets start selling fast, these outlets want to be restocked in a matter of days. Customers aren't going to wait two weeks for an item to come into stock when they need it *now*! For a given model of dehumidifier, DHM knows that the lead time to procure the raw materials and components, and get the whole product through fabrication and assembly, is about eight weeks. So as summer approaches, DHM has no option but to rely on forecasts from its sales department.

Just as we saw in the previous environment, the longer the lead times, the less accurate you can expect the forecasts to be. Knowing full well that the forecasts will be inaccurate, you accumulate safety stock at all levels: raw material stocks, component and subassembly stocks, and finished goods stocks. Also, as a result of forecast errors, you do not use some inventory as expected. So once again inventories start to swell up for both these reasons.

Now comes the punch. You receive orders for products that were not in the forecast. They are needed very soon. Some of these orders are from core customers that cannot be ignored. Perhaps Wisconsin is experiencing an unexpected heat wave and WalMart wants dehumidifiers for all its stores in Wisconsin. Again we have the call from the VP-sales to the VP-manufacturing. Again we have the heroic feat by manufacturing. And again the "rush job" gets out in a short lead time, but the other jobs are delayed beyond their regular quoted lead time.

In DHM's case it might find, after a while, that products being made with an 8-week lead time are actually taking 10–12 weeks to come out of the shop. "What's the point of us giving you a forecast of what we need 8 weeks from now if we get it 12 weeks from now?" the VP-sales asks the VP-manufacturing at a meeting. The company is losing sales because the right items do not reach the stocking point on time, and the retail outlets turn elsewhere to get what they need. Worse, by the time the items reach the warehouse they have missed the peak demand, and either they just sit in inventory or they have to be sold at a large discount. Either way DHM's profitability takes a hit.

When this situation reaches a crisis, you get a similar top management meeting. Once again, after much finger pointing, DHM also finds no clear indication of inefficiency; in fact, it too finds high utilization of machines and labor, and many departments working overtime to try to clear the backlogs. So DHM decides that these longer lead times are a more accurate indicator of the best performance that it can achieve.

So the sales department is told that from now on it must provide a 12-week forecast instead of the 8-week one. The longer time horizon means you can expect an even less accurate forecast. Hence you get even more inventory, and of course even more unscheduled jobs. Once again, you soon find the regularly scheduled jobs are unable to achieve even these longer lead times—and so the spiral unfurls for the make-to-stock company.

The Response Time Spiral for an Engineer-to-Order Company

One might expect that in a company where each order is custom-designed and manufactured, the impact of each job could be clearly predicted and there would be no Response Time Spiral. You will find no such utopia. Some of the worst Response Time Spirals and late deliveries that I have seen were at engineer-to-order companies. The driver in this case is the customization process itself, and that originates in the engineering department. In an engineer-to-order company, the process begins with designing the customized product, creating the detailed engineering drawings, and then releasing those drawings to downstream departments. These downstream departments typically include the materials department, which needs to procure raw materials and components; the manufacturing engineering department, which creates the process routings and NC programs; and the planning department, which plans when each of the fabricated parts will be manufactured. The planning department also plans the dates when the subassemblies and the final assembly are to be performed, and prints and releases the corresponding job tickets to the shop floor.

Since such a company usually handles a number of custom orders at any given time, it too needs to plan its downstream resources. So when the order is first received, the engineering department and the planning department meet to predict when the engineering drawings will be released to the downstream departments, when each of those departments will complete their part of the work, when the first manufacturing operations will begin, when subsequent manufacturing departments will receive portions of the job, and so on. These predictions are combined with predictions for other jobs, so that each department's load can be forecast and reasonable schedules given to each department. For example, if you have only two experienced welders who can weld thick stainless steel plates, you don't want to assign a lot of thick stainless steel parts to the welding department for completion in the same week. Taking into account many such constraints, the engineering and planning departments create a plan for the new order.

This plan is, unfortunately, short lived. Engineering itself is not immune to the usual enemies of planning—there are the hot jobs that come in with top management pushing them; there is rework that appears from a downstream department and needs to be rushed through because it is late; there are clarifications to customer specifications that take time to iron out; there are miscalculations in the complexity of a given customer requirement—all of these chew up unexpected amounts of time. As a result, the drawings arrive at downstream departments later than originally forecast.

Now the cycle of expediting starts in the downstream departments and continues all the way through manufacturing, because this order is quite likely to be late. A typical comment I hear from manufacturing managers at engineer-to-order companies is as follows.

> *Manufacturing Manager*: "We quote our customers 10 weeks for delivery. Five of those weeks are meant to be for engineering, materials, and process planning. I'm supposed to have the other 5 weeks for fabrication and assembly. But many times I don't even see the first job ticket until 2 weeks before the due date! Sometimes I don't even get the job until *after* it's due! And of course, since I'm the last person in line, I'm caught holding the evidence when the job is late, and so I get all the blame."

Again, inventories accumulate, even though this is supposed to be an engineer-to-order company. The reason is that some parts or subassemblies are manufactured—or they may arrive from suppliers—but others are not yet done due to delays in upstream departments, so the final product cannot be assembled. Expediting to get those upstream departments to deliver is again the rule. This further disrupts any plans those departments might have. After a while, the company experiences more and more late deliveries.

Once again you see from the quote above that the first department to get the blame is manufacturing. Again you have the high-level meeting, the display of utilizations and efficiencies, and the final conclusion that quoted lead times need to be longer. But it is not just manufacturing that is affected by this decision; all the upstream departments, such as engineering, now insert more safety time into their estimates of how long it will take to perform a given job. The Response Time Spiral has attacked the engineer-to-order company too.

OTHER POLICIES THAT PROMOTE THE RESPONSE TIME SPIRAL

In reality, few companies exist in the three "pure" environments just discussed. An engineer-to-order company, for example, may fabricate commonly used components based on a forecast. A primarily make-to-stock company may still manufacture a few rarely used products on a make-to-order basis. There are many secondary policies at work in manufacturing companies that promote the Response Time Spiral. The five most common are discussed here.

1. *Always making a minimum quantity.* Companies often have policies that state the minimum batch size of a shop order for a given part. Typically, these policies arise from attempts to reduce the cost of setting up for each job. For example, each time a given type of shaft is to be machined, the company may require that a batch of 20 be produced. Even if the company is make-to-order and they receive an order for 5 end-items that need only 5 shafts, the company's planning system will still require that an order for 20 shafts be released to the shop floor. They will use 5 of these and the other 15 will go into stores until (hopefully) they are needed for a similar order.

2. *Building ahead to save on setups.* This is related to the previous item, but the perpetrator is different. In the previous case it is usually a policy embedded in some planning software, while in this case it is a shopfloor supervisor. As an example, a supervisor in the lathe department sees an order for 20 bushings that need to be turned. She knows that it takes more than two hours to set up a lathe to do this job, because the tolerances required are high. She also notices that in the planned schedule that was printed out this week, eight weeks hence she can expect another order that will need 40 of the same bushings. There is already enough bar stock in her department, so she decides to turn all 60 this week.

3. *Making components to forecast.* In a similar vein is the policy where, even though there are no firm orders in hand for a make-to-order or engineer-to-order company, components are made ahead to a sales forecast.

4. *Building ahead to keep machines and people busy.* In most manufacturing companies there is pressure to ensure that people are busy working on something. Nothing makes senior management more nervous than walking along the shop floor and seeing numerous machinists sitting idle next to their equally idle and expensive machine tools. Therefore, when department supervisors experience an unexpected lull in work, they will rummage through the printed schedule for the next several weeks (or even months) to find work that they can assign to their people and machines.

5. *Making to customer's forecast.* Some companies that are suppliers to other manufacturers may consider themselves make-to-order as far as their customers go, but their customers may be building to a forecast. For example, consider a company that makes calculators. Based on marketing forecasts for a new model of calculator, this

company has its own "build schedule" (final assembly schedule). This may generate an order to a vendor that provides the faceplates for the calculator. This vendor, not being in the calculator business, only makes customized faceplates to specific customer orders. So it considers itself a make-to-order company. On the other hand, its schedule is really being driven by the forecast of its customer.

Recognizing that your own company engages in these practices, and they seem to make sense, you may well ask, "Why are these policies bad?" The first and simpler answer is: Each of these policies uses up resources in anticipation of future needs. That takes away resources from customers whose needs must be fulfilled today. If you could have devoted all that capacity to fulfilling the current needs, you might not have been late. This is, however, a tough argument to sell, because one gets embroiled in issues of efficiency. For example, the counter arguments would be, "If we built everything in small lot sizes and never economized on setups, we would use up all our capacity doing setups and hardly produce any parts at all," or, "If we didn't make these small components ahead of time to forecast, we wouldn't be able to assemble the final product to order in the lead time that we quote." Should you build ahead and use up capacity for the sake of efficiency and having items on hand, or only use capacity when you are sure you must, so that you use it for the right items?

To truly answer this you must dig for deeper reasons why all the policies above are bad. That requires understanding the roots of the Response Time Spiral.

ROOTS OF THE RESPONSE TIME SPIRAL

The Response Time Spiral is a direct outgrowth of scale- and cost-based strategies. Starting with the industrial revolution and culminating in today's huge consumer demands for manufactured goods, the promise of modern production has been to make seemingly luxurious goods available to the average buyer. The epitome of this was Henry Ford's 1913 invention of the moving assembly line at his new Highland Park plant in Detroit. When combined with other key elements of mass production, such as interchangeable parts and simplicity of assembly, Ford's new plant brought the cost of a Model T Ford down to an incredible $260 in 1927—literally an order of magnitude lower than the price of cars made a decade earlier by other manufacturers who were still using craft production techniques. In the past, only the richest elite could conceive of buying a car; now the automobile was within reach of the large middle class.

For decades, the question that successful producers of manufactured goods had to answer was, "How do we cut costs?" Over time, this obsession with cutting costs led to a progression of developments. The key to keeping costs down, as formalized by Frederick Taylor and demonstrated so effectively by Henry Ford at his Highland Park plant, was to break jobs down into specialized and very simple steps. Each step could then be done by a person with very few skills. This allowed the organization to keep its wage bill low. The degree to which Ford achieved this job specialization is truly amazing:

> ...the assembler on Ford's mass-production line had only one task—to put two nuts on two bolts or perhaps to attach one wheel to each car. He didn't order parts, procure his tools, repair his equipment, inspect for quality, or even understand what the workers on either side of him were doing. Rather, he kept his head down and thought about other things. The fact that he might not even speak the same language as his fellow assemblers or the foreman was irrelevant to the success of Ford's system.[5]

It is no surprise then, that such an assembler required little training. The low investment in training meant that wages could be kept very low. If a worker didn't like the wages or the job, it hardly took long to train a replacement—and replacements were plenty, since many laborers were happy to leave the farms to work in the factories and dream of striking it rich in the city.

> With this separation of labor, the assembler required only a few minutes of training.... As a result, the workers on the line were as replaceable as the parts on the car.

It is worth noting that division of labor to achieve economies of cost had been practiced in manufacturing for almost a century prior to Taylor and Ford. As far back as 1840, this practice was well enough established that the famous historian and political scientist Alexis de Tocqueville noted that "when a workman is engaged every day upon the same details, the whole commodity is produced with greater ease, speed, and economy."[6]

At the same time, de Tocqueville was quick to note the detrimental effects of such work on the human being:

> When a workman is unceasingly and exclusively engaged in the fabrication of one thing, he ultimately does his work with singular dexterity; but at the same time he loses the general faculty of applying his mind to the direction of the work. He every day becomes more adroit and less industrious; so that it may be said of him that in proportion as the workman improves, the man is degraded. What can be expected of a man who has spent twenty years of his life in making heads for pins?

Division of Labor in the Office

Lest you think such "Tayloristic" policies were confined to the shop floor, and that the office side of operations remained more "civilized," a few examples should show that in this century we learned to practice this division of labor just as well in the office.

- Many companies have a data entry department staffed with clerks whose job is to enter data from paper forms into a computer. A company that receives a large number of mail orders might have a department with order entry clerks who open the mail and enter the order information into computers. Notice that these clerks usually do not service the customer order in any other way. They simply take the information from the written form and type it into the computer. This means people working in this job need only have a minimum education requiring basic reading and typing skills.
- Before the spread of personal computers and word processing software, companies had pools of typists whose sole job was to take handwritten drafts of documents and type them to produced finished letters, memos, papers, and so on. This job only required good typing skills, along with basic reading and writing knowledge.
- Companies that perform a lot of engineering design activities typically have people whose job is just to do the detailed drafting of the drawings after the designer has created the concept. Even though CAD systems are available to most designers, many companies still relegate the detailed drawings to more junior draftspersons. Such people require less training and need to have less experience than designers.

Formation of Departments

So the modern enterprise learned how to use division of labor to keep its wage bill low for both the shop floor and the office. And in order to ensure that employees were doing their specialized jobs effectively, the enterprise needed to be organized in departments, each performing a specialized function. In this way, all the people in the department could learn from each other how to do this function in the most effective way ("the one best way" as Taylor would always say).

The formation of these departments led to a new requirement. In any one department, each person knew his or her own task, but no one understood the overall picture of how to satisfy a customer's requirements. If an order entry clerk made a data entry mistake that resulted in inconsistent information, and a designer was unable to create a product design based on that information, who would track down the problem and resolve the situation? It was soon realized that departments would need managers, who would have additional skills and training, and be responsible for communicating across the organization. Thus, while the responsibility of doing a particular task was that of the department worker, the responsibility of satisfying the customer was accomplished (when needed) by the managers.

The manager's job did not stop there, however. With the relentless emphasis on cost reduction, it soon became clear that the first measure of a manager's own performance would be the efficiency with which he or she ran the department. If last year the order entry department had entered, on average, 50 orders per day per clerk, this year the pressure would be on from senior management to achieve an average of 55. If the drafting staff was each putting out 10 drawings per week, it would be given a target of 12, and so on. In this environment, ask yourself what would happen if the following sequence of events was observed.

Scene: *The order entry department of Signs-by-Mail Company. Of five clerks, only three are typing orders, two others are reading the local newspaper. Walter, the manager of the department, is sitting at a desk in the same room. The president of Signs-by-Mail, Lee, enters the room, takes in the scene, then goes over to Walter's desk.*

President: "Walter, how come two of your employees aren't working on entering orders today?"

Walter: "Yes, Lee, I know. I'm sorry, but we just didn't get that many orders today."

President: "Hmmm..." [*Leaves the room*]

Scene: *The same department a week later. Two clerks are typing orders; three others are sitting in the coffee room chatting. Walter is at his desk. Lee enters the room, reviews the situation, then goes over to Walter's desk.*

President: "Walter, it isn't break time right now. Why aren't your people working?"

Walter: "Ah, yes. Once again we have a lull in incoming orders, so I gave them permission to take an early break."

President: "Hmmm..." [*Leaves the room*]

Scene: *The same department two weeks later. Three clerks are typing orders; two are talking to each other. It is clear that their in baskets are empty. Walter is at his desk. Lee, enters the room, evaluates the situation, then goes over to Walter's desk.*

What do you think happens next? When I ask this question of managers, after describing this set of scenes, the first answer I get is, "The president tells Walter to fire two of the clerks." However, after some more thinking, people say, "The President tells Walter to fire two of the clerks, then he fires Walter!" Why? Because it is clear Walter is not doing a good job of managing his department. He should know by now that five clerks are too many for the level of work his department gets. He is wasting company resources by keeping all five on the payroll.

In fact, few department managers would allow the above scenario to occur. Of course you might expect that cunning managers would ensure that their employees feign work if the president appeared unexpectedly. However, this is not at all necessary. There is a better way, as most managers quickly find out.

The Art of Keeping People Busy

The first lesson that managers learn is how to make sure their people are always busy. This does not mean that they create work. Rather, through experience they make sure they have just the minimum number of resources to keep functioning and at the same time ensure that everyone has work. Figure 3-4 clarifies this by showing the number of hours of incoming work for the order entry department of Signs-by-Mail Company for each week of the previous year (solid line). Also shown are two other lines. The upper line shows the peak level of work received; in order to complete that work in that same week, the department would need four clerks. The lower line represents the average level of work received over the year. This is equivalent to the time available from about three clerks. If the department manager staffs up to four clerks, the department will typically finish each week's work without accumulating a backlog to the next week. However, this means that there will be times (such as week 20) when several clerks are idle. On the other hand, if the manager staffs the department with only three clerks, then during some weeks (such as 14–16) there will be work that cannot be completed. A backlog will start to accumulate, and will eventually be worked off in weeks 18–21 (see the shaded areas and the arrows in Figure 3-4). With this policy, since the work coming in to the department equals the capacity (on average), there will almost always be work to be done by the clerks. This also means that there will almost always be a sufficient backlog to get them through low periods. So it becomes the norm for each department to operate with a backlog of work.

This art of keeping people busy, maintaining high machine utilizations, and maximizing efficiencies was carried even further in manufacturing—one might even say it was refined to an exquisite art. In order to make items on a large scale and to reduce cost, companies did all the operations of a given type in one department. So the milling for all products was done in the milling department, which enabled the company to economize in several ways. First, by pooling the demand of all the products, it could smooth out the ups and downs of production. This allowed them to minimize the number of machines and laborers, and just the right number of resources could be utilized close

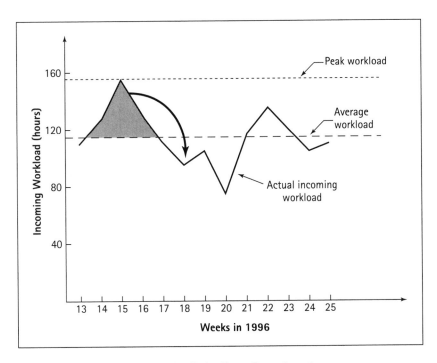

Figure 3-4. Work Arriving at the Order Entry Department

to 100 percent all the time. So far the argument seems the same as before. However, there was more to be gained. With many different products requiring similar components, the company could build a large batch of those components economically by using one setup and store those components for later use. Third, it could train its labor with a minimum number of skills, just to run the operations in that one department. With the skill levels kept low, hourly rates and total payroll stayed low.

This kind of organization is none other than the logical evolution of Henry Ford's assembly line ideas from 1913, but adapted for more general fabrication operations. As the arena of manufacturing competition evolved past the 1970s, however, this organization developed several disadvantages.

Ills of the Cost-Based Organization

Although the manufacturing system perfected by Henry Ford was a breakthrough in its time, this same cost-based organization began to display many problems in terms of today's competitive needs:

- Products requiring many operations suffered long and tortuous routes through the factory.
- Often, even before release to the factory, these orders had equally tortuous routes through numerous office departments.
- All departments, whether office or shop floor, lost sight of their customers. The emphasis, because of resource utilization and efficiency concerns, was to ensure there was a backlog of work at each step. Obviously backlog is the antithesis of responsiveness to the customer. To see how quickly the backlog effect can pile up, consider a customer order that needs to be processed by five office departments and five shopfloor departments. Suppose that each office department has a 2-day backlog, and each shopfloor department has a 1-week backlog. This means that you must allow 35 days—7 weeks—of lead time just for queue time for this order. Adding actual processing time, say a total of 3 days across all the departments, gives us a total lead time of almost 8 weeks for an order that only has 3 days of real work.
- There was another way that the organization lost sight of the customer. If a given customer order needed resolution of difficulties across departments, the proper way to communicate problems or special needs was up through the hierarchy of managers, instead of directly from one task worker to his or her counterpart in the other department. For example, if a design engineer noticed for the fourth time in a month that a given data item had been interpreted wrongly by a data entry clerk, and as a result information in the computer was wrong, the engineer would not pick up the phone and call the data entry department. Instead, she would bring this to the attention of her manager, who would call the manager of the data entry department. They would resolve the problem and then communicate the solution back down to their respective employees. Again the reasons for this hierarchical communication were manifold. One, the task workers were trained in individual tasks and had little or no expertise in the total customer requirements. So such expertise had to come from the managers, who had a broader training. Two, the task workers' efficiencies would be degraded if they

spent a lot of time problem solving, and this would reflect poorly on the department manager. Thus the manager preferred to do such tasks. Three, sometimes problem resolution involved "political" issues of turf and resource allocation in the organization. Such issues were beyond the purview of the task workers. The result of this "up–then across–then down" communication structure was that even routine issues could take a long time to resolve, and all the while the customer was waiting to receive the product.

- Low skill levels led to low quality and to quality by inspection. Rework and scrap became the norm. This was exacerbated by efficiency measures that motivated workers to make parts fast at minimum quality. In fact, it became a game to see how much they could compromise the quality before the parts were rejected by the inspector.

- Since each department produced a high variety of products using general-purpose machines, setups tended to be long. This, coupled with the desire to minimize handling across the long interdepartmental distances, resulted in a preference for working with large batches. Lead times became correspondingly long: Large batches not only take a long time to be completed at each step (since a large number of parts need to be completed), they also create a long wait for jobs in queue behind them. I will give a precise quantitative discussion of this in Chapter 7; however, a simple example serves to drive home the point. If a housing for a pump requires an hour of milling, then a batch of 40 such housings can occupy a milling machine for an entire week of a one-shift operation. If this job has just been started, there are three jobs in queue with similar requirements, and a new job arrives at the milling machine, that new job could wait a month before its turn.

- The proclivity to use large batches had another negative effect. Rather than improving quality through repetition and doing things the "one best way," quality-related costs soared. The reason for this combined human error with the long lead times between operations. Despite numerous inspection steps, some features were not earmarked for inspection, or despite the best efforts at inspection some defects still went undetected, until several departments later, a part would not fit an assembly or could not be machined properly. At this point it was often discovered that a machine setting had been incorrect in an earlier operation and the entire batch was defective. Instead of scrapping one or two parts, dozens or even hundreds had to be scrapped or reworked. Worse, little was learned from this experience. The earlier operation might have been done months ago in a department at the other end of the factory, and there was no point trying to track down who did the operation so they could learn from the mistake.

- The irony is that the department that originally machined the bad part might have shown very high efficiencies in the same month that it was turning out these bad parts. Such discrepancies between measured performance and what was actually good for the company were seldom detected.
- In a similar vein were the undetected discrepancies when components were made to stock and stored for later use, but then were not needed because of market or engineering changes, and had to be discarded or sold at a loss. The same performance reports showed that these components were made very efficiently, even though the company eventually lost money on them. The incentive was for manufacturing managers to make large quantities of such components regardless of their eventual profit potential.

Looking at these features of the scale- and cost-based organizations that evolved over time, you realize that all the elements were in place for the Response Time Spiral to unfurl outward beyond control. Not only did lead times grow longer and longer, but quality suffered too. The greatest insult of all, though, was yet to come. Remember that the driver of this organizational strategy was the desire to minimize cost. Instead, companies found their costs *escalating* because of the amount of rework, expediting, excess inventory, obsolete parts, and the multitude of other non-value-added activities. Figure 3-5 summarizes this whole cycle of effects.

ELIMINATING THE RESPONSE TIME SPIRAL

It is clear from the preceding discussion that to improve responsiveness you have to combat the Response Time Spiral. In fact, implementing QRM requires *eliminating* this spiral—there is no way to control it. As a matter of fact, many companies did try to control this spiral when it first started to get out of hand. In the late 1960s and early 1970s, large companies with complex bills of materials and thousands of component parts were seeing their manufacturing spiraling out of control and their

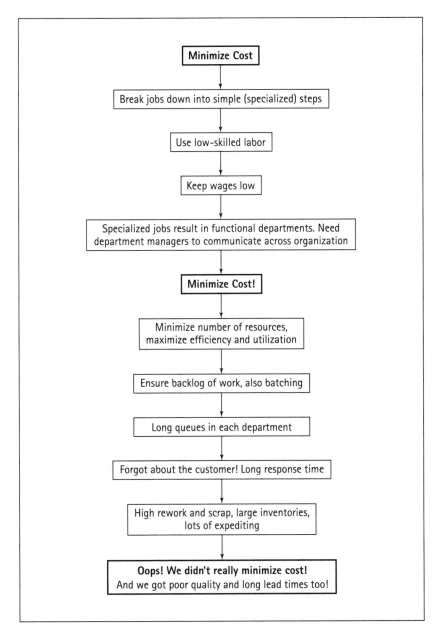

Figure 3-5. The Result of Obsession with Cost Reduction

planning functions completely overwhelmed. Material requirements planning (MRP) systems running on mainframe computers were promoted as a means to manage this complexity. Although computers certainly enabled companies to store all the complex data and make vast numbers of computations, they did little to diminish the Response Time Spiral. Instead, the MRP systems were more fodder for the spiral to feed on—but insidiously! Inside the system, in the database and in the calculations taking place, was buried a new source of energy that would eventually burst forth and make the Response Time Spiral worse than it ever was. I will discuss the reasons for this in Chapter 8.

In the meantime, as companies realized that MRP alone would not solve the problem of late deliveries and expediting, the next savior perceived by industry was computerized shopfloor scheduling and control. "If only we knew where each job was at each moment, and if the computer could tell us what to do next to help keep jobs on time, we would never be late," thought many manufacturing executives. Sophisticated and expensive shopfloor scheduling programs, along with improved MRP systems (now called manufacturing resource planning, or MRP II), became all the rage. Yet the Response Time Spiral continued to blossom relentlessly, defeating even the best computers and scheduling systems.

In the late 1970s and early 1980s most corporations saw their lead times increase severalfold as a result of this Response Time Spiral. I personally had the misfortune to witness the quoted lead time for a company in the aerospace industry go from 6 months to 9 months, then to 15 months, and finally to 24 months as the spiral unfurled at this company.

Killing the Response Time Spiral involves taking time out of the whole manufacturing system. You need to understand that the spiral derives its energy from time. The more time you put into the system, the more it feeds on this time and gets stronger. To weaken it and eventually kill it you must take time out of the system. Instead of asking, "How much lead time should we allow for this activity?" we should ask, "Why does this take so long and what can we do to enable it to be completed in less time?"

This search-and-destroy mission to eliminate time is not confined to manufacturing, nor even to the order fulfillment cycle that spans order processing functions in the office. You need to look at the entire organization and also the whole supply chain outside the organization, even your customers and your customers' customers. You need to review all activities including:

- *Marketing:* Campaigns and policies prior to generation of leads.
- *Sales*: Policies relating to promised delivery dates and quantity discounts.
- *Estimating and quoting:* Activities that work with sales leads in an attempt to get a firm order, including inside sales, engineering, and cost estimating.
- *Order processing:* Office tasks performed after a firm order is received, including order entry, cost review, engineering, process planning, scheduling, and release to manufacturing.
- *Materials:* Policies such as make versus buy, supplier selection, reorder points, and reorder quantities.
- *Manufacturing:* This includes fabrication and assembly activities.
- *Shipping and distribution*: Policies such as how full each truck should be, or truck routing, can be critical.
- *The customer:* A serious QRM program will even change the behavior of customers, for their eventual benefit.

In this book you will learn how to take time out for each of these activities, activities that are concerned with supplying the current spectrum of products. To implement quick response in new product introduction, you will also learn how to eliminate time from activities such as:

- Design
- Prototyping and testing
- Manufacturing ramp-up

However, taking time out of various activities requires a major change in management thinking from a *cost-based* approach to a *time-based* approach. As mentioned earlier, virtually every management system that is in place today has to do with measuring cost. Very few systems measure time, and if they do, the numbers are not much used. Consider two examples.

Example One: How Much Time Did it Take?

Ask any department manager, "Can you tell me how much your department spent last year, broken out into the major categories of expenditures?" The manager will immediately reach into a file cabinet and pull out a report with all the cost information on it. Now ask the inside sales manager, "For the last ten customer requests for quotations (RFQs), how much time did each RFQ spend in each activity that was performed in each department? Each time an RFQ was waiting at some department for information or for other reasons, what was it waiting for and for how long?" Hardly any companies that I know would be able to give you an accurate answer to this question. Sure, they may have some rough guesses, such as, "We usually respond to a request for quote in five days," but ask for firm data and you'll find no one measures it. Here is the proof: Management is mainly concerned with cost. *If management was truly concerned about time it would measure it.*

Example Two: Promoting Numbers on Responsiveness?

In 1996 I was invited to give a seminar to the management of one division of a publicly owned company in the packaging industry. They told me that short delivery time was essential in their industry and all their divisions were working hard at reducing their lead times. Prior to the seminar I received the annual report for this company. This report contained numerous descriptions of the company strategy and many statistics. None of these, however, mentioned anything about responsiveness. The report mentioned aiming for above average profit, product innovation, and efficient manufacturing processes. Similarly the statistics were almost all financial. *If management and stockholders really believed that responsiveness would be significant to the future of the company, why were they not measuring and reporting this?* Only a few companies, such as Hewlett-Packard, actively promote numbers on their responsiveness. The majority of companies are like the one whose report I reviewed. They may give lip service to such items as speed, but when it comes to annual reporting they (and their stockholders) only want to see the bottom line in financial terms.

The implication of these examples is not good. It means that cost-based thinking drives most of our firms. As you will see throughout this book, changing from cost-based to time-based thinking is not easy. Yet it is precisely such a change, in each and every policy throughout the organization, that is essential for successfully implementing QRM. The more you understand the source of each cost-based principle and its QRM counterpart, the more likely you are to convince your colleagues. The first step toward this was showing them one big pitfall of cost-based thinking—the Response Time Spiral itself.

Main QRM Lessons

- Implementing QRM does not mean speeding up existing operations. It requires finding whole new ways of completing a job, with the emphasis on short lead time.

- The "whole new way" for QRM requires a new organizational structure. Present organizations are designed to minimize cost, not minimize time.

- The result of cost-based organizations is the Response Time Spiral, which causes lead times to grow longer and longer. At the same time, quality suffers and costs, far from being minimized, start to escalate.

- Secondary policies that promote the Response Time Spiral are listed, such as always making a minimum quantity and building to a customer's forecast.

- You cannot control the Response Time Spiral, you have to eliminate it, which is accomplished by taking time out of the system. You need to rethink all operations, not just manufacturing. From marketing to order entry, materials procurement through distribution, and even the customers' policies—all need to be redesigned for minimization of response time.

- Taking time out of the system is easier said than done, however, since existing manufacturing and management systems are designed to reduce cost, not time. Implementing QRM is therefore going to fly in the face of many existing organizational norms.

Part Two

Rethinking
Production and
Materials Management

4

Reorganizing Production

IN PART ONE I focused on setting the stage for QRM, providing the scenery that is the mental preparation for the actual QRM drama that is about to unfold. In this part, I will detail the production needs and key ideas behind implementing QRM for delivery of existing products. By *existing products* I mean products that make up your present offering to customers; they are in your current catalog, or lie within a defined envelope. These products may involve some degree of customization or design, but do not require radically new designs or technology.

QRM's aim is to reduce lead time from receipt of order to shipment of the completed product. This involves all the processes in this cycle, including inside sales, order entry, engineering and process planning (if product customization is necessary), manufacturing planning, materials procurement and preparation, order scheduling, fabrication and assembly processes, testing, packaging and shipping. Although I'll occasionally refer to this entire set of operations, in this and the next two chapters I will focus on the shopfloor side of these operations.

As I stated in Chapter 3, to implement QRM it is necessary to find whole new ways of completing a job, with the primary focus on minimizing lead time. Specifically, I discussed why fine tuning or speeding up the existing methods would not work, and also described in detail the drawbacks of the traditional organization. The whole new way of reducing the lead time for existing products requires major restructuring.

SEVEN KEY PRINCIPLES FOR RESTRUCTURING YOUR COMPANY

There are seven key principles that depart from cost-based manufacturing strategy that will help you restructure your company for QRM. Without these principles in place you will not be able to reduce your lead time for existing products.

1. You must change the organization of tasks, procedures, equipment, and processes from a functional basis to a product-oriented basis. All the resources needed to complete a given product are located close to each other. Such product-focused groups are also called work cells, cellular manufacturing, or simply cells. Note that this shift to cells includes both shopfloor and office operations.
2. You must transform the structure of your organization from hierarchical, with many levels, to flat, with many teams.
3. You must shift human resources from being trained in narrowly defined tasks to being cross-trained in a number of operations.
4. You must change the management of processes from top-down control of individual processes in each department to ownership of the entire delivery process by product teams.
5. With principles 1 through 4 in place, you must replace complex centralized scheduling and control systems with simpler, local planning and scheduling procedures.
6. With the operations now closer to each other you no longer need to process an entire batch at each operation before moving it. You must move the parts from one operation to the next as soon as they are completed, or in transfer batches that are much smaller than the whole production batch. This means that there is little work-in-process (WIP) in the cell.
7. With the right organization in place, teams now must run smaller and smaller production batches while at the same time improving quality and reducing waste of all types.

However, implementing the above "cellular" changes is far from trivial. Although the ideas of cellular manufacturing have been promoted for more than two decades, and there has been much press about "cellular manufacturing," some companies have been disappointed in the results and have even torn out the cells and reverted to the old organization. I've also found in my seminars that there is still a great deal of reluctance to implement cells. Key reasons for the failures have been misinterpretation of the basic concepts, and maintaining traditional organizational policies after the cells are put in place. Some of this reluctance stems from the persistence of beliefs such as this: "Relayout is an expensive, disruptive road to establish flow discipline. It's not necessary 90 percent of the time!" And, "One should try to increase the throughput velocity of the old system before starting to relocate equipment." [1]

As you learned in Chapters 2 and 3 you can barely increase the "throughput velocity of the old system" any further—all the past attempts at efficiency improvement tried to do just that. Plus it is the very structure of the old system that creates the problems.

Even if there *is* a desire to implement cells, managers still struggle with fundamental issues of how to properly implement them. Similar findings have been reported in a study of Midwest manufacturers conducted through the Center for Quick Response Manufacturing,[2] and also in the U.K., where a recent paper describes implementing cells as "this often difficult and traumatic process."[3] Again, the reason for this resistance and/or trauma has been misunderstanding or misinterpretation of some basic ideas of cell design and implementation.

To successfully implement cellular manufacturing and create this new QRM organization you must ensure four things:

1. Thoroughly understand the seven key principles for restructuring your company.
2. Understand how to change current practices to accommodate the new organization.
3. Be aware of the many worker and management concerns about the change, and be prepared to address these concerns.
4. Support the new organization with consistent policies in all parts of the company.

This chapter and the next two focus on the first three points. The last point deals with a number of areas of company operation and will be covered in several subsequent chapters.

THE MANUFACTURING CELL—CREATING THE PRODUCT-FOCUSED ORGANIZATION

For reasons of economies of scale and cost firms traditionally have been organized with all the order entry done by order entry clerks in the order entry department, milling done by mill operators in the milling department, welding by welders in the welding department, and so on. Instead, you need to organize all the process components necessary to deliver a finished product (or a family of related products) into one department, often called a *cell*. Since there are a number of different definitions of cells, I will make my own definition concrete. By "manufacturing cell" I mean the following:

> A manufacturing cell consists of a set of (usually dissimilar) machines, in proximity to each other, arranged according to product routing to minimize part movement (often a U-shaped layout is used to minimize worker travel time). The cell is operated by a team of multiskilled workers who are cross-trained to perform several operations in the cell, and who take complete responsibility for quality and delivery performance. The cell is dedicated to producing a family of products that need similar operations, the aim being to complete all the operations for the products within the cell. This means all the resources required to complete the operations should be available within the cell.

I'll highlight six key points in this definition, and then discuss several issues in more detail.[4]

1. The aim of the cell is to start with raw material and end with finished product, with all operations being completed in the cell. This may not always be possible or feasible, and several remedies exist in that case.
2. Machines are dissimilar. This is in contrast to the traditional organization where each department has similar machines—the turning department has only lathes, for example. I've been in the situation where a factory manager proudly displays a "cell," but it is simply a number of lathes (or other identical machines) bunched together with a team of operators running the department. This is still a functional layout; the only change is that the operators are working as a team.[5] Such an organization may result in some quality improvements and other procedural improvements, but will do little to reduce lead time. Reduction of lead time requires cutting across traditional functional boundaries.
3. All the resources are located close to each other, again in contrast to a functionally organized factory where jobs need to go long distances from one operation to the next. Some companies have

attempted to create "virtual cells" where equipment is earmarked as belonging to a cell but not relocated. I am not an avid fan of this partial approach.

4. In contrast to the age-old efficiency principle of division of labor, you have a multiskilled work force performing various operations.
5. Instead of having a hierarchy of managers and task workers, you give ownership of the cell's performance to the team of workers.
6. The cell is dedicated to a set of products, which means that its resources are not diverted to making anything outside that family.

These six principles may sound straightforward, but their application is more subtle than may appear at first glance. In several instances, I have seen management compromise one or more of the principles while attempting to implement a cell. For example, there may be reluctance to engage in cross-training because it would result in higher wages and thus higher labor cost. However, each of the six principles is *equally* critical to the success of cellular manufacturing. Misapplication, or ignoring one of the principles altogether, can lead to failure.

Hypothetical Example: SteelShaft, Inc.

SteelShaft, Inc., produces steel shafts for motors and generators. SteelShaft serves a wide range of markets ranging from shafts of standard dimensions made in high volume, to custom shafts made in lot sizes of five or even just one. SteelShaft is facing lead time pressure, particularly in the small quantity custom market. It is quoting lead times (which include the custom design as well as manufacturing) of 10 weeks but is still experiencing late deliveries. You can see all the ills of the Response Time Spiral in operation at SteelShaft. At the same time, its competition is quoting 8 weeks and has a more reliable delivery record. Market information from its customers and its field sales force lead SteelShaft's management to believe there is a substantial market opportunity for a company that can achieve a significantly shorter lead time, perhaps to the point of capturing more than 50 percent of the market. To regain competitive advantage, SteelShaft's management thus wants to shock the market by advertising a lead time of three weeks for small quantity customized shafts.

In such a case, you might organize the operations of inside sales, order entry, design engineering, process planning, materials planning, scheduling, and shop ticket printing into one office cell dedicated to the small quantity customized market. In fact, you would be well advised to consider shortening the order capture cycle too, which involves estimation and quoting. I will discuss the design and operation of this type of office cell, as well as the issue of quoting, in Chapter 11. This office cell would be followed by a manufacturing cell, which takes the shop tickets and performs material preparation, sawing, rough turning, finish turning, face milling, grinding, inspection, packaging and shipping. One might go further and combine all of these office and shopfloor operations into one cell. Some companies have done just that. For now, let's assume that there will be two cells, and we will focus only on the manufacturing cell portion of this reorganization. By showing the routing for custom shafts before and after cell creation Figures 4-1a and 4-1b highlight the change in manufacturing organization under the cell concept. Notice that the flow is simpler, and parts travel directly between machines instead of between departments. Also, in keeping with cellular manufacturing principles, the inspection operation has been completely eliminated–the machine operators and the packaging operator are responsible for ensuring quality of the finished product.

QRM versus JIT Cells

Based on JIT or flow manufacturing principles, many managers think that cells are best suited to the situation where a known set of products is made and demand is reasonably predictable. This is *not* a prerequisite of successful cells. Note that the two cells I've identified for SteelShaft are focused on customized products–each order could be completely unique. The key, however, is that there is a similarity between the set of products chosen. They lie within a predefined envelope of design requirements, office tasks, and manufacturing processes. As an example, the cells could focus on customers that need shafts made of 8640 stainless steel only, with upper and lower limits on their length and diameter, and requiring custom features that are in a prescribed set, such as slots, holes, chamfers, splines, and keyways that lie within specified dimension and tolerance limits. However, orders for different customers need not have the same routing through the cell. For example, while one order for shafts may follow the route in Figure 4-1b, another

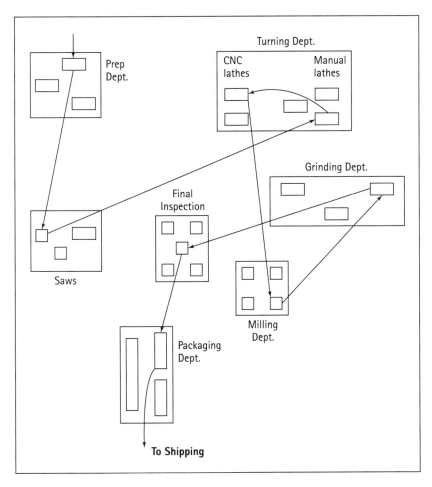

Figure 4-1a. Custom Shaft Routing Before Cell Creation

order might follow the route in Figure 4-1c. Notice that the order in Figure 4-1c skips two of the operations, but also, it backtracks in the cell—a "no-no" in JIT cells. Hence we see that QRM cells may differ from those in standard JIT or flow manufacturing. QRM cells are designed to be more flexible and do not need linear flow. This has implications for the way QRM cells are managed and operated: specifically, issues such as capacity planning, work force planning, and mater-

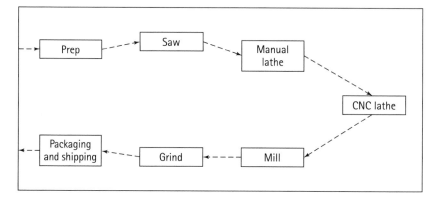

Figure 4–1b. Custom Shafts Cell and Flow of a Typical Product

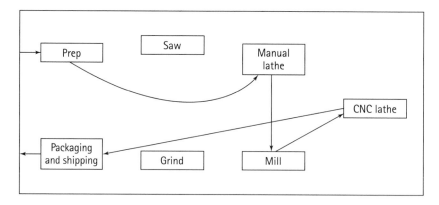

Figure 4–1c. Flow of a Different Product Through the Custom Shafts Cell

ial planning and replenishment, which I will address in more detail in later chapters. To drive home that you can create cells for custom parts with unpredictable demand, and to further differentiate QRM from flow or JIT manufacturing, I will describe in Chapter 6 how one company has been successful in using cells for the production of one-of-a-kind orders for aftermarket parts.

If the number of processes is so large as to make a cell unwieldy, you split it into a few smaller cells. In this case, it is important to make each cell responsible for an identifiable subproduct, with a clearly defined handoff to the next cell. You can go further by capitalizing on the

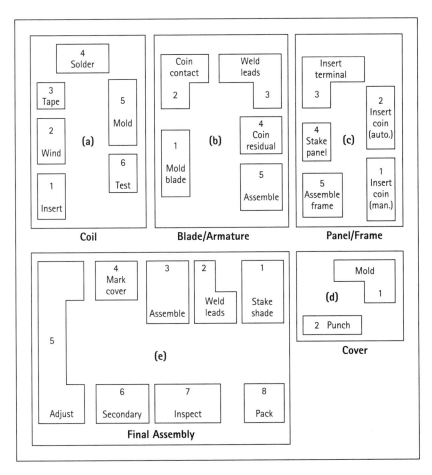

Figure 4-2. Set of Cells for Production of Electromechanical Relays

U-shaped layouts for all the cells to ensure a good flow of materials to the downstream cells as well. Figure 4-2 demonstrates this with the layout for a series of cells that produce an electromechanical assembly.[6]

STAFFING AND TRAINING OF CELL WORKERS

You staff each cell by a team of employees, and give this team ownership of the entire process in the cell. This means that the team has complete say on decisions such as which people run which machines, or who can take time off, subject to overall company policy.

It is also imperative that the team members in each of these cells, whether white or blue collar, be cross-trained so that each person can perform several of the operations that are in the department. This is not just so you can pay lip service to the popular notion of cross-training. There are several good reasons for investing in this multiskilling:

- *Cross-training ensures that the cell can operate even if some people are absent because of sickness or vacation.* In the traditional organization, a department such as milling might have a half-dozen milling machines and an equal number of operators. If one or even two people were not at work, the department still chugged along, albeit at a lower pace. Plus, with the large backlog of work, it wasn't noticeable if on some days the pace was slower than others—with a lead time of several weeks, a day here or there didn't faze anyone. However, in the cellular organization, there may be only one machine for each operation, as you saw in Figure 4-1b. In addition, there is little or no inventory between operations. In such an environment, if only one worker is trained to operate each machine, then the whole cell comes to a halt whenever any person is away from his or her work. Hence the criticality of cross-training: The cell can continue to operate because that person's work can be reassigned or shared by others. The cell may produce parts at a slower rate, but it will keep producing instead of being completely shut down.
- *Such an organization of work enriches each person's job and helps to motivate them.* The detrimental effect of traditional manufacturing organizations on the human being was noted more than a century ago. To repeat part of the quote from de Tocqueville, "When a workman is unceasingly and exclusively engaged in the fabrication of one thing ... he loses the general faculty of applying his mind to the direction of the work ... the man is degraded."[7] The QRM organization reverses this trend. Baxter Healthcare Corporation reported a 37 percent reduction in employee turnover when it implemented cells in several Mexican plants.[8] Even in traditional companies with a history of adversarial labor relations, the new organization can have a powerful impact, as illustrated by the case study of Beloit Corporation that follows. Instead of being dedicated to one step of production, the worker

is now responsible for a completed product to be delivered to an identifiable customer, whether internal or external. In the traditional organization, not only was a worker confined to one step of manufacturing, but he or she also had little visibility of the customer. For one job, the customer might be the gear hobbing department at one end of the factory, and for the next job it might be the grinding department at the other end of the plant. In addition, given the large batch sizes and long lead times at each operation, the next operation might not even commence for weeks. So if any quality problems were detected at a later operation, there was little memory of how the present operation was done, and little chance for learning and improvement. The situation with a cell is just the opposite.

- *Cross-training results in many productivity improvements.* As team members take turns working on different operations, many opportunities for improvement are suggested. This is the greatest benefit of cross-training, and over time, it results in substantial productivity gains. This benefit results from the synergy of three factors in a cell: (1) ownership of the whole process, combined with (2) cross-training and performing multiple operations in the process, along with (3) the proximity of all team members, which leads to frequent communication and feedback in case of any problems. As stated by Deborah Davis of Baxter Healthcare Corporation, regarding the company's experience with cells in its Mexican plants:

> Cells create responsibility centers where none existed before. When responsibility centers are operating, procrastination, finger-pointing, and excuses fade; the stage is set for conversion to a culture of continuous improvement.[9]

She also reports a 32 percent reduction in defects at those plants after the cells were implemented. Compaq Computer found that cells used for the assembly of its computers in Scotland and Texas raised labor productivity by 51 percent, and NEC reports almost the same figure.[10] HUFCOR, a manufacturer of accordion doors and operable partitions in Janesville, WI, experienced a 99 percent reduction in rework with a cell that makes carriers for the partitions.[11] In fact, underestimation of these tremendous productivity improvements is one of the key reasons why companies have difficulty in justifying and implementing cells.

In Chapter 16, I'll discuss additional details concerning creation of the team-based organization. Now I'll demonstrate all the above benefits by a case study of Beloit Corporation, a hundred-year-old company with a traditional organization, a strong labor union with rigid job classifications, and a history of adversarial management-labor relations.

A Flow Tube Manufacturing Cell at Beloit Corporation[12]

Headquartered in Beloit, WI, Beloit Corporation is a global manufacturer of systems and machinery for the pulp and paper industries. Founded in 1858, it now employs more than 7,000 people worldwide. In 1990, Beloit began experiencing increased competition from both small domestic manufacturers and larger foreign companies with faster delivery times. Concerned about the company's delivery performance, Jim Schneider, Beloit's vice president for materials management and logistics, and Jim Riihl, planning manager, approached me to help them promote QRM concepts throughout the company. As a result of several years of training on QRM, along with involvement with the Center for Quick Response Manufacturing, the company implemented many lead time reduction projects.

One effort focused on the manufacture of flow tubes. The groundwork for QRM policies having been established in Beloit through the training seminars, this particular project was implemented completely by Beloit's staff.

The starting section of a paper-making machine is called a headbox. Here, the pulp flows out of a reservoir onto a web (around 30 feet wide). After several hundred feet of pressing and drying operations in the rest of the machine, the pulp laid on the web will become a sheet of paper. For the paper to have consistent quality, it is imperative that the pulp be laid out properly to begin with. This is accomplished by hundreds of flow tubes in the headbox that discharge the pulp onto the web.

One of the functions of a flow tube is to create turbulence for improved fiber dispersion before the pulp is laid on the web, which results in better paper quality. Traditionally, flow tubes had a round cross section, like any other pipe. In the early 1990s, as the result of an extensive joint development program by Beloit Corporation and Mitsubishi Heavy Industries, Beloit introduced the Concept IV-MH headbox. In this revolutionary design, the flow tubes changed in cross section from round at the input to rectangular at the discharge point.[13] Included partway in the tube was a sudden increase in diameter. With all these features, Beloit found that it could achieve directional flow distribution, establish a uniform high velocity profile, then reduce pressure and create a uniform velocity and consistency profile, leading to a good basis weight profile and fiber orientation on the web.

Although the Concept IV-MH headbox was functionally superior to its predecessor, Beloit experienced long lead times and quality problems in the manufacturing of the flow tubes. The company had a functional organization, and the production of the tubes involved thirteen different operations, including three trips between plants that were miles apart. Although the stated lead time for tubes was four weeks, often the final assembly of the headbox would be held up because of late delivery or defective tubes, and the usual cycle of expediting would ensue.

In 1993, Jodi Servin, an industrial engineering student then working as a co-op at Beloit (she is now an employee there), wrote a report analyzing the flow tube production and recommending the formation of a cell for the tubes. Beloit management accepted the recommendations, machinery was purchased, and the cell was put in place in Beloit's Blackhawk (IL) facility in March 1995.

The first anecdote of interest in this case study involves the process of finding workers for the cell. Negotiating with the bargaining unit of the union prior to forming the cell, Beloit management wanted a new classification of operator that could perform all the operations in the cell. Previously, operators with three different job classifications were required to complete all the operations on the tubes. Jodi's report had indicated that only two operators would be required, but they needed to be fully cross-trained to perform all the tasks. The bargaining unit was concerned that workers might not be willing to learn all these skills for the pay scale that management was offering. However, facing the possibility of losing jobs to the competition, the union agreed to try the cell with one job classification as an experiment, and the job was posted for workers to apply. Then came the first QRM lesson for everyone: Despite the union's concerns, a substantial number of workers at Blackhawk bid to work in the cell.

The second anecdote strikes at the heart of our preconceived notions of labor in factories. The previous anecdote shows that traditional union perceptions are not in keeping with the times. This one goes equally against many managers' conception of union workers trying to make as much money for as little work as possible. As is typical in a union environment, the cell jobs were allocated based on union seniority. One of the people who got the job was not a machinist, but a man who had spent his decades of work at Beloit in the packing and crating department. He was more than 50 years old, and in order to work in the cell he would have to devote a considerable amount of time learning to operate all the machines. This surprised me, so I

(continued)

decided to talk with him in person. He was not far from retirement, I remarked to him, and he would have good benefits coming to him after all his years of seniority. Why did he want to change jobs and put in so much effort this late in his career? His reply could have been taken straight out of de Tocqueville's book: "My brain was getting rusty with the work of packing and crating. I wanted to do something where I could use it." A second cell worker had a similar response: "As long as I can keep learning, I'm staying at this company, else I'm out of here."

The third part of this story concerns the performance of the cell. The increased operator ownership did indeed lead to a number of quality and productivity improvements. Defective material was detected earlier in the process, resulting in less rework and less work being put into bad parts. Over a period of time, through knowledge of all the processes and their interactions, the operators came up with numerous ideas for improving the processes. Implementation of these ideas increased their productivity and raised the quality of the tubes. Finally, lead time dropped substantially and the headbox assembly department found the cell very responsive to its needs. Two numbers show the extent of the cell's achievements. First, by early 1996, the lead time had dropped from the original four weeks to three days.[14] Second, the cumulation of all the improvements, coupled with the learning by the operators, has resulted in a doubling of the production rate, without any changes in staffing level. *In short, lead time has decreased 80 percent and, simultaneously, productivity has almost doubled.*

All this, of course, has been accomplished in a company which, by all descriptions, would be considered a traditional organization with a labor union that has been in place for a long time.

PLANNING, SCHEDULING, AND CONTROL WITH CELLS

Since these terms can have different definitions at different companies, I'll clarify what I mean by them. By *planning* I mean the allocation of resources in the longer term, for example, anticipating the number of people and machines that you may need to meet production in the next quarter, next year, and so on. *Scheduling* refers to which jobs to run this week or today, on what machine, in which order, and who will work on which tasks. *Control* is the function of ensuring that you meet deadlines and are capable of reacting to unexpected circumstances, such as a machine breakdown or a bad component part, and taking actions to rectify such a situation.

As you transform a factory into a number of product-oriented cells you will be able to replace the complex, centralized planning, scheduling, and control systems with simpler, local scheduling procedures. In the traditional functional layout, a job would visit a multitude of departments. As far as each department, such as milling, was concerned, the jobs appeared from somewhere to milling and disappeared somewhere else after they were done. Milling did not have responsibility for, or even much visibility of, the overall product routing and delivery schedule. Its goal was simply to do a good (read this as "efficient") job of milling any parts that arrived. In such an organization, each department needed to be told by a central scheduling system which jobs to work on next.

On the other hand, in the cellular organization there is high visibility within each cell of where the orders are and the current status for each one. In this case, it is not necessary for a central system or manager to schedule each piece of equipment in the cell. The team operating the cell is given the delivery schedule for the cell's products, and it takes responsibility for deciding actual equipment schedules, labor hours, and priorities. Now the function of the central system becomes assigning overall delivery schedules, ordering and allocating material, and coordinating between cells if necessary. In Chapters 7, 8, and 9 I'll discuss the tools that enable the cell team to take responsibility to operate this system, a system that is far simpler than one that needs to schedule every piece of machinery for every order in the factory.

As many companies that have installed complex scheduling systems have found out, the detailed schedules produced by such systems often do little to help the shop floor because changes caused by expedited jobs, machine failures, or material shortages make the schedule obsolete before it is even printed. As a result, supervisors ignore the detailed schedules and decide on priorities via an informal communication network. Their decisions make the printed schedule even less usable, since assumptions made by the scheduling system are no longer true. In this atmosphere, the centralized scheduling system has no chance to succeed, and the enterprise reverts to running by manual expediting of hot jobs.

In contrast, in the cellular factory, the central system allows for manual scheduling of jobs within each cell. With its simpler task of coordination, supported by the fact that the cells consistently achieve their lead time and delivery targets, the central system can maintain its effectiveness. Not only is this system more effective, but it also consumes fewer resources—another source of reduction in overhead costs from implementing QRM. In a survey of 36 firms that have installed cells, 80 percent stated that the cells had simplified production planning and control procedures. In fact, significant reductions in effort and expense were reported, ranging from 25 percent to 75 percent.[15]

CELLS FOSTER CONTINUOUS IMPROVEMENT

Finally, as each cell begins to operate in this organizational environment, they can focus their efforts on how to run smaller and smaller batches in order to make the cell even more responsive. These efforts are enabled by four characteristics of the cellular organization.

1. Each cell is now responsible for a limited range of products. Now you can tailor the setups on each of the machines more closely to this family, and make setup times much shorter than they were in the functional layout.
2. The proximity of all machines in the cell encourages the use of smaller transfer batches (leading to overlapping operations)—this will be illustrated in detail in Chapter 7.
3. With the cross-trained group of operators working as a team to deliver the product, additional setup reductions and many other quality and process improvements will be realized over time.
4. As the cell experiences occasional delivery problems from machine failures, the team will be motivated to implement preventive maintenance and other policies to reduce machine down times.

Compared with the traditional cost-based and efficiency-focused organization, these worker-led improvements are so dramatic that they often surprise managers who have been in the company for a long time. A recent study reports that such improvements comprise "the most frequent positive unexpected outcomes" of implementing cells. As one manager stated: "I didn't realize that you will get continuous improvement with the same operators in the same cell day after day."[16]

From several of the preceding points you can see that the kaizen (continuous improvement) philosophy[17] would provide a concrete framework for encouragement and implementation of numerous employee-generated ideas for productivity improvement in cells. Thus, creation of a kaizen mind-set in the organization assists the successful implementation of QRM.[18] It also helps in the rapid introduction of new products.[19]

As a result of all these improvements, products that were made in lot sizes of 100 will soon be made economically, rapidly, on schedule, and with high quality, in lot sizes of 20, or 10, or even 1. The resulting reduction in lead time will be phenomenal. It is not unusual for a firm to transform a product that has a two-month lead time and is stocked in large quantities into a make-to-order item with a one- or two-day manufacturing lead time. Or, in SteelShaft's engineer-to-order environment, they might find that they can complete the office operations in a week and the manufacturing in three days, thus achieving an even more aggressive lead time than targeted. The combination of lot size reduction and kaizen efforts will also make a significant impact on quality. One study of 114 manufacturing firms showed that when lot size reductions are supported by other continuous improvement methods, there is a roughly one-to-one relationship between the reduction in lot size and the reduction in quality rejects. In other words, a 75 percent reduction in lot size is accompanied by a 75 percent reduction in scrap and rework.[20]

A case study of CISCO Precision Components serves to illustrate the incredible level of productivity improvement that you can obtain through the combination of all these effects.

Productivity Multiplies by 250 percent at CISCO

CISCO Precision Components, a sheet metal stamping company, is a division of Central Industrial Supply in Grand Prairie, Texas. CISCO has dozens of punch presses up to a capacity of 275 tons. The company has developed a heavy emphasis on assembly of the components it manufactures. One particular item involves the production of eight different metal stampings, and the assembly of all eight (A–H) into a single movable component. Historically, each component was held in inventory and passed in large,

(continued)

skid-size quantities from station to station. In other words, several hundred pieces of component A were assembled to the same number of component B, and stacked neatly in a container. The container awaited transfer to the next station where component C would be attached, and then each of these assemblies placed into another large container for transport to the next station down the line, ending with the attachment of component H. At each station, large containers of the partial assemblies were stored to ensure that the station would have enough product for the day's activity. If any single component was unavailable for any reason, the affected station was closed and effort focused on one of the other stations to build up its inventory. CISCO estimates that it carried more than 8,000 components in various stages of assembly at any time. About 25 people worked in the assembly area, with an output of 1,200 pieces per day.

In 1996, David Bibb, plant manager of CISCO, attended one of my two-day seminars on QRM. Upon his return to Grand Prairie, CISCO management decided to implement the principles of QRM. The results have been nothing short of astounding.

They have reduced inventory virtually to zero, with an annual savings of $85,000. More impressive is the increase in productivity. By the summer of 1997, a 12-person cell was producing 1,500 pieces per day. This is an average of 125 pieces per person—2.5 times the original figure of 48 pieces per person. Also, true to QRM principles, the growth of the company has been so fast that every displaced worker has been reassigned to other manufacturing tasks.

Order of Implementation Is Critical

It is important to note that you *must* sequentially implement the various steps above. For instance, it would be counterproductive to attempt the batch size reduction step before the others. In the early 1980s, many Western managers visited Japanese factories and came back with the idea that a lot size of one was the goal. They immediately cut all the lot sizes in their own factories without implementing any other changes (these factories were still organized in a functional layout). The result was huge bottlenecks and even worse delivery performance. Even with setup reduction in place, the functional layout offers limited gains. The large amount of material handling needed to move pieces discourages the running of small batches anyway, and the lack of a product-oriented team (as in a cell) means that other improvements in quality and setups are not obtained.

Indeed, this is why I am not a fan of virtual cells. Even if a team is created for such a cell, without the proximity of team members and the constant communication between them, the gains will be limited. What it boils down to is this: The very fact of locating the machines and people together sends a clear signal to the whole company that you are committed to this QRM journey and there is no going back to the old ways.[21]

Summary of Benefits of Cells

- Simple and clear product flow, leading to high visibility of jobs and ease of control.
- Reduction in material handling, which not only cuts down on time and cost, but can also reduce the defects caused by frequent handling and movement.
- Job enrichment, leading to increased worker satisfaction.
- Ownership combined with cross-training and frequent communication, leading to continuous improvement efforts, which reduce non-value-added activities such as setups and down times, and also improve productivity through process improvements.
- Better quality and reduction of rework.
- Decentralization of detailed scheduling and control, leading to simpler central systems that have a greater chance of success in their tasks.
- Ability to run small batches, which, combined with proximity of operations and transfer batching, result in short lead times and low WIP.

As a result of all these benefits companies have seen productivity increases of 50 percent or more, and floor space reductions of 30 percent.[22]

Although the benefits of cells can be substantial, reorganizing the company into cells is not a trivial process. The next two chapters deal with steps for implementing cells, methods for dealing with a number of management and employee concerns, and issues related to manufacturing processes and equipment.

Main QRM Lessons

- The organization of tasks, processes, and equipment must change from a functional basis to a product basis. Such a product-focused approach is also known as cellular manufacturing.

- Despite years of publicity about the cellular approach, there is still incomplete understanding of the basic concepts, resulting in limited success or even ineffectual implementations.

- A precise definition of a manufacturing cell, and the key departures of the cellular approach from cost-based strategy.

- Why cross-training is essential for the success of cells.

- A case study of a successful cell at a traditional 100-year-old company with a strong union.

- The effect of cells on planning, scheduling, and control.

- Why cells foster continuous improvement.

5

Structured Methodology for Implementing Cellular Manufacturing

GIVEN ALL THE ADVANTAGES of cells detailed in the previous chapter, you may be motivated to explore this approach right away. "Which products should I begin with?" you might ask, "and which operations should we put in the cells?" This chapter gives you a procedure for implementing cells. This is only part of the story, however. To ensure success you need to follow the full road map for QRM implementation in Chapter 17. The reason for covering a key set of issues here, though, is that it reinforces an overall understanding of cellular manufacturing concepts.

Although there has been a lot of discussion about cells in the literature, much of the knowledge is fragmented in different research papers, or else it is anecdotal and known only to people who have wrestled with many implementations. Also, in my discussions with managers and employees, I find that existing literature does not adequately cover all their concerns. In addition, no source seems to adequately combine discussion of the management and technological issues that you need to tackle. You often derive the best cells when you consider organizational, technological, and employee issues simultaneously. So in this section of the book I've attempted to put together in one place information that accomplishes three aims:

1. A thorough description of the concept of cellular manufacturing, its benefits, and why those benefits are obtained (Chapter 4);
2. A complete methodology for implementing cells, along with answers to the key concerns of managers and employees (Chapter 5);
3. How to rethink many organizational and technological issues, in order to enable effective implementation of cells (Chapter 6).

SEVEN STEPS TO SUCCESSFULLY IMPLEMENT CELLS

Figure 5-1 summarizes the structured methodology for implementing cellular manufacturing.

Step One: Begin With a Market Opportunity or Threat

Right here, at this first step, is where many efforts stumble. Company managers attend a seminar, decide that cells are the way to go, and begin creating cells based on product routings without thinking about the customer. In many cases, such cells are doomed from the start. The result is that the organization shys away from cellular manufacturing as a whole, losing the ability to use a powerful QRM tool.

Like any effective manufacturing strategy, cellular manufacturing must begin with thinking about a specific customer need or complaint. Is there a segment of the market demanding shorter lead times? Or a segment of customers complaining about repeated late deliveries? Are you losing market share to competitors that have shorter lead times? Is your cost too high and your quality too low as well? In this context, focusing on an internal customer is acceptable too. In the Beloit Corporation example, the customer for the flow tubes cell was the head-box assembly department. In the SteelShaft example, they targeted a particular market segment based on both customer needs and the competition's ability.

On the other hand, one company went through a lot of effort to create cells and reduce its lead time, and the market just yawned. It turned out that this company's customers had other suppliers with much longer lead times, and the lead time for this company's product had never been an issue in that market. A related example is of a company that went for the right market segment, but didn't spend enough effort putting in place a marketing strategy and training its field sales force about the opportunity. Lead time dropped but sales did not come in at a sufficient rate, and the company had to lay off some workers temporarily. The message from this is twofold: (1) given the energy that you will put into creating cells, it is essential that you focus on a market where your improvements will be received with great enthusiasm, and (2) you must have your marketing and sales people involved with this program from the start so they can plan ahead to exploit the results.

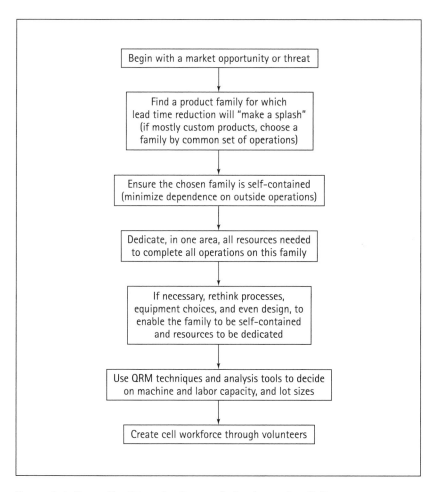

Figure 5-1. Seven Key Steps for Success in Implementing Cells

Step Two: Find a Product Family That Will Make a Splash

Having identified the general opportunity, home in on a product family where a significant reduction in lead times will make a splash, that is to say, create a highly noticeable change in sales or customer satisfaction or both. In the case of SteelShaft, it is clear that taking a majority of market share will be the requisite splash. In the Beloit case, the signifi-cant reduction in the number of problems and delays experienced by the assembly department was the planned splash, but several other remarkable achievements emerged from the cell too. But why do you need to make a splash in the first place? Why not just go for any improvement opportunity? Because implementing QRM changes many established traditions in the company. Some employees and managers will support the changes, but there will be many skeptics sitting on the fence, and some that are opposed to the changes, perhaps vehemently. It is imperative that your first few implementations be not just moderate successes, but *huge* ones, so that no one can deny the benefits that have been achieved.

How does one define a family if many of the products are custom made and/or there is no obvious grouping of products? If a company makes many types of shafts or many types of housings, then it is clearer how to define a family. What if the company makes a large number of custom-engineered products? As I've noted before, you can apply QRM to such a situation. *The key is to focus on the common set of operations needed by the products.* A company that makes all kinds of spare parts for machines might find within this large variety a set of parts that can be machined from bar stock, and only need the operations of sawing, turning, milling, and grinding. The example of Beloit's aftermarket parts division, in Chapter 6, is along these lines. A few brainstorming sessions that combine the market opportunity view ("need to make a splash") with the process commonality view will usually lead to candidate families of parts. A company that assembles printed circuit boards was able to create an effective cell by focusing on small-dimension boards needing only top and bottom surface mount components, with no through-hole or manually inserted components.

Step Three: Stepwise Quantitative Approach to Find Candidate Families

When even the brainstorming technique is not yielding good candidates, or there is disagreement about the candidate families, the quantitative procedure described here can provide effective results. This procedure consists of the following seven steps:

1. *Obtain a table of some recent representative period's sales by part category, in dollars.* (The period could be the last year, or last

quarter, or a forecast for an upcoming period. It will be simplest if this table is in a computer spreadsheet.) By *part category* I mean a classification that is by part type rather than detailed part number. For example, even if every housing made for every customer is different, you could still have categories called *small housings*, *medium housings*, and *large housings*. In fact, sales targets are often set by such categories, so they already exist in most companies. If the parts sold have many components, it may be necessary to bring the table down to the component level. Any aggregation you can do in this first step to reduce the number of detailed parts, using common sense or existing categories in use at your company, will make the procedure simpler.

2. *Sort this table by descending order of sales; parts with the highest dollar sales should appear at the top.* Examine the first dozen or two dozen part categories. Is there a good candidate family in these? If so, then you are done. If no obvious candidate shows up, or the volumes are too small to justify a cell, then go to the next step.

3. *List the main processes needed for each category in a new column in the table, called Process Set.* Use a letter for each major process in your company, and then list the processes in alphabetical order, not in the order they need to be done. Thus a part that needs processes F, C, and S would have the code "CFS" in this column. If there are more than 26 processes, you can add the digits 0–9 to get up to 36, or further, you can use a letter-digit combination for each process, e.g., A0, A1, A2,...,B0, B1.... In this case, a part that uses the processess of R2, D2, C3, P0, would have its entry in the Process Set column as "C3D2P0R2" (remember the alphabetical rule). This coding option will give you the ability to represent 260 processes, which should be enough. If not, you can expand the above method in obvious ways. However, remember that I'm trying to represent the major processes and not to make minor distinctions, so chances are, you should not need to go much above 36 in the first place.

4. *Now sort the table by the Process Set column.* Total the sales for all parts that have the same Process Set, and generate a new table with just two columns, Process Set and Total Sales (for each Process Set). Sort this new table by the Total Sales column, again in descending order, and examine the top dozen or two Process Sets for likely candidates. If you find one or more candidates, then stop.[1] If not, then go to the next step.

5. *Consider if some Process Sets can be combined.* For example, CHOP and CHOPS might represent parts that follow a similar route except that some of them need a grinding operation. As another example, BET and BENT could be parts that need three of the same operations, but some of them require a slot to be milled, while others have no milling. If there are obvious (common sense) combinations you can make, do so and create a new process code for each set of processes that have been combined. Now repeat the totaling and sorting operation.

6. *Repeat this procedure (combine, sort, examine)* until you find a suitable family with sufficient volume.

7. *If, after repeated applications of this procedure, no family presents itself clearly, ask yourself if your business is too fragmented.* Should you try to create more focus to what you do in your factory? The message from the above procedure, in such a case, is that it is time to examine this issue *prior* to attempting QRM.

Even individually made handcrafted products yield to such analysis. Figure 5-2 shows examples of common routings extracted for a company that makes pottery items for direct marketing. Given the craft nature of the business, many of the company's managers were skeptical that cells would be a possibility. However, as Figure 5-2 shows, we found several candidate families that represented a large percentage of the company's sales. Finally, rethinking product design and product options simultaneously with cell design can provide effective families of products and cells.[2]

Step Four: Choose a Self-Contained Family

This step ensures that the product family chosen is self-contained, meaning there is minimal dependence of the cell on operations outside of the cell. To pave your way to success you need this step. Companies often stumble on this step, as seen by this story.

A company put in a cell that required a heat treatment operation part of the way through the cell operations. Heat treatment for the whole company was done in one central furnace. This operation was often backed up and parts were not returned to the cell on time. The cell, in turn, was late in its deliveries. Unfortunately, there was little the cell team could do because the heat treatment operation served

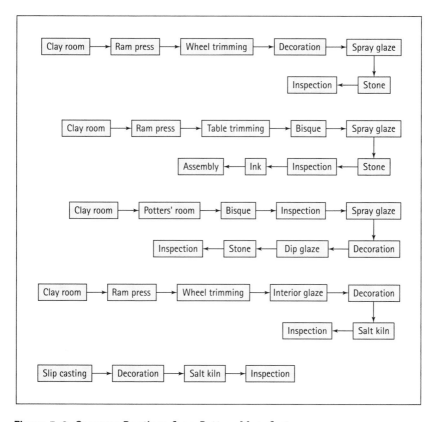

Figure 5-2. Common Routings for a Pottery Manufacturer

many other areas of the plant, and some of those areas had more clout than the cell team. Another effect of these delays in the middle of the cell was that the quality and process improvements resulting from feedback between cell operations were not being realized. At the same time, management had dedicated additional machines to the cell, and also established higher pay grades in view of the higher operator skills, so the cell costs were higher than the previous production costs. After several months, management concluded that the cell was a failure, and it decided to outsource the parts.

In reviewing this situation note that the problem lay partly in the fact that the cell never got a fair chance to prove itself. Put another way, it's not fair to ask the cell team to play cards if they only get dealt from a partial deck. Give them full control over all their operations. But what if most of your products require an operation such as heat treatment, plating, or another process that is done in a central facility or by a subcontractor? This issue is not as insurmountable as it seems. In the next chapter we'll discuss several ways you can get around this obstacle.

Step Five: Create the Physical Cell

Next, plan to locate all the resources needed to complete your operations in one area of the plant, laid out to minimize handling between operations. If it is difficult or prohibitively expensive to locate the resources together, see the suggestions in the next chapter. In planning the layout[3] and floor space requirements, how do you know exactly how many machines are needed? And how many operators? What if the product family is quite broad? Do you put all the parts in the cell or only some of them? How does this affect the number of machines and operators? In Chapter 7, I'll show you how to answer these questions. using simple cell planning tools.

In an effort to make the family self-contained, and in deciding if resources can be dedicated, be sure to reexamine *all* the process decisions for this family, including process choices, equipment choices, and even product design—this point is discussed in considerable detail, with numerous examples, in the next chapter.

Step Six: Education on Manufacturing System Dynamics

In planning and operating the cell, it is vital that both management and workers understand the relationships between (1) setups, lot sizes, and lead times and (2) capacity utilization and lead times. Basic education on these points can make the difference between success and failure: Inaccurate or misguided perceptions of lot sizing and machine/worker utilization can lead to incorrect decisions on numbers of machines and lot sizes to be used in the cell, resulting in cells with long lead times. This is not a rare problem; a study of several companies showed that managers typically have an incomplete understanding of these issues.[4] Chapter 7 is devoted to this topic.

Step Seven: Call For Volunteers

Finally, create the cell work force through volunteers. Begin this process by giving everyone in the company an overview of QRM and why it is imperative for your company to embark on the QRM journey. Then describe some of the key elements of implementing QRM, including cells. The reason for beginning with these basics is that if you implement some of the organizational changes needed for the cell without explaining them to everyone, there may be some mutterings, rumblings, or even noncooperation from others in the organization who don't understand what is being done.[5] Continue by giving everyone a good vision of what cellular manufacturing entails. Those attracted by the

---

From Failure to Success with Cells at Deltrol Controls

Deltrol Controls, a manufacturer of electromechanical relays and solenoids, located in Milwaukee, WI, first attempted a cell in the early 1990s. Its management had read about the advantages of cells and the theory behind them, so it designed the cell and decided which workers would be in it. Dubious about the cell to begin with, the workers were unable to gel into a good team and the productivity of the cell was worse than the stand-alone operations previously being used. The cell was disbanded after just a few months, and the company abandoned the ideas of cells for a while.

   In the fall of 1994, under continued market pressure, company management approached the Center for Quick Response Manufacturing for help with a second attempt at cells. A study conducted by the Center showed the potential for creating a cell for a group of products known as general-purpose relays.[7] A specific cell design was proposed by the study, along with additional principles for cell operation. The recommendations were further refined by Deltrol staff, after which they were accepted by Deltrol management, and implementation began in fall of 1995. This time they involved the workers much earlier in the project. Also, partway through the implementation, Frank Rath (a colleague from the Center) and I were asked to conduct a half-day workshop with the workers, supervisors, and manufacturing engineers who were involved in the project. The workshop helped to further educate employees on the concepts of cellular manufacturing, as well as to address their specific concerns and allay some of their fears. By the fall of 1996, production lead time had dropped by over 60 percent, and on-time delivery was at an all-time high of 99 percent.[8]

vision will volunteer and will be your best bets for success of the cell.[6] Results of two different cells at Deltrol Controls underscore the importance of this procedure.

This completes the description of the seven core steps to implement cellular manufacturing. For complete success, however, you need to supplement these steps with organizational policies discussed in the next few chapters. Specifically:

- Capacity and lot sizing (Chapter 7)
- Material and production planning (Chapter 8)
- Material control and replenishment (Chapter 9)
- Creating teams as well as their performance measurement (Chapter 16)
- The overall approach to a QRM project (Chapter 17)

While considering the possibility of cell implementation, management needs to know how to address several concerns.

CELLULAR MANUFACTURING IMPLEMENTATION CONCERNS— HOW TO OVERCOME THEM

Despite the overwhelming advantages of cells, many companies are still reluctant to implement them, or else struggle with many issues of how to implement cells, including the issue of justifying that the investment in cells will pay back. In dealing with hundreds of managers at QRM seminars, and through cell implementation projects at many companies, I've identified the key concerns. I should emphasize that in many other

Table 5-1. Concerns About Cellular Manufacturing Implementation

- Will dedicating equipment or workers to a cell reduce the utilization of these resources?
- Will multiskilled workers be less efficient than the specialized workers in place today?
- Will the labor union resist this change?
- What will happen to supervisors?
- Will dedicating resources to cells reduce our flexibility?
- How can we cost justify the cell when it will have underutilized resources and higher pay scales?
- How should we deal with operations that must be done outside the cell?

books and seminars on cells, the focus has been on the *mechanics* of implementing the cell—identifying suitable part families, how to create the layouts, and so on. However, the concerns that I will focus on here are primarily those voiced by management and workers (see Table 5-1). Some of these concerns are addressed in other places, but without enough emphasis on really understanding such concerns and providing specific solutions. (The concern of dealing with outside operations, or out of cell operations, is sufficiently complex as to warrant being the topic of Chapter 6.)

Dedication May Result in Lower Resource Utilization

Invariably, when managers discuss cells or review a recommendation for a specific cell, the first concern voiced is: Will dedicating equipment or workers to a cell reduce the utilization of these resources? To make this concrete I'll present two hypothetical situations for Metal Products Company (MPC). See what you would recommend in each case.

Hypothetical Situation One

MPC is a small midwestern firm that makes sheet metal cabinets. The company is exploring cellular manufacturing for a line of custom-designed cabinets, where management feels if they cut lead times in half there is an opportunity for market growth. An industrial engineer with cell experience at a previous job has proposed a cell with four items of equipment: a manual shear to cut the sheet metal; a CNC turret punch to punch holes and slots, or notch the parts; a press brake to bend the metal; and a welding booth where the pieces are welded into a final product and finishing operations such as grinding and cleaning are also completed. She would staff the cell with a team of four cross-trained workers; typically each would run a given operation on a given day. While planning the cell, using last year's custom cabinet orders as a benchmark, it appears that on average each operation would have a total time requirement of about 50 minutes per cabinet (for all the pieces in a cabinet, and also including setup times), except for the CNC turret punch, which (in part due to its short setup time) needs only 25 minutes, on average, per cabinet. Based on some simple modeling and past experience, the IE estimates production lead time for this cell will be under four days, from receipt of shopfloor order to completion of all cabinets in the order (in this niche market, usually just one to five cabinets are ordered in a batch). The IE puts together a proposal supported by the director of manufacturing, and together they present it to MPC's president and senior staff.

MPC's marketing director predicts that with this shorter lead time, the demand for cabinets should reach about 2,000 cabinets per year. At this volume, and operating with one shift (about 2,000 hours a year), the shear, brake, and welding booth should be utilized about 80 percent of the time, but the CNC punch will only be busy 40 percent of the time. When the president, who is also the CFO and the owner of MPC, finishes listening to the proposal he complains, "In order to implement the cell, you want me to purchase that $200,000 CNC machine—but more than half of my investment will lie idle, which means that darn cell will waste over $100,000 of my money!" Then he adds, "Not only that, but about half the time one of those operators will also be sitting around waiting for work. In these times, including benefits, that's going to cost me more than $25,000 a year."

If you were the manufacturing director, what would you do now? Spend a little time thinking about your answer. Jot down your suggestion, so you can review them against the discussion later. Now I'm going to make the situation a little more difficult. The cell imbalance cannot be anticipated; it occurs after the cell is implemented.

Hypothetical Situation Two

I'll use the same situation for the Metal Products Company, however, in this case, the benchmark from previous orders shows that the average time requirement for all four operations will be around 50 minutes per cabinet. Thus the cell will be reasonably "balanced" and it will use all four equipment items over 80 percent of the time. The president likes the proposal. He approves the cell and it is implemented.

After the first few months of operation, the cell team has made improvements in all the operations, and the average time for three of the four operations is now 40 minutes per cabinet. In the case of the CNC punch, however, the team realized substantial reductions in setup time, and the average time requirement per cabinet has dropped to 20 minutes. So the punch is idle about 50 percent of the time, as is the person who is operating it. This development has not gone unnoticed. The president, who frequently walks down to the cell to see how his new investment is doing, has observed on several occasions the CNC punch and the operator sitting idle. "Half of my $200,000 investment is idle," he complains to his director of manufacturing, "which means that darn cell is wasting $100,000 of my money! Plus I'm paying a person for sitting around half the time. Why couldn't you anticipate this when you planned it? You'd better figure out what to do with that person's time right away. Then I want you to propose something that will give me a better return on that $100,000 that's being wasted."

Again, if you were the manufacturing director, what would you do? Once more, spend some time thinking about the situation, because thinking through this exercise is important to your understanding of QRM. First, think about solutions to the operator problem. Then propose some solutions to the problem of the 50 percent idle machine. As before, note down all your ideas, and please don't go on until you have thought through both examples in detail.

So What's the Problem?

Well, did you spend a lot of time thinking through solutions to this problem? If you did, then to put it bluntly, and with all due respect to the senior executives reading this book, you've been suckered. Why? First the president of the company said there was a problem. Then I asked you to assume the role of manufacturing director and solve this problem. You then spent some time solving the problem. *But is there really a "problem" here?*

The cell is very successful and MPC is grabbing market share like you wouldn't believe—lead time is even lower than predicted. Customer satisfaction is at its all-time highest. Plus, with all the improvements that led to reduced cycle times, the cell team is turning out 25 percent more cabinets than originally estimated. Also, through improvement in quality and reduction in rework, material costs have decreased as well. In fact, the total cost of production is 30 percent lower than before. Finally, marketing now believes there may be a much larger market out there than it first anticipated, and MPC is considering going to a second shift for the cell. Now this is the type of problem an owner would *like* to have!

Then why did I sidetrack you with two hypothetical situations? Because I know most of you still think in traditional paradigms: An expensive machine with low utilization is bad; a worker who is partially idle is a waste of money, and so on. I conduct this exercise routinely when I give QRM seminars, and invariably, senior managers come up with a number of solutions to the "problem." Very seldom does someone question me, "Do we really have a problem?" I go through this exercise precisely to show the audience that it's not easy to get rid of our traditional mind-set. Here are three of the most commonly proposed solutions that managers give when confronted with this "problem." You may have thought of a similar set of solutions.

- *Let's double the production rate of the cell.* The idea here is that if you assign twice the production quantity to the cell, you can end up with acceptable utilizations all round, as follows. You plan for the cell to have two shears, two brakes, and two welding booths, but only one punch. You place 7 workers in the cell. Since in the previous case you were using about 3.5 operators it is easy to see that with double the production you will keep about 7 people (and the 7 machines) busy. This will work, provided you can find enough suitable products (those that fit the cell's capabilities and flow pattern) to double the total quantity produced. In some cases, management may not believe that the demand exists for this volume of products. There is another drawback to this solution—it almost doubles the total investment in the cell. Management may not be willing to put out that kind of money for an idea that hasn't proven itself as yet. One more point: when the idle time cannot be anticipated, as in situation two, you will not know ahead of time that you should plan for twice the number of machines for some operations, so this solution will not be anticipated and you will still be stuck with the idle time "problem."
- *Let's eliminate one worker.* This solution involves having the other cross-trained workers moving between their machines and the CNC punch, so they share the work on the punch and no operator is required on that machine. This appears to be an acceptable solution to the worker idle time problem, because now the remaining 3 operators will be busy more than 90 percent of the time. However, this solution has two drawbacks. One, the load on the 3 operators, which is approaching 100 percent of their available time, may lower the capacity of the cell or substantially increase its lead time

through queuing and backlogs (see Chapter 7). Two, you still have a $200,000 machine that is idle about half the time, so we haven't tackled that issue. I'll address this in the next proposed solution.

- *Let's bring in some other jobs that need punching.* Since MPC makes a lot of other cabinets, perhaps the idle time of the punch can be used up by assigning the punching operation of other jobs to the cell. Manufacturing managers frequently propose this solution. We have to realize that this solution is the beginning of a slippery slope that will take us all the way back to the eras of scale and cost. Let's see what its effect will be on the cell by imagining that MPC implements this solution.

Hypothetical Situation Three

This is a continuation of situation two. In order to appease the president of MPC, the manufacturing director has decided that some of the higher volume cabinets that MPC makes will have their punching operations done at the CNC punch in the cell. For a few weeks this looks like a good idea. Now the punch and its operator always have work to do. After a few months, though, everyone realizes something is wrong with the cell. Inventory levels in the cell are three times what they were before this change, the lead time has also tripled, and few additional improvements are being realized in the cell. Compared with the original pace of improvements that occurred when the cell was first implemented, it seems that the kaizen in the cell has completely evaporated. Actually, it is worse than that: quality has degraded. Now more defects are occurring; some go undetected until the welding has begun so they have to scrap entire cabinets. In sum, lead times are long and customers are complaining, quality is bad, and costs are going back up: What has happened to MPC's cell?

Before you read further, try to solve this mystery. Can you explain all the happenings in the cell? This is a test of your understanding of the concepts in this chapter as well as previous chapters. To determine what has happened in the cell, we interview Shirley, a skilled punch operator who had been with the company ten years before she volunteered to be in the cell team earlier this year.

"The additional cabinets that they have me working on," Shirley explains, "are for the Navy. They have to withstand severe conditions on board warships and so they are a heavier gauge than the ones that normally go through the cell, and they need a substantially different set of tools loaded in the turret of the punch. Also, several other settings need to be adjusted on the machining table and other places. This means that every time I have to work on the Navy's cabinets, it takes me almost three hours to set up the punch. So after a couple of weeks of doing this, the team decided it only makes sense for me to do the Navy cabinets once a week, all in a row. Typically, I work on the Navy jobs for two days straight, then come back to the regular cell jobs. If I did the Navy jobs more often, I'd spend so much time setting up I'd have no time left for production!"

Shirley continues with her explanation. "We also realized that if we used the original cell policy of having no between-machine inventory, and transferring pieces one at a time down the cell, then Bill (the press brake operator) would be idle the whole time that I made the Navy cabinets. We'd just be 'robbing Peter to pay Paul'—fixing my idle time but creating idle time for Bill. That problem was easy to fix because my cycle time is faster than Bill's for the custom cabinets that belong in the cell. I just work ahead and build up a couple of days of inventory for Bill, then he can work that off while I do the Navy jobs. Then there is the counterpart to this too, and that is Joe, the shear operator. Likewise, Joe's cycle time being slower than mine, he continues working while I do the Navy jobs. This provides me with enough inventory that the punch and I can remain busy when I'm done with the Navy jobs!"

Does this sound familiar? What we have created is a set of "mini" functional departments, with a pile of inventory between them. All the ghosts of scale/cost strategies come back to haunt us in this cell—long lead times, lack of immediate feedback, resulting in poor quality, rework, expediting ("Shirley, you *have* to interrupt that Navy job so we can get this custom cabinet order out on time!") and the rest. By trying to improve the utilization of the punch you have started sliding down that slippery slope back in time. This points out the need for one other rule in implementing cells:

....

QRM Nugget: For success in cell implementation, it is imperative to maintain "cell integrity."

....

Everyone, from the cell team to the manufacturing director and even the president, must understand that for the cell to be effective it has to work on a consistent family of parts, and the product flow has to be maintained. Any attempts to disrupt the flow in the name of efficiency must be fought off. Keeping such a consistent product focus is called cell integrity and the peril of not maintaining *cell integrity* is that the whole cell implementation can fail.[9] This is why you should avoid the cell having to share its machines with other areas, and conversely, having the cell depend on machines in other areas—the latter is frequently necessitated by processes that require large batch machines such as heat treat furnaces. I'll discuss how to resolve this in the next chapter.

Although I could mention other proposed "solutions" to the original MPC problem, I've made my main two points, which are:

1. While implementing QRM concepts such as cells you must continue to be wary of being trapped by the traditional mind-set.
2. A focus on utilization can lead us back on the path to the obsolete scale- and cost-based strategies. I will have a lot more to say about the pitfalls of focusing on utilization in Chapter 7, after I have laid some more groundwork on the topic of manufacturing system dynamics.

At the same time, I should emphasize that I'm not ignoring the need for companies to achieve a good return on investment (ROI) since it is a legitimate concern of an owner or CFO that a partly used machine may lead to a low or even negative ROI. The approach to resolving this issue is discussed later in this chapter.

Will Multi-Skilled Workers Be Less Efficient?

This concern is rooted in our Tayloristic mind-set that a specialized worker is the most efficient in performing a given operation—a person who has twenty years of experience on lathes should be more efficient in turning parts than someone who has had to spend time learning about lathes, grinding, milling, surface treatment and other operations. Wouldn't the specialized person produce a better quality product as well? I've visited some U.S. companies where manufacturing managers have advanced just such an argument against cells. This argument ignores five important points.

1. Modern manufacturing technology has changed the way that you can impart quality and features on a product. A lot of what used to be craftsmanship is replaced by training on how to use computer-controlled equipment—the craft is embedded in the NC program.

2. Interaction effects between operations have a high leverage on cost and quality. What this means is that there is great opportunity for improvement not by optimizing a given operation but by optimizing the interaction effects. Often, quality defects are caused by a sequence of operations, and eliminating the root cause involves determining this interaction effect. Since this interaction effect is highly product-dependent, cells, by their product focus plus physical and human organization, are ideally suited to discover such improvement opportunities. Indeed, the basis for the success of JIT was not so much in eliminating the cost of inventory, as in forcing companies to figure out such interactions—the absence of WIP forced tighter coupling of operations—and then to improve their processes.

3. It is well known that specialized workers do not have an incentive to maximize quality. What really happens is that measurement and reward systems encourage them to maximize production at the expense of quality. In contrast, the organization of the cell gives workers pride of ownership, and quality becomes one of their goals. As illustrated by the earlier quote from the flow tube cell worker, job enrichment in a cell leads to higher satisfaction and better quality work results.

4. Specialized workers tend to prefer long runs and accept long setup times as the norm. Cross-trained workers promote flexibility and quick response.[10]

5. As shown by the performance achieved by the flow tube cell, teams of cross-trained workers in a cell, brainstorming on a regular basis, can implement improvement leading to higher productivity than was present in the individual specialized operations.

What you are doing through all this, in contrast to the mass production view of creating a specialist to optimize individual operations, is creating a team that specializes in a focused product family. Focus is key, because training people to do every possible combination of operations could take a lifetime. However, training workers to do the focused set of operations required by a family of products is achievable in the space of weeks or months at most. *The choice of product family also limits, clarifies and hence simplifies, the scope of the cross-training effort.* The following principle summarizes these points.

> QRM Nugget: In designing cross-training programs to support cellular manufacturing, do not aim to train workers on all the skills needed to run each operation. Rather, focus the cross-training efforts on the specific skills needed by a team for processing a given product family through all its operations.

Related to this issue is the practice at some companies of having one classification of specialized workers who only do setup, while production is performed by another classification of workers. Again, the reason is one of cost—the setup workers are higher skilled and paid more, but since they're only needed occasionally, less skilled and lower paid workers can do the bulk of the production. At companies with such an organization I am usually asked, "Should the setup people be in the cell?" The answer, ideally, is that you should train the cell operators to do all the operations including setups. If this isn't possible in the short run, the cell can begin by having one or more setup people dedicated to the cell as part of the cell team, and goals can be set for both the setup people to train the others on how to do the setups, and the workers to train the setup people on performing all the production operations. Without such an organization, the cell team will lose ownership and the usual problems will result. If the setup person is not included in the cell team, there could be finger-pointing whenever there are quality or delivery problems.

Will the Union Resist a Cellular Organization?

Companies with unions are usually concerned that going to a cellular organization will require a lot of work in modifying and renegotiating union contracts. This often delays management's decisions for implementation. The biggest issues are (1) breaking down job classifications (or defining new, broader classifications), and (2) simultaneously defining pay scales for the new classifications. However, there is evidence to show that such fears often loom larger in management's mind as a result of traditional thinking than as a result of unions per se. A study of U.S. companies showed less reluctance overall to put in cells in companies with unions, as compared to those without unions.[11] Even a company such as General Motors with a long history of adversarial management-union relations found ways to redo job classifications in its Saturn and NUMMI ventures.

The key to resolution of these issues is understanding and accepting that management is as much part of the problem as is the union, and both sides need to work together for resolution. It was, after all, management's overexploitation, underpayment, or unreasonable termination of employment during the initial industrial period that led to the growth of the unions. Although unions continue to provide safeguards against such situations, a major reason for their existence continues to be the individual worker's concern with long-term job security. Unions relish the numerous job classifications because they help to protect the total number of jobs. In such an environment, management needs to recognize that if it can provide job security in other ways, the union may be willing to broaden job descriptions. One way to provide such security might be an agreement not to lay off the cell workers during the (say) three-year term of the new contract. If management believes in the new organization, it should be willing to back up its beliefs to the fullest.

If your company is in a market where shorter lead time is critical to survival, you can use this argument to help sway the union: "We can put in the cells and have jobs for everyone, and if you can help us be successful we'll have even more jobs than today, or we can stay the way we are and eventually all of us will be out of a job!" It is important that neither management nor the union perceive cells and other QRM techniques as a way to downsize. On the contrary, QRM implementation should help grow the company and increase the total number of employees.

As a final point, one shouldn't immediately assume a union will be opposed to the creation of a worker team. In recent years some unions have even taken the initiative to propose teams at their factories—beating management to the punch. I'll deal with this in Chapter 16 under the topic of teamwork.

What Will Happen to Supervisors?

When teams are formed many supervisors see loss of control and possibly loss of their job. This is a sensitive issue and can even lead to supervisors opposing cell formation through putting forward numerous arguments why the cell concept won't work for their area. In general, the QRM organization is flatter than a traditional one, so there will be fewer layers of management between the shop floor and the president. However, people who have worked as supervisors have a number of skills useful to the organization and there are ways to keep such people in the organization. Teams need to have champions in the organization, people who will evaluate their needs and then press upper management to provide the requisite resources. The supervisor's knowledge of the company operation, skills in cost justification, and past experience dealing with management could be useful to the team workers who may never have done such tasks. A supervisor could become such a champion and could be used by a number of shopfloor teams.

Training is another application for supervisor skills, such as in quality methods or machine operation. Continuous training is a fundamental part of a QRM organization, and an experienced supervisor could be a valuable resource for this. In other words, instead of being in charge, the supervisor must see himself or herself in the role of providing support to the team or teams.

Another alternative for the supervisor is to become a member of one of the teams. In the new organization, there are more people who "do" the work and fewer people who "instruct or manage." The supervisor may be the right person to work a complex machine, or to be in a cell team responsible for high precision parts, where that person's skills will be needed.

The company could also encourage the supervisor to acquire the skills necessary for a white collar position. For instance, if SteelShaft is trying to reduce its office processing time for estimation, quoting, and order processing, who better to be on the team than someone who has been responsible for seeing that the shafts are made correctly on the shop floor?

Will Cells Reduce Your Flexibility?

The advantage of the functional layout is that it enables you to make a wide variety of products with varying demand levels. Since you pool all the resources of a given type, ups and downs in individual product demand get averaged out at the functional departments dealing in joint demand. Of course, this has all the drawbacks too, such as a large backlog to ensure even utilization, and resulting long lead times. But the organization is flexible in the sense that it can accommodate not only varying demands but products with very different process routings.

On the other hand, a major principle of cell design is that you dedicate resources to a family of products—I've commented on several occasions on the importance of cell integrity. In such a situation, managers worry about loss of flexibility, specifically due to events such as:

- A surge in demand that surpasses the cell capacity.
- A fall in demand that leaves the cell idle.
- A change in product design that requires new processes or eliminates existing processes.

So management may see the transition to cells—involving initial costs of analysis, cell design, moving equipment, and training the teams—as highly risky investment if the cell becomes obsolete in a short time. There are several solutions to this problem.

- You should not deny the presence of risk in any manufacturing venture, and the importance of planning to minimize this risk. Every time a company launches a new product, sets up a new facility, or even buys one machine, it is taking a risk in terms of assuming the market is there to justify the investment. Cells are no different. You must begin with analysis of the market need—remember the very first step I listed for implementing cells. Up-front analysis of the market and a good deal of planning for alternatives such as those below are likely to alleviate future problems.
- Choose cell space, equipment capabilities, and material handling so as to allow minor variations in product family and process to be made easily.
- Make the cell work force flexible so that it can move between cells based on load.
- Design cells with overlapping capabilities so you can redirect particular orders from one cell to another based on backlog. For example, a company could put in three cells making shafts, one dealing with shafts under 0.5″ diameter, the second with shafts from 0.5″ to 2.5,″ and the third with shafts larger than 2.5.″ But a better solution might be to have the first cell be able to machine shafts up to 1,″

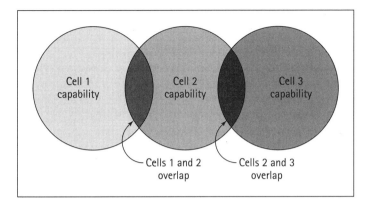

Figure 5-3. Cells with Overlapping Capabilities

the second cell from 0.5″ to 3,″ and the third cell from 2″ up (Figure 5-3). The cell teams would then have considerable flexibility to allocate incoming work in order to keep lead times short and to keep all cells busy.

- Change your mind-set. Unlike the traditional factory where one invests in a machine and expects to see it operating in the same spot for 25 years, you should know at the outset that cells are not permanent. One needs to continuously rethink and reevaluate all the cells. Baxter Healthcare found that "a good cell, like a good factory, is flexible. The best machine cell is one in which the machines are easily uncoupled from utilities, easily detached from machine-to-machine transfer devices, and easily moved."[12] Companies should rethink their choice of equipment and the way it is secured to the floor and to utilities. It may be better in the long run to invest in more flexible machines even if they are more expensive. Some companies have found they can put machines on wheels or other devices, enabling them to be moved easily. Setting up utility lines (electricity, water, pressurized air) that you can access at several points and easily hook up or disconnect will assist as well.

COSTING AND JUSTIFICATION OF CELLS

There are two financial issues that you need to resolve regarding cells: (1) how to cost-justify the decision for cells, if management needs such a justification before proceeding, and (2) how to measure costs in the

Table 5-2. Alternative Calculations to Justify a Cell

(a) Unit Cost with Current Operation (Traditional Calculation)

Operation	Setup Time (hr.)	Run Time per piece (hr.)	Average Time per piece (hr.)	Labor Rate per hour	Labor Cost per piece
Shear	0.80	0.15	0.158	$12.00	$ 1.90
Punch	1.50	0.20	0.215	16.00	3.44
Brake	1.20	0.25	0.262	15.00	3.93
				Total Labor Cost/pc	$ 9.27
				Material Cost/pc	52.05
			Overhead @ 400% of Labor		37.08
				Total Cost/pc	$ 98.40

Note: The average time per piece assumes an average lot size of 100, thus it includes one setup for every 100 pieces.

(b) Unit Cost with Proposed Cell (Traditional Calculation)

Operation	Setup Time (hr.)	Run Time per piece (hr.)	Average Time per piece (hr.)	Labor Rate per hour	Labor Cost per piece
Shear	0.10	0.135	0.145	$18.00	$ 2.61
Punch	0.15	0.180	0.195	18.00	3.51
Brake	0.20	0.225	0.245	18.00	4.41
				Total Labor Cost/pc	$ 10.53
				Material Cost/pc	52.05
			Overhead @ 400% of Labor		42.12
				Total Cost/pc	$104.70

Note: It is predicted that setup times will be greatly reduced due to similarity of parts in the cell, to the times shown above, plus run times can be reduced by 10% through optimization of tooling and operations for the focused set of parts. With the shorter setup times, the cell will run with average lot sizes of 10, thus the average time per piece includes one setup for every 10 pieces. However, since labor is cross-trained, all operators will be paid at a higher rate of $18/hr. Material and overhead is based on standard rates. The unit cost appears to be higher, thus the cell cannot be justified.

(continued)

cell once it is up and running. In both cases, the issues you must deal with relate to the detailed example I gave at the end of Chapter 2, concerning overhead allocation as well as estimating the potential benefits of cells. Therefore, for effective cost-justification of a cell it would be better to have agreement from management to engage in nonstandard overhead allocation procedures, as well as to estimate productivity improvements in the future. Without these, the cell may not appear financially viable.

Table 5-2 illustrates two alternative calculations for justifying a potential cell. In part (a) of the table, the current cost of production is

Table 5-2. Alternative Calculations to Justify a Cell *(continued)*

(c) Unit Cost with Proposed Cell (Calculation Incorporating QRM Arguments)

Operation	Setup Time (hr.)	Run Time per piece (hr.)	Average Time per piece (hr.)	Labor Rate per hour	Labor Cost per piece
Shear	0.10	0.135	0.123	$18.00	$ 2.21
Punch	0.15	0.180	0.166	18.00	2.99
Brake	0.20	0.225	0.208	18.00	3.74
				Total Labor Cost/pc	$ 8.94
				Material Cost/pc (see note)	47.89
				Overhead (see note)	33.04
				Total Cost/pc	$ 89.87

Note: In addition to the improvements in (b), it is estimated that scrap will be reduced by 8% and rework by 7%. Thus, while the estimated setup and run times remain the same, the net labor time needed for the same quantity of production will be 15% less, and the average time per piece is therefore reduced by 15% from the values in (b). The scrap reduction also means material costs will be lower by 8%. Finally, it is argued that overhead shouldn't be assigned as a multiplier, but rather, adjusted from its current value. Certainly, overhead shouldn't be higher than its current value of $37.08/pc, and it ought to be lower based on elimination of activities. By quantifying anticipated reductions in overhead activities such as material handling, scheduling, and expediting, it is expected that overhead will be reduced by $4.04/pc, giving the value of $33.04/pc. It is seen that the unit cost is now $8.53 lower than its current value. Assuming the cell will run two shifts, for 240 days/year, and the bottleneck operation (the brake) will not run beyond 80% capacity, this allows production of 14,769 pcs/year, with an annual cost savings of over $125,000/year. It is estimated that the cost of setting up the cell and training the workers will be around $230,000. Hence with this approach, the cell is seen to provide a payback in under two years.

calculated. Then in part (b), you have the first calculation for the proposed cell. This computation takes into account the new estimates for setup and run time in the cell, but it "goes by the book" in all other aspects, and you see that the cell is not viable. The second approach, in part (c), uses QRM arguments about productivity improvements as well as in regard to overhead, as explained in the notes to the table.[13] This calculation shows the true potential of the cell. This potential would not be realized, and the company would be the loser for this, if it stuck by the first calculation and did not invest in the cell.

Main QRM Lessons

- Steps for successful implementation of cellular manufacturing.

- How to address implementation concerns such as: Will there be lower machine utilization? Will there be lower efficiency when you replace specialized workers with cross-trained ones? How to deal with the union? Or with supervisors? Will cells reduce your flexibility?

- Effective strategies for cost justification of cells.

6

Creative Rethinking for Cellular Manufacturing

THERE IS A SIGNIFICANT CONCERN in moving toward implementation of cells, namely, how to incorporate processes such as heat treatment, sand blasting, and plating in a cell when typically these processes are done in one large facility shared by the whole plant. A similar concern is voiced regarding operations that are subcontracted outside the company. These issues can become insurmountable obstacles to implementation. Or, if they are not properly resolved at the outset they can be stumbling blocks after implementing a cell, as illustrated in the previous chapter with the case of the cell failure caused by the heat treatment operation.

In this chapter I'll deal with this issue of outside operations and show that through creative rethinking there are many ways of solving this difficult issue. In order to do this, however, we need to effectively combine management and technological ideas in unique ways. I'll do this by detailing examples of each idea as well as presenting new ideas such as time-slicing and time-sliced virtual cells.

Once you understand the methodology you can extend the concepts to solve similar problems in cellular manufacturing as well as apply the same approach to rethinking many other details of product and process issues in cells. To deal with the issue of outside operations I need to introduce another key principle of cell implementation.

....

QRM Nugget: A cell is not a mini-replica of your existing operations. When planning a cell, rethink everything you do to make it more amenable to cell operations. Battle conventional product costing approaches and notions of efficiency to implement solutions that maintain cell integrity.

....

Many companies stumble in cell implementation because they assume they will need the *identical* machines, to do the *identical* operations, in the *identical* sequence as they are done today. Instead, when a cell is being planned you have a rare opportunity to rethink every procedure and every decision for the family of products being considered. Not only should you reexamine your choices of machines, but also your choices of processes and sequence of processes. This rethinking is not confined to manufacturing. You can reexamine the product design, the materials used, or even step back further and look at customer needs to see if you can modify some specifications.

Once you break out of the mold of your current way of doing things, you will find a host of solutions concerning outside operations. The principles behind these solutions are listed in the Main QRM Lessons at the end of this chapter. I will use examples to illustrate application of these principles and this way of thinking to create successful cells. This is a unique departure from typical discussions on cells that focus either on the basic definition of a cell along with part and machine selection using routings, or on the mechanics of cell layout, machine selection, material handling, automation devices, and the like.

CHALLENGE CONVENTIONAL CHOICES

The underlying approaches to challenging the conventional way of doing things are to combine operations, eliminate operations, or bring these operations into a cell. Right off the bat these approaches require that you get away from conventional cost-based methods of process design or conventional organizational choices of who does the work. First, by rethinking your choice of material you can often eliminate an operation. Also, by getting away from cost-based methods of choosing the lowest cost machine for any operation, you can often combine operations or bring them onto machines in the cell. Finally, by challenging conventional union or organizational rules about work assignment, you can also bring operations into a cell.

In its companywide efforts at reducing lead times, Beloit Corporation has used all of these approaches, and its work provides us with concrete examples to assist in firming up these ideas.

Use of New Process Design Principles at Beloit Corp.

In its efforts to drastically reduce lead time for aftermarket parts, Beloit Corporation found an effective way to eliminate the casting operation for many of these parts. At Beloit, castings typically had a long lead time, often several weeks, while customers for aftermarket parts wanted delivery in days since usually a machine was down at their plant. When Beloit created a cell for aftermarket parts, and then found that most of the lead time was spent waiting for this "out of cell" operation, the company decided to replace the use of castings by bar stock whenever a customer had an urgent need, and when this was feasible (in their case they decided to do this for rotational parts up to 21″ diameter). The cell Beloit created is truly a QRM cell, as defined in Chapter 4, in that it machines one-of-a-kind parts each with its own unique routing through the cell.[1]

A second strategy used in the same cell was to bring traditionally non-cell operations into the cell. The cell was originally dependent on forklift operators to bring parts and other material to the cell. These operators had a different job classification than the cell workers, so they did not belong to the cell team, and lack of needed material was causing some delays. Similarly, another classification of workers had traditionally done the inspection at Beloit. Again, the cell team found that waiting for inspection was causing substantial delays. Rather than accepting this as necessary, or going to management for help, the team approached the other union workers directly and negotiated with them to bring their jobs into the cell. The cell team is now responsible for all operations including material handling and inspection.[2]

In another instance, mild steel parts used as structural components in a paper machine needed to be painted to prevent corrosion and also to provide a good appearance to the customer. After some initial welding to fabricate the parts, they required three coats of paint with drying time in between, and then needed to be attached to other parts of the machine. In addition to the lead time for painting and drying, this involved moving large parts from the weldry to the paint booth and back again. Following a lead time

(continued)

reduction seminar, Beloit's estimating department was motivated to reexamine the choice of processes for these parts. They proposed that Beloit use stainless steel with glass-blasted finish instead, which would satisfy the criteria of both corrosion resistance and good appearance. They could accomplish the glass blasting in the vicinity of the weldry, while the paint area was relatively far away. Although conventional costing of one of the parts projected a somewhat higher cost for the glass-blasted stainless option, the estimating department argued that because of the lead time benefits and other indirect savings Beloit should adopt the glass-blasted option.

Following QRM seminars to the whole company, Beloit's industrial engineering (IE) department has picked up on this thinking and has been encouraging process designers to think of "time-based" routings in addition to the usual cost-based approach. Some housings, for instance, had operations being performed on boring bars, after which they were moved to a different location for drilling on manual drills. On examining the drawings they found that they could do drilling operations on the boring bars. The reason they were not was that with a standard cost approach the manual drill is a cheaper cost center than the expensive boring bar. Transferring to the drill, however, requires material handling, waiting in queue for a drill to be idle, as well as creating a new setup at the drill with possible loss of accuracy. Both lead time and quality suffer, and indirect costs go up. As part of its move toward making smaller batches of parts with shorter lead times, Beloit now usually does the drilling operations as part of the setup on the boring bars. Also, Beloit's IEs now routinely examine whether they should use a routing with shorter lead time instead of a routing with lowest standard cost.

These examples also show the cumulative effect, over time, of QRM education in a company. After investing in a series of QRM seminars over one year to expose all its employees in Beloit, WI, to QRM principles, Beloit has found staff and workers proposing more and more ideas for lead time reduction. More important, many of these ideas question established beliefs and traditional ways of doing things and are having a fundamental impact on the way the company performs work.[3]

Regarding the strategy of replacing castings with bar stock, you might ask, "Why do companies not do this routinely?" The answer is, for larger quantities of production, the unit cost per part is lower when you use castings, compared with the amount of machining that you need to do starting with bar stock. However, when applying quick response to a small quantity of parts, using bar stock is the better approach. Remember, however, that if you move from castings to bar stock, your standard costing system will probably show that your unit costs have gone up, and you may have a battle to fight. This is related to the discussions on cost in previous chapters, and I'll tackle this issue again in Chapter 16.

Another example of rethinking material choices concerns the use of prepainted blanks. Often, painting is an operation that needs to be in a central facility because of environmental constraints. In such cases it is worth exploring prepainted material, especially since coatings available today are capable of withstanding forming and other operations.[4]

USE TECHNOLOGY THAT ENABLES A SMALLER-SCALE PROCESS IMPLEMENTATION

Applying this principle has a lot to do with changing an organization's mind-set. Typically, large batch processes such as furnaces, painting lines, plating lines, and sandblasting booths are designed that way for efficiency. A large furnace, for example, can be more energy efficient than a small one. Simple physics shows us this. Heat loss is proportional to surface area, which goes up as the square of the dimensions while volume goes up as the cube. So if you double a furnace size, you get eight times the volume but only four times the heat loss. If you could fit one workpiece in the smaller furnace and heat loss cost $20, now you can get eight pieces in the bigger furnace and heat loss costs $80, which is only $10 per piece. The larger the furnace gets, the greater is this "efficiency." Considering that most factories use a larger furnace there is a built-in mind-set that to be efficient you must use large furnaces.

Once you break out of this mold you can find many alternatives. Just such an alternative has been exploited by an industry you don't often associate with manufacturing, but it is a production process nevertheless—that of bread making.

You can exploit the same idea in more conventional manufacturing processes. Many equipment manufacturers supply small ovens that you can place in a cell. In the same vein, instead of a large, high-capacity wave solder machine shared by an entire circuit board assembly factory, you can put several small wave solder machines in individual cells. Small machines for glass beading and sandblasting exist. Similarly, you

---

How the Baked Goods Industry Is Reexamining Scale Operations

For decades it was assumed that mass-produced bread, made in large ovens, would eventually replace all the neighborhood bakeries. In the last couple of years, though, the market for freshly made "gourmet" breads and pastries has been exploited by several franchises setting up neighborhood bakeries in urban areas. These bakeries offer freshly baked French baguettes, various flavors of organic or "natural" breads, muffins, and other pastries. In contrast to the large ovens used by the mass producers, these neighborhood bakeries use many small ovens baking different items based on today's demand—much less "efficient," but more responsive to the consumer. The good news for the bakeries, of course, is that the customers pay higher prices than they would at the supermarket, and yet are happy doing that—a win-win situation for both the producer and the consumer.

---

can replace a conveyor-operated paint line designed for high-volume production with a single robot-operated spray paint booth in the cell. You might even go to an old-fashioned manual paint booth. A bicycle manufacturer that put in cells for custom bikes capitalized on this by saying in its advertising that the bikes were individually hand-crafted.

An obstacle to this approach is that there are environmental or safety constraints on many items of equipment, making it prohibitively expensive to put the environmental control equipment or safety devices in each cell. This might be the main reason for having one common facility in the factory, or using a subcontractor that devotes its facility to this type of operation and can afford the environmental controls. Again, to overcome this problem, try to rethink *everything*, using all the principles in this section. Switching from a large oil-fired to a small electrical furnace might eliminate the problem. Changing the materials may solve the problem: Plating might be required for punched parts because you are using mild steel and after punching them you create opportunities for corrosion, while if you used stainless steel you would eliminate this problem. Rethinking processes or design might do it, too.

Consider several solutions to the problem of parts used as structural fasteners that need to go from a cell to a plating operation and then back to the cell (see Figure 6-1). The plating is done at an electro-deposition machine. In order to think this through, you should realize that most surface treatment is done for one of two reasons: (1) to improve

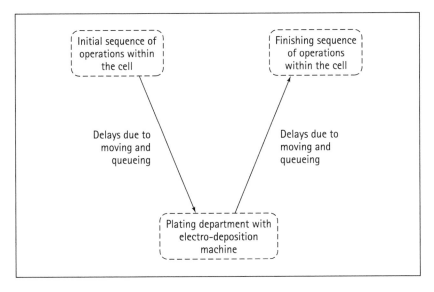

Figure 6-1. Routing for Fasteners Through Plating Operation

corrosion resistance and/or (2) to improve external appearance. The first question to ask is can you achieve the desired property via choice of material? In many cases, corrosion resistance can be achieved this way. As examples, we have the addition of 1 percent tin to 70/30 and 60/40 copper/zinc alloys to create admiralty brass and navy brass; addition of small amounts of copper and chromium to low- and medium-carbon steels, which allows a rust film to form that itself becomes protective and at the same time is attractive (called Cor-Ten); and the addition of

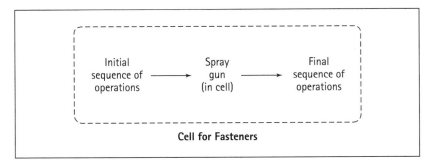

Figure 6-2. Improved Routing Using Sprayed Metal Coating

Rethinking Operations Sequences in the Apparel Industry[6]

The clothing industry found that in implementing Quick Response they needed to alter several conventional paradigms for the production of tailored clothing. Modern textile mills are based on the principle of large batch operations that first weave the cloth, then dye it, then print it. After that they have cutting and stitching of the tailored items. Using very large, high-speed machines, they can perform the weaving, dyeing, and printing operations very efficiently. The penalty, of course, is long lead times. In fact, in the tailored clothing industry these lead times were approaching not months but even years!

The first paradigm that was questioned was the one of dyeing before making the finished item. With the general public's fickleness about color—the rate with which colors become suddenly popular or equally unfashionable—having a stock of dyed material of the wrong color can be a liability. Equally bad can be the lost opportunity for sales because the right color isn't there. The Benetton clothing chain began a new trend by holding undyed natural wool sweaters in stock and dyeing them based on demand patterns. Although the dyeing process for sweaters is less "efficient" than the continuous process used for unstitched wool, it allows the lead time from color commitment to retail delivery to be a few days. In the traditional process, this lead time was almost a year, requiring enormous foresight (or enormous risk) when making decisions on what colors to manufacture.

The second paradigm shift promoted by the Tailored Clothing Textile Institute has been to stitch first and print later. The reason here is different. It enables you to use smaller machines for printing, and these machines can be part of a cell. Of course, several technological challenges had to be overcome to enable printing on stitched material, but having mastered the technology they can now bring the additional benefits of cellular manufacturing to bear on clothing production.

chromium and nickel to low-carbon steels to form stainless steel.[5] A second solution is available in the case where appearance is not a major consideration, but thick pore-free characteristics are more important. In such situations, sprayed metal coatings may be a solution. This is appealing because you can use spray guns in the cell itself (see Figure 6-2).

CHANGE THE SEQUENCE OF OPERATIONS

Conventional sequences of manufacturing operations are sometimes in place because of technical process requirements, and at other times because of attempts to maximize efficiency of the process. Challenging the second reason can lead to new operations that are better for quick response, as demonstrated by two recent changes in the apparel industry.

Often the exercise of questioning the sequence of operations has beneficial side effects in addition to redesign of the process. The U.S. Air Force's McDonnell Douglas C-17 Globemaster III is one of the heaviest transport planes in the world. At the same time it has set a total of 22 world records, including those for payload to altitude, time to climb, and short takeoff and landing. McDonnell Douglas is in the process of implementing an impressive 46 percent cost reduction program for these aircraft, and no design or production decision will remain unexamined in this quest. One decision already reexamined involved the procedure for installing fasteners—the C-17 has some 1.4 million of them. The original specifications called for installing them with a wet sealant. The mechanic had to get fasteners and sealant from stock, apply sealant to the hole and the fastener, check for squeeze-out of the sealant, and clean off the excess sealant with a solvent. To streamline the process, a McDonnell Douglas team developed a process to bake on a dry sealant that completely covers the fasteners. The new process eliminated several instances of non-value-added time spent by the operator: getting the fasteners from sealed stock kits, getting the sealant, checking for squeeze-outs, and cleaning the excess sealant. The baking, in this case, turned out to be easy to do after the assembly process. The nice thing was that tests showed the precoated sealants also had superior corrosion protection and increased joint fatigue life. As an added benefit, the coating reduced the force required to install the fasteners and reduced operator fatigue.[7]

The use of prepainted blanks, described earlier, is another instance of changing the sequence of operations. Instead of forming and fabrication followed by painting, we have painting followed by forming and fabrication. Of course, this has been made possible only recently because of coatings that can withstand the later forming and fabrication processes. Similar to this is the emerging use of tailor welded blanks (TWBs). These consist of two or more pieces of flat sheet (of different materials) joined together before forming, to provide customized properties in the finished part. TWBs eliminate the need for subsequent spot welding or other reinforcement operations.[8]

ASK, WILL THE OPERATION STILL BE REQUIRED?

Sometimes the very fact of implementing a cell, with its shorter lead time and better quality processes, eliminates the need for an operation. However, you must keep alert to find these opportunities. Just such an opportunity was discovered by a process improvement team at General Electric's Wilmington, MA, plant. The routing for thermocouples made at the plant included two of the typical "out of cell" operations, sandblasting and baking. In questioning the need for these processes the team discovered that materials spent considerable time sitting on the factory floor between various steps. In one case this produced excessive oxidation of the parts, which meant that sandblasting was necessary prior to the welding operation, and in another case it allowed the powder in swaged leads to become damp, causing electrical defects if assembled, thus necessitating baking in an oven for several hours before assembly. In improving the process flow so that the waiting time between steps would be less than a day, the team decided they could eliminate both operations.[9]

USE TIME-SLICING AT THE SHARED RESOURCE

If, after all the brainstorming is over, it is still felt that the operation is needed and can't be done in the cell for the sake of cost, environment, safety, or other reasons, there are still some approaches that you can apply to minimize the downside risk to the cell. One is to use "time-slicing" at the shared resource. Remember that a major goal of creating the cell is to let the team have ownership of each job from start to finish. Loss of this ownership is the primary cause of problems with an out-of-cell operation. One way to restore this ownership is to let cell teams "own" the shared process for certain blocks of time. Consider a factory where three cells plus all the remaining operations share a furnace that has a four-hour cycle. Based on capacity required for each cell, you might devise the schedule in Figure 6-3. During the time periods allocated to each cell, the cell team is responsible for bringing parts over and running the furnace. This time-sharing schedule gives each team a predictable amount of capacity that it can turn into a reliable and short response time for heat treatment.

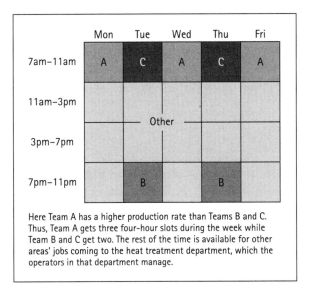

	Mon	Tue	Wed	Thu	Fri
7am–11am	A	C	A	C	A
11am–3pm			Other		
3pm–7pm					
7pm–11pm		B		B	

Here Team A has a higher production rate than Teams B and C. Thus, Team A gets three four-hour slots during the week while Team B and C get two. The rest of the time is available for other areas' jobs coming to the heat treatment department, which the operators in that department manage.

Figure 6-3. Time-Slicing at a Heat Treat Furnace with a Four-Hour Cycle

For this to be successful, however, it is imperative that the company be disciplined in making this capacity available to the cell teams with no exceptions. It takes only a few instances of the plant manager grabbing away capacity for hot jobs, for the team to decide that it doesn't really "own" that period of time. On the other hand, if a team decides that it does not need its time slot, it can "trade" this with other teams, perhaps acquiring a different time slot for the future. This seems like a progression backward—to the ever-changing schedules of the traditional organization—however, it is not likely to degenerate into the expediting scenarios of that organization. The reason is, instead of managers imposing schedule changes, which then have many ripple effects, this transaction is conducted between two teams who have control over the remaining portions of their routings, and can think through the implications of such a trade. If conducted properly, this decentralized transaction can be a win-win situation for both teams.

Another issue arises with multiple people sharing one piece of equipment. There can be denial of responsibility for maintaining the equipment in good working order. In addition, substantial training may be needed to properly operate a specialized machine. In such cases, a high degree of training and cooperation is required between teams of users, or else the company can continue to employ a skilled operator whose job is to be a resource to each team when it "owns" the equipment, but also to keep the equipment in good working condition and maintain quality of the processes.

IMPLEMENT TIME-SLICED VIRTUAL CELLS

You can carry the time-slicing concept even further when a facility has a lot of expensive equipment that cannot be replicated for each cell. It also applies to factories that have very large pieces of equipment, such as rolling mills, that cannot easily be moved and dedicated to individual cells. I call this approach *time-sliced virtual cells*, to distinguish it from other virtual cells.

The virtual cell concept normally refers to cells where equipment is not moved together to form a cell, but is earmarked for use with a given family of parts.[10] Thus a turning department may have five lathes, three of which are used for general jobs, one dedicated to the bearing housing team, and one dedicated to the shafts and spindles team. These teams similarly have equipment earmarked for them in other departments.

My extension of this concept involves dedicating to each team *slices of time* on each item of equipment rather than dedicating items of equipment to teams. Although I am not normally an advocate for virtual cells, in the case where there is only one machine of each type in a company, and the machine is either very expensive or very large, my extension of virtual cells may be the only hope the company has of moving toward a QRM strategy in the short term. A good example of this is the fabrication of printed circuit boards (PCBs), that is, the substrates onto which components are placed or inserted (also called printed wiring boards or PWBs). Consider the production of single-sided boards using the following processes:

- cutting and preparation
- photoresisting
- exposing
- developing
- etching
- stripping
- drilling and deburring
- testing
- solder coating

Since several of these steps involve expensive resources, a company that makes PCBs may not find it economical to form cells. Also, it is not evident that the equipment needed for one set of PCBs differs from that needed for another set. On the other hand, the company does recognize that they serve different focused markets, each with its own characteristics in terms of board specifications, degree of customization, and lead time. One way to meet this need is by forming teams of workers for each market and giving them time slices on each piece of equipment. This time-slicing could be time-phased as shown in Figure 6-4, allowing each team to progress through the facility in a round-robin fashion.

Why bother with this, you might ask? You might as well have functional departments. Actually, no. There is a significant difference. The ever-important attribute of full ownership has been retained in this

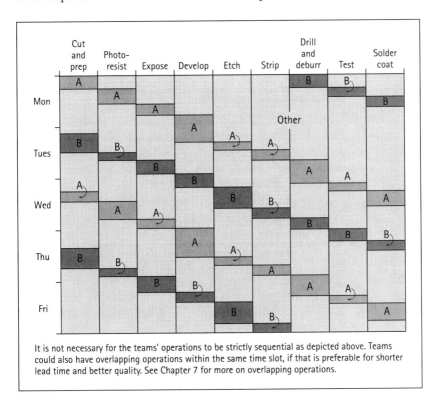

It is not necessary for the teams' operations to be strictly sequential as depicted above. Teams could also have overlapping operations within the same time slot, if that is preferable for shorter lead time and better quality. See Chapter 7 for more on overlapping operations.

Figure 6-4. Time–Sliced Virtual Cells for PCB Fabrication

setup: Each team is responsible for all the operations done on a board, from start to finish. The actual assignment of time slices at shared operations can be done quite easily using the procedure and calculations shown in Table 6-1.

Another question that arises when people are presented with the concept of time-sliced vertual cells (TSVCs) is: Isn't this the same as developing detailed schedules using a sophisticated finite capacity scheduling system? In fact, it is not; there are significant differences (see Table 6-2). First, a centralized scheduling (CS) system generates a detailed schedule of every task at every resource, while for the TSVCs the central management only makes strategic decisions on the sizes of time slices assigned to each team. The product teams decide what to run within their slice; that is, they have full ownership of their slice. The organization structure is also very different. With the CS approach you still have

Table 6-1. Production Analysis for Assigning Time Slices

Product	Volume Allocated for This Month
Family A	102
Family B	140
Family C	38
Other Families	95

Analysis for Operation: VT Lathe	Team A	Team B	Team C	Other	Total	Avail. Hrs.
Weekly hours needed	16	22	10	18	66	80
Minimum slice (hours)	5	6	4			
Actual slices assigned	5+5+6	7+7+8	5+5			

"Volume Allocated" is a management decision based on priorities for the current month. From these volumes, the number of hours needed each week is calculated for each resource that will be time-sliced. The above example shows one resource, a vertical turret lathe. This is a large lathe that is too expensive to duplicate and is being shared by several operations. The "minimum slice" denotes the smallest time slice that it makes sense to assign (based on setup time and typical lot sizes for each family). Based on the hours needed and the minimum slice we can get the actual time slices to be assigned, shown in the last row. The detailed sequencing of time slices across different operations is not too critical provided the slices are frequent enough. If we try to sequence across operations the complexity will multiply. Instead, we rely on the frequency of time slices to ensure that when teams need multiple shared operations they can get to one of "their" slices reasonably soon at any operation they need.

a functional structure, while with the TSVC you can have a product-focused organization. In the CS approach, if there is a problem at one operation, such as an operation taking longer than expected or material arriving late, this creates a ripple that affects all products and eventually all other operations. In the TSVC approach, while there will still be real-world problems, the ripple effects are usually confined within each team's slices. Finally, as we examine the operation philosophy of a CS system, you'll find it conflicts with almost every QRM principle, while the TSVC allows you to apply all the remaining QRM principles (other than machine dedication).

You can put the TSVC concept in perspective with other manufacturing system structures by viewing all these alternatives on a continuum (Figure 6-5). At one end of the continuum you have traditional functional departments, and at the other are perfect cells. In a way, you could think of the line in Figure 6-5 as asking the question: How far are you willing to go toward implementing QRM?

You can also use the TSVCs to deal effectively with the issue of *prototyping*. Companies sometimes have to supply a prototype for the customer to check, before they receive a firm order. Production of these prototypes is often under time pressure, and the quantity is small, so it can play havoc with a company's production schedule. The ideal way to do prototypes is to dedicate a whole cell to them. If you cannot do this because the cost of a cell is prohibitive relative to the volume of prototype production, then you can use time-slicing to provide the prototyping team with access to all the required equipment at predefined times.

Table 6-2. Comparison of Centralized Scheduling with Time-Sliced Virtual Cell

Centralized Scheduling	TSVCs
Centralized schedule generation	Center only makes periodic strategic decisions on allocation of slices. Teams generate schedules and have full ownership of them (within their slices)
Functional organization	Product-focused organization
Changes create ripple effect across all products	Ripple effect usually confined within each team's time slices
Conflicts with most QRM principles	All remaining QRM principles can now be applied

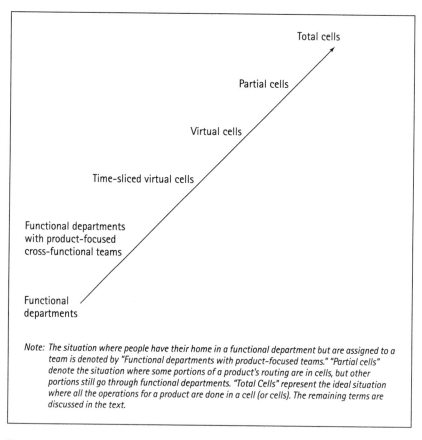

Note: *The situation where people have their home in a functional department but are assigned to a team is denoted by "Functional departments with product-focused teams." "Partial cells" denote the situation where some portions of a product's routing are in cells, but other portions still go through functional departments. "Total Cells" represent the ideal situation where all the operations for a product are done in a cell (or cells). The remaining terms are discussed in the text.*

Figure 6–5. Advancing Toward a QRM Organization

MAKE THE RESOURCE FACILITY BEHAVE LIKE A SUBCONTRACTOR

If you cannot implement time slicing, then one way to proceed, for an internal shared facility, is to establish a subcontractor relationship between this facility and the cell. This requires the facility manager to plan ahead for capacity for the cell team, quote lead times to it and view the obligations to the team as typical customer-supplier contractual obligations.[11] If the shared facility is already an outside subcontractor, then you should delegate the day-to-day subcontracting of the cell parts to the cell team, instead of the purchasing department. (The purchasing department would still have responsibility for negotiating the contract and fine-tuning the legal and commercial terms.)

SPLIT INTO TWO CELLS

Finally, a less effective approach to use when an outside operation is required is to divide the cell into two cells, one preceding and one following the out-of-cell operation. Suppose a bearing housing visits seven operations: CNC lathe, drill, boring bar, heat treat oven, cylindrical grinder, packing, and crating. The organization desires to create a

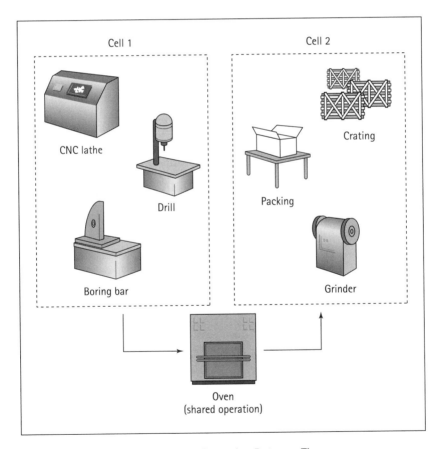

Figure 6-6. Two Cells with Outside Operation Between Them

cell for these bearing housings, but is unable to implement any of the preceding suggestions, and so it feels a cell would not be feasible. It might still be worth implementing two cells, one with the CNC lathe, drill, and boring bar, and the second with the grinder, packing, and crating steps (Figure 6-6). This way, each cell will still have ownership of its segment of operations from start to finish, and you can set lead time goals for each team without them being frustrated about the inability to achieve targets due to factors outside their control. However, the organization should not stop at this. It should pursue interteam meetings to assist in quality across the two cells, and continue to pursue ways of eventually incorporating the heat treat in the cell's operations through all the techniques described above.

In the last three chapters I've shown how you can reorganize production for quick response and given you specific techniques to support the transition to cellular manufacturing. Proper implementation of cells, however, requires more than this—it requires a whole new attitude toward two additional key concepts in manufacturing management, namely capacity utilization and lot sizing. These are discussed in the next chapter.

Main QRM Lessons:
Principles for Dealing with "Out of Cell" Operation

- Rethink material choices to eliminate operation (battle conventional product costing).

- Use smaller-scale implementation of process technology in the cell (battle conventional notions of efficiency).

- Rethink process choice (particularly if smaller equipment is available for alternative process).

- Rethink design to modify feature or functionality, to enable one of these solutions to be used.

- Use time-based routings instead of cost-based routings to determine equipment choice (again, battle conventional product costing).

- Change sequence of operations.

- Bring operation into cell by changing company norms or policy.

- Eliminate operation completely—may not be required with cell processes in use.

- Use combinations of above strategies. One strategy alone doesn't solve the problem, but a combination provides a solution.

- If none of the above are possible, then use time-slicing at the out-of-cell operation.

- If most of the machines needed cannot be dedicated, create a time-sliced virtual cell.

- Split into two subcells (before and after).

7

Capacity and Lot-Sizing Decisions

ALTHOUGH RESTRUCTURING YOUR MANUFACTURING along product-focused lines is a big step toward quick response, you will not tap its full potential until you also make the appropriate changes in the way you determine capacity and lot sizes. To review the traditional approaches to capacity and lot sizing and discuss their negative impact on responsiveness, I'll begin this chapter by introducing some basics of manufacturing system dynamics. A current criticism of the way manufacturing management courses are taught is that managers fail to develop sufficient intuition about manufacturing system dynamics. In contrast, stock analysts learn about stock market dynamics, and good financial managers learn how to determine which key variables are influencing balance sheet performance. It makes sense that to be an effective manufacturing manager you need to understand the basics of manufacturing system dynamics. This chapter will help you develop a basic understanding of these dynamics without having to delve into much technical detail or mathematics. The overview will also give managers a basis for appreciating why they need to change the traditional ways of making decisions. Once this is covered I'll introduce new ways of making decisions about capacity and lot sizes.

But this chapter is not only for the manufacturing manager, it is also aimed at anyone who works in any part of a company whose business involves manufacturing. The reasons are threefold.

1. *QRM involves changing the way the whole company operates.* No department in the company will remain unaffected by the implementation of QRM. For effective transition to QRM operation,

all managers should understand how they need to reconfigure *all* parts of the company so that there is companywide support for this transition.

2. *Policies often cross-departmental boundaries.* In such cases, managers from different parts of the company need to be on board. To make these first two points concrete, consider the following situation that I encounter frequently when working with companies implementing QRM. A manufacturing manager attempts to change a lot sizing policy based on the principles discussed in this chapter. However, the materials manager and the accounting manager, both of whom are still driven by the traditional principles, resist the change. Eventually, the suggestion to change the policy is shot down in a joint meeting where the president hears the arguments from all three, and is convinced by the materials and accounting managers that costs will go up if they implement the new lot sizing policy.

3. *The basic principles of system dynamics also apply to office operations.* To appreciate some of the points I make later when discussing quick response in office operations you'll need an understanding of these system dynamic principles.

DO YOU HAVE GOOD INTUITION ABOUT MANUFACTURING SYSTEM BEHAVIOR?

An experienced manager might argue that it's not necessary for managers to learn the basics, since they've already developed a good intuition for the system behavior based on their work experience in manufacturing. Strange though this may seem, it is my observation that such an "experienced" intuition is not well developed. I'll give two specific examples to illustrate this point.

Example One: A Seminar in Madison, Wisconsin

In 1995 more than 150 managers from various departments of manufacturing companies attended a seminar in Madison, WI. The companies ranged in size from small family-owned to very large public companies, and they covered a wide spectrum of industries. The following question was asked at the seminar:

On a press in a factory, a given job takes a setup time of 2.8 hours and a processing time of 5.2 hours. What will happen to the lead time of this job through the press if the setup time is reduced by 50 percent?

Test your own intuition by trying to answer this question. Can you quickly come up with a rough guess? Attempt to give an answer before reading any further.

Now comes the startling fact. Of the more than 150 managers present at the seminar, not a single one gave the right answer, or even an answer that was close. In fact, not a single one had the right *approach* to answering the question. I will answer this question later in the chapter after I've presented the tools to answer it.

Example Two: The QRM Quiz

This example draws from the second question in our QRM Quiz in Chapter 1. Think about this question again.

---

2. To get jobs out fast, we must keep our machines and people busy all the time.

                                       ❑ True   ❑ False

As you may recall three-fourths of the 400-plus managers in U.S. industry who answered this question believed this principle to be true. Through anecdotal data and more examples you will soon learn why question 2 of the QRM quiz is false—and why it results in longer lead times, and eventually, through other interactions, in higher costs. However, in my work I have repeatedly encountered the belief that you must keep machines and people busy all the time to get jobs out faster. I'll present two of my experiences here to drive the point home.

Tale of Two Equipment Manufacturers

During 1995 I visited two equipment manufacturers. One was a world-class manufacturer of large machines, the other a small manufacturer of industrial equipment. Both companies were losing sales to competitors with shorter lead times, and both had been trying for some time to cut their lead times. The lead time reduction programs at both companies had not been effective, and so they had asked me to advise them on what to do next. At both the companies, I was told the same story: "The President believes in 100 percent utilization of our resources—he never likes to invest in more machines or people, until it is clear that all our resources are being used well beyond capacity."

At one of these companies, the manufacturing manager asked me several times during my visit: "Now explain to me again why 100 percent utilization is bad for us."

What does this tell us about how well we are doing in developing intuition about manufacturing dynamics? As you can see from these responses, from manufacturing managers to presidents of companies, basic intuition about manufacturing dynamics is still missing. The only way you can understand why 100 percent utilization is bad is if you have a clear understanding of the basics of manufacturing dynamics.

FACTORS INFLUENCING LEAD TIME

Measures of utilization alone are not sufficient to predict the lead time performance of a manufacturing system. Consider a flexible machining center (FMC) that machines a variety of parts. Material arrives at this FMC from initial operations at the rate of one part every 10 hours. We will observe this FMC for 30 hours of operation, and our observation period starts with the first arrival of material. I will now construct four possible scenarios for this period of observation.

Scenario One: Resulting Lead Time of Eight Hours

Suppose the FMC has a machining time of exactly eight hours for each part. It is clear that the FMC will have a utilization of 80 percent, there will be no queue time for any parts, and each part will have a lead time of exactly eight hours at this operation (see Figure 7-1). (Sometimes "lead time" is used to signify the planning time used for releasing products to the shop floor, but here, by *lead time* I mean the actual time from arrival of raw material to departure of finished part. Queue time denotes the time spent from arrival of material to the time it begins its turn on the machine.)

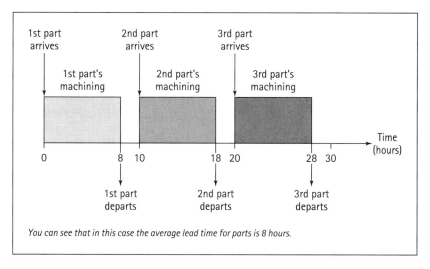

Figure 7-1. Performance When Machining Time Is Exactly Eight Hours

Scenario Two: Resulting Lead Time of Nine Hours

Now consider the situation where the average machining time (over three parts) is still 8 hours but actual times vary from part to part. As you see for the specific machining times illustrated in Figure 7-2, the utilization is still 80 percent, but now the average queue time is one hour and the average lead time is 9 hours.[1] This illustrates my first point, that the same utilization at the same machine can result in different queue times and lead times.

Scenario Three: Resulting Lead Time of Eight and Half Hours

In this situation the machining time is fixed at 8 hours per part, but the delivery of parts is variable. The average time between deliveries is still 10 hours but actual delivery times vary. As we see for the specific delivery times illustrated in Figure 7-3, the utilization is again the same 80 percent, but now the average queue time is half an hour and the average lead time is 8.5 hours.[2] Once again, we've shown that utilization alone is not sufficient in determining lead times.

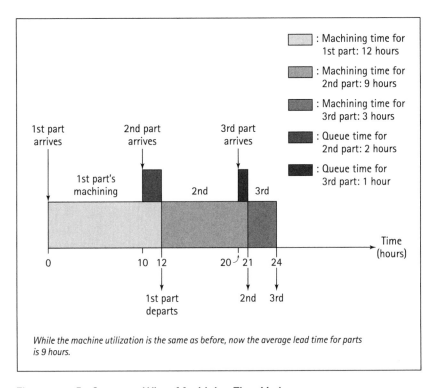

Figure 7-2. Performance When Machining Time Varies

These examples illustrate that at least two other variables affect queue time (and hence lead time), namely the variability in processing time and the variability in material arrival time.

Scenario Four: Resulting Lead Time of Eleven Hours

Finally we consider the interaction between utilization and variability. Consider the example in Figure 7-2, but now suppose that the machining times are an hour longer for each part, for an average of nine hours per part. The FMC utilization now increases to 90 percent. If there were no queue time, the average lead time for each part would be the same as the average machining time, or nine hours. However, Figure 7-4 shows the average queue time becoming two hours and the average

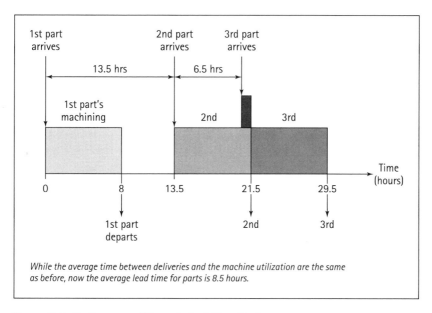

While the average time between deliveries and the machine utilization are the same as before, now the average lead time for parts is 8.5 hours.

Figure 7-3. Performance When Arrival Time Varies

lead time 11 hours.[3] Note that just a 10 percent increase in utilization has doubled the queue time. The lead time has deteriorated as well.

Why do I say this is an example of the interaction between utilization and variability? Suppose I made the machining times exactly nine hours each in Figure 7-1, what would happen to the *queue* time? There is no change. Without variability, you don't see any effect. In the real world, where there are many unpredictable events (tools break, material arrives late, a setup takes longer than expected), variability is always present. As I've shown in these examples, you need to consider the variability in both machining times and arrival times. I will now make the above illustrations concrete via some quantitative relationships. If you are not a math whiz don't worry. These relationships don't require much sophistication in mathematics.

A BASIC FORMULA FOR LEAD TIME FOR THE SINGLE WORK CENTER

Since variability affects lead time performance in the single work center example, we need a way of quantifying variability. A good measure is

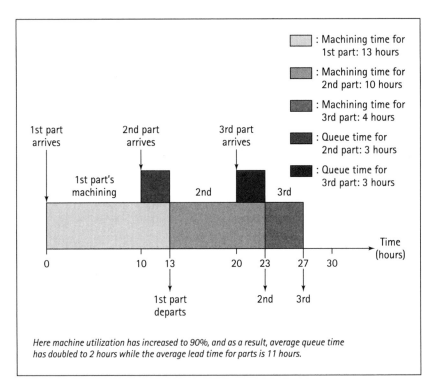

Here machine utilization has increased to 90%, and as a result, average queue time has doubled to 2 hours while the average lead time for parts is 11 hours.

Figure 7–4. Performance When Utilization Increases

already available from statistics. If you look at a sample of machining times for a number of parts, the *mean* quantifies the average machining time for a part, while the *standard deviation* is a way of quantifying the spread of machining times.

Using these terms, I'll define some concepts more precisely for our work center and generalize our previous example somewhat by talking about jobs that are entire *lots* of parts and focusing on the performance of each job. These input parameters are defined as:

TJ = mean time to process a job (including setup time and process time for all pieces in the lot)

SJ = standard deviation of time to process a job

TA = mean time between arrivals of jobs to the work center

SA = standard deviation of time between arrivals of jobs to the work center

These data are sufficient to predict the performance of the work center with reasonable accuracy. I'll focus on five main performance measures:

U   = utilization of the work center (fraction of time it is working on a job)

Q   = average number of jobs in queue

WIP = average work-in-process in the system (number of jobs in queue plus any on the machine)

QT   = average queue time for a job (time from arrival of job to when it begins being processed at the work center)

LT   = average lead time for a job (time from arrival of job to its completion)

The first performance measure, utilization, is simply the ratio of the average time it takes to do a job versus the average time between arrivals (see Figure 7-1 for a simple instance):

$$U = \frac{TJ}{TA}$$

It is important to note here the difference between my definition of utilization and one often used in manufacturing. Managers frequently use the word utilization to denote what you might call *productive utilization*, namely the amount of time a work center is making product. This excludes times such as setup time or down time (when the machine is broken or being repaired). My definition, more commonly used in discussion of manufacturing system dynamics, denotes *all the times the work center is occupied for some task*, be it setup, processing, or even repair. The reason is, during these times the work center is not available to do another job. This explains why in TJ (and hence in U) I include setup time as well.

Let's assume that U is less than 1 (i.e., less than 100 percent). This is because if it is greater than 1, then jobs are arriving faster than you can process them and you will build up an ever increasing backlog, never to be worked off.[4] Let's also assume that we are observing the work center for a long enough period so that there is representative data for the inputs and representative behavior for the performance. While this notion can be made mathematically more precise, I'll not try to be more technical about this issue and appeal to the intuitive nature of the statement in the preceding sentence, which is sufficient for my intent in this chapter.

The remaining performance measures are given by a branch of applied statistics called *queuing theory*. Their derivation is advanced, but you will still find the formulas to have simple intuitive explanations. First I'll define some intermediate terms which will make it easier to interpret the formulas:

VRA = variability ratio for arrivals = SA/TA
VRJ = variability ratio for job times = SJ/TJ
V   = total variability = $VRA^2 + VRJ^2$
M   = magnifying effect of utilization = $U/(1 - U)$

The first two ratios above are simply a way of measuring how large the standard deviation is, in comparison with the mean. Suppose I asked you, "Is a standard deviation of 1 hour acceptable for this job?" Your answer would depend on how long it took on average to do the job—if it normally takes one minute, then this standard deviation is unacceptable, while if it normally takes 100 hours, then it seems quite acceptable. Thus it is not the absolute value, but the ratio above, that is a better measure of the degree of variability in a system.

With these terms we can state the first formula, which is for the average queue time:[5]

$$QT = (1/2) \times V \times M \times TJ$$

Since each job needs to queue and then be processed, the average lead time is the sum of the average queue time and the average processing time:

$$LT = QT + TJ$$

There we have it, a simple formula for lead time.

EFFECT OF UTILIZATION ON LEAD TIME FOR THE WORK CENTER

To illustrate use of this formula, consider the case where you are planning capacity for a work center by deciding how many jobs you will assign to that work center. The more jobs assigned, the higher its utilization; thus you can consider utilization as a proxy for number of jobs being assigned. What about the lead time performance of the work center? Suppose it takes one shift, or eight hours, to process a job, on average, with a standard deviation of two hours (giving a VRJ value of 0.25, which is quite good in manufacturing situations). Suppose also that jobs arrive based on customer orders that can be considered quite unpre-dictable. In this case applied statistics tells us that a good guess for VRA is 1. Figure 7-5 then shows the behavior of lead time as a function of utilization. At very low utilization, there is almost no queue time (as you would expect, since at low utilization, arriving jobs usually find the work center idle) and the lead time stays close to eight hours. However, as utilization approaches 100 percent, the lead time starts to grow rapidly. For example, at 90 percent utilization, the lead time is over 46 hours—almost six shifts for a one-shift job. This rapid growth of lead time with utilization occurs because of the behavior of the term M, which I aptly named *magnifying effect of utilization*.

You can also see from the formulas above that increasing variability in either arrival times or job times will cause the lead time to increase. The graph in Figure 7-6 illustrates this behavior. For our work center, at 70 percent utilization and low total variability, the lead time might be just 9 hours. However, with higher variability, the lead time could be 18 hours even though the utilization remains the same. It is not my inten-tion for managers to remember these formulas and compute the actual

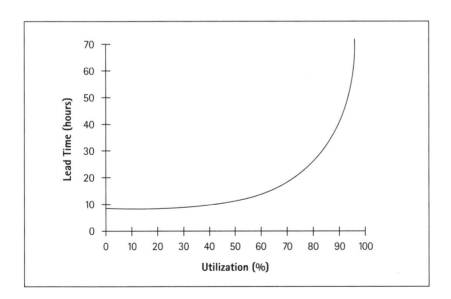

Figure 7–5. Effect of Utilization on Lead Time for the Work Center

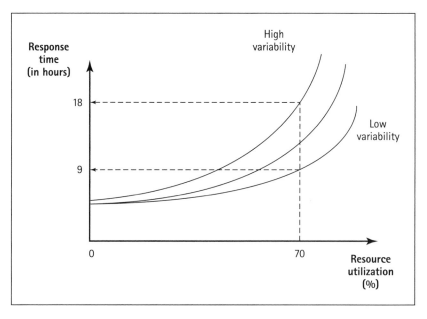

Figure 7-6. Combined Effect of Utilization and Variability on Lead Time

numbers. Rather, I would like the shape of the graph in Figure 7-6 to become part of your intuition, along with some more observations made below, based on this and related graphs.

The shapes of these graphs should not be that surprising if you stop to think about your everyday experiences. Visit a supermarket on a weekday morning (low utilization of checkout counters), and compare your waiting time at the checkout with that on a weekend afternoon (high utilization of checkout counters). In terms of variability effects, consider a McDonald's restaurant where one customer arrives every two minutes exactly. Chances are, no customer ever has to queue. Now let the same restaurant have to deal with a schoolbus full of kids on a field trip, and then no arrivals for a long time. You can make the data such that the average utilization of the servers is the same in both cases. Yet, in the second case, most of the kids will have to wait for service.

A final example shows the interaction between utilization and variability. Suppose a machine that has high utilization breaks down and takes a long time to be fixed. By the time it is repaired, a large backlog of work has already built up due to the high utilization expected of this machine, plus, even as this backlog is being worked off, new jobs continue to arrive at a high rate. Thus the machine develops a long queue that takes a very long time to dissipate, and lead times are very high for arriving jobs. So much is clear, but what does breaking down have to do with variability? In terms of our formulas, the down time is equivalent to a very long processing time for a job that is occupying the machine, and it results in a very high value of VRJ, the variability ratio for jobs. This means our formulas would also predict that a machine with high utilization and occasional long breakdowns will have very long lead times.

We can summarize the above discussion with the QRM principle that should replace the belief in quiz item number 2:

> **QRM Principle: Plan to operate at 80 percent or even 70 percent capacity on critical resources.**

Many managers react to this principle with the statement, "We can't afford to do that. We will be wasting our investment in expensive resources and our costs will go up!" From the preceding analysis, though, you see that the idle capacity ensures short lead times. As a result, the series of dysfunctional interactions resulting from the 100 percent utilization policy and present long lead times will go away,[6] and costs will go down over time.

Yet another way to justify maintaining some idle capacity is to see it as a *strategic investment* that will pay for itself many times over in increased sales, higher quality, and continued lowering of costs in the future via the productivity improvements described in Chapter 4.

IMPACT OF LOT SIZES

You can extend the simple formulas above to get an appreciation for the effect that lot sizing decisions have on lead times. Suppose now that for the work center for which we derived the lead times formula, the total quantity of products you are to make during a given time period is fixed, but you can choose the lot sizes with which to make the products.

The False Security of High Utilization

A Midwest electronics company with a functionally organized factory ran its critical resources at 90 percent utilization. But because of defects, rework, and scrap, the net yield (proportion of good product at the end of the line) was 70 percent of the total material that was being pumped through the machines. Despite repeated attempts at quality improvement, the traditional operating structure combined with long lead times and delayed feedback loops limited quality improvement to where 70 percent was the best yield this company could achieve.

A company with very similar products, and also located in the Midwest, implemented cells. In keeping with QRM principles, this company ran its cells at 80 percent of capacity. As a result of the improvements the cells were able to achieve a net yield of 90 percent (as described in Chapter 4). Let us compare the total output of the two companies:

Net output for traditional company
$$= 0.90 \times 0.70 = 63 \text{ percent of capacity}$$
Net output for QRM company $= 0.80 \times 0.90 = 72$ percent of capacity

In addition, the QRM company also had lower overhead costs, for all the reasons discussed in Chapters 2 and 3, and so, per unit cost of installed capacity, the QRM company had lower overall costs than the traditional company. These are the counter-arguments to the statement that "our costs will be higher."

This might model the following real-world situation. A component manufacturer serves a number of customers that have assembly plants. These plants need similar components all made at one work center, with some minor customized details that are different. Because of the detailed differences, each customer's components require their own unique setup of the machine. Over a year, you know roughly the total number of components required by each customer, but you are not sure when each customer order will arrive. Also, you can influence the lot sizes ordered by the customer through business negotiations. How does the decision on lot sizes affect the lead time of the work center?

Traditional thinking might suggest that the most efficient solution would be to set up the work center once for each customer and produce a year's worth of product. This would minimize the time and cost spent in setups during the year. However, you might end up carrying a lot of inventory over the year, so the next thought might be to optimize the trade-off between cost of setups versus cost of inventory. Further analysis based on this thought would lead to the well-known EOQ (economic order quantity) formula for setting lot sizes. I'll have more to say on EOQ later. Suffice for now to say that none of the thoughts in this paragraph has considered responsiveness at all.

Instead, let us see if we can apply our newfound knowledge about lead times to this problem.[7] Let:

D = demand during the period (total quantity of parts to be shipped)

H = number of hours worked during the time period

L = average lot size of each order (to be decided by management)

From these you can define some useful quantities relating to the time between arrivals of orders. If customers only ordered individual pieces (that is, always a lot size of one) there would be D orders during the total H hours, so the time between orders, on average, would be H/D hours. On the other hand, with a lot size of L for each order, the total quantity of D would result in D/L orders over the period. So the time between orders would now be, on average, $H \div (D/L)$ which you can write as $L \times H/D$. To summarize this, let:

TA1 = time between arrivals if orders were for one piece = H/D

TA   = time between arrivals when average lot size of L is used = $L \times$ TA1

Note that TA has the same meaning as before, when we gave the formula for lead time. Now consider the work center. The time to complete a whole job (order) is the time to set up for the order, plus the time to complete all the pieces in the order (which equals the lot size multiplied by the time per piece). Thus if

TSU = time to set up for an order

TJ1 = time to make one piece (after the setup is done)

then you can write

TJ = TSU + $L \times$ TJ1

where TJ also has the same meaning as before (time to complete a job).

Impact of Lot Sizes on Utilization

Since we have expressions for both TA and TJ, you can now determine the utilization of the work center for a given average lot size L. Earlier we showed that U = TJ/TA, so we have:

$$U = (TSU + L \times TJ1) / (L \times TA1)$$
$$= TSU/(L \times TA1) + TJ1/TA1$$

Let us observe here that if orders were always for one piece, and if there were no setup needed, then the time between arrivals would be TA1 and the time for a job would be TJ1. In this case, the utilization of the work center would be

$$U1 = TJ1/TA1$$

so you can write the previous equation for U as

$$U = U1 + TSU / (L \times TA1)$$

Remember that all these quantities, except L, are given for this situation. Only L is a decision variable; the rest are fixed. So let us graph the behavior of U as a function of L (see Figure 7-7).

Begin by looking at the right side of this graph. When L is very large, U approaches U1. This is to be expected, since with large L you perform very few setups and almost achieve the utilization of the system with no setups. Moving to the left, you see that as L gets smaller, U increases (you are doing more setups and using more capacity for them), until an interesting point is reached: At the value LMIN U equals 1, or 100 per-

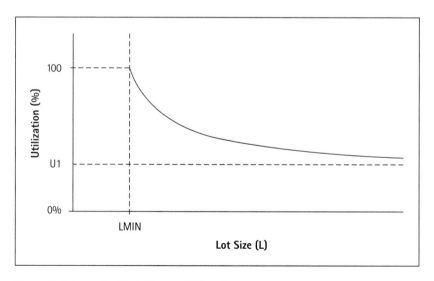

Figure 7-7. Impact of Lot Size on Utilization

cent. You cannot make L any smaller without a disaster for your business—an ever-growing backlog of orders. LMIN represents the smallest lot size that you can make without running out of capacity.

Impact of Lot Sizes on Lead Time

So far you have only calculated the utilization, not the lead time impact of lot sizes. To estimate the lead time, you also need variabilities. For the purpose of this example, and to keep formulas simple, I will assume that the variability ratios VRA and VRJ both equal 1. In this case, the total variability, V, equals 2 and the formula for queue time[8] conveniently reduces to

$$QT = M \times TJ$$

This means that the lead time of a job is

$$LT = M \times TJ + TJ = TJ \times (M + 1)$$

If we put in the value for M in terms of U, you get the simple formula

$$LT = TJ / (1 - U)$$

You now have everything in place to analyze the behavior of lead time as a function of the lot size decision (see Figure 7-8). First, consider what happens when L is very large, so that U approaches U1 as previously explained. Then the denominator above can be considered equal to $1 - U1$, essentially a constant, for all values of L sufficiently large. Since the numerator, TJ, equals $TSU + L \times TJ1$, for large values of L you see that the lead time, LT, increases almost linearly with L. This behavior is seen on the right side of Figure 7-8. Right away, this behavior shows that traditional notions of efficiency work counter to responsiveness—in order to increase efficiency a work center must have fewer setups and spend more time in production, that is, run large lot sizes. However, our analysis and the graph in Figure 7-8 show that the result of this is very long lead times.

Now let us gradually reduce L (i.e., start on the right edge and then move to the left in Figure 7-8. For a while, U is still close to U1 and the lead time drops almost linearly with L, for the reasons in the previous paragraph. Then U starts increasing significantly (refer also to Figure 7-7, to enhance your understanding of this argument), so the denominator in the formula for LT starts to get smaller. This magnifies the value

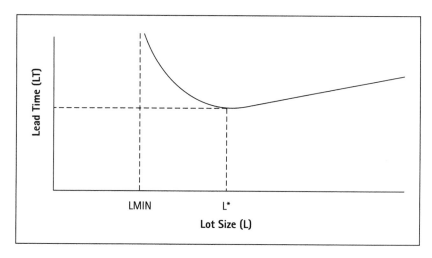

Figure 7-8. Impact of Lot Size on Lead Time

of TJ and lead time doesn't drop quite as fast. Eventually, U approaches 1 and as it does so, the denominator (1 – U) becomes very small. Dividing TJ by this small quantity produces a very large lead time. So lead time starts to increase steeply as L gets even smaller. Finally, as L approaches the value LMIN described in Figure 7-7, it becomes impossible to service all the customer orders and the lead time grows without limit. This discussion is summarized by the U-shaped curve in Figure 7-8.

Somewhere between the increasing lead time on the right and the increasing lead time on the left is the point at which lead time reaches a minimum. The lot size that achieves this minimum lead time is marked as L* in Figure 7-8. If responsiveness is the goal of the company then its lot size policy should be to operate at a lot size of L* on average. Note that this value of L* bears no relation whatsoever to the "optimal" lot size determined by the EOQ formula.

You might ask, if we know the demand over the year, why not minimize the cost of setup and inventory by using the EOQ formula? And anyway, who cares about responsiveness, since we know the forecast demand for the period and can always have the product ready ahead of time so it can be shipped? Unfortunately, this argument ignores all the issues related to the Response Time Spiral—errors in forecast, changes in

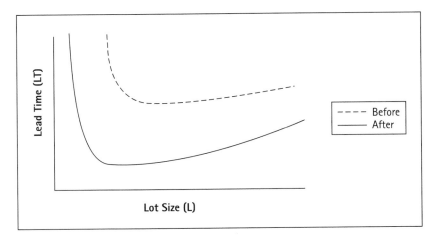

Figure 7-9. Impact of 50 Percent Setup Reduction on Lead Time

what the customer needs, hot jobs, expediting, and so on, along with all the extra costs associated with this behavior, which are not quantified in the EOQ formula. These and other issues related to EOQ, will be discussed after I describe a few more details of lead time behavior.

IMPACT OF SETUP REDUCTION ON LEAD TIME

Suppose you implement a setup reduction program at the work center in the previous example, resulting in a 50 percent cut in setup time. How will this affect its lead time? You are now, finally, in a position to answer the question posed at the beginning of this chapter. Using our formulas, you can cut the value of TSU in half and plot the new graph. Figure 7-9 shows the graphs for lead time before and after the setup reduction. (The "before" curve is the same as the one in Figure 7-8.)

You see that the impact of the setup reduction program depends on the company's lot sizing policy—it is minimal if large lot sizes are in use, and if they do not change these lot sizing policies after the reduction in setup time. On the other hand, it can be significant in some cases where utilization is very high prior to setup reduction. So the answer to the question posed earlier is, "it depends." Specifically, the amount of reduction that you will see in lead time depends on:

1. The utilization of the work center.
2. The lot sizes currently being used.
3. The variability of the order arrivals and job times.
4. Whether the lot sizes are changed after the setup reduction is in place.

You might say, "Whoa! That's not fair. You didn't give us all this information about the system to begin with." Guess what? It *is* fair. This is how questions occur in the real world. All the data isn't handed over on a plate, and so managers need to know the approach to answer such a question. If any of the 150 managers at the seminar had known the right approach to answering this question, they would have either said, "It depends on the utilization and other variables," or, "You need to tell us more about the existing lot sizes and utilization for us to answer this." But no one made this point, and the few that tried to answer the question guessed some absolute numbers.

There is another significant point to make here. Setup reduction alone is not the best strategy for lead time reduction. Usually it takes a combination of setup reduction *and lot size reduction* to get the maximum reduction in lead time. Figure 7-10 shows a company operating at a lot size of LB (for "L before"), resulting in a lead time of LTB ("lead

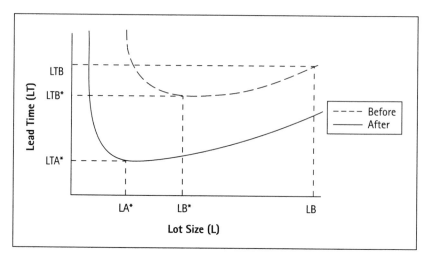

Figure 7-10. Combined Impact of Setup Reduction and Lot Size Reduction

time before"). Even if it optimizes its lot size for minimum lead times, choosing the value LB*, it will only improve its lead time slightly, to LTB*. However, after implementing a setup reduction program, if it also revises its lot size to LA* ("A" for "after"), it can achieve a significant drop in lead time to LTA*.

In dealing with factories in differing segments of industry I've found companies almost always using much larger lot sizes than necessary, perhaps as the legacy of an EOQ computation the company did many years ago. This means there is usually an opportunity to reduce lead time by just reducing lot sizes. However, they can realize an even greater opportunity by combining a setup reduction program with lot size reduction—it is not unusual to see lead times drop by 80 percent in this case.[9]

Unfortunately, most managers only see setup reduction as a way of increasing capacity, and fail to target this combined strategy. For instance, in a study of fourteen lead time reduction projects, we noted:

> Management and operations staff at four of the companies who either identified setup reduction as one of their main objectives for participating in the study, or who found out during the initial phase of the study that setup reduction was a key variable for significantly reducing lead times, did not have a good understanding of the important relationship between setup and lot size reduction and reducing lead times. The most common belief among the above group was that they were wasting too much valuable production time in setup, and by reducing setup times they would create additional manufacturing capacity which would automatically reduce their lead times. *They did not, however, understand that the more powerful way of reducing lead time is not just to reduce utilization via setup reduction, but to simultaneously reduce lot sizes.* They did not anticipate that the end result is often more frequent but shorter setups with smaller lot sizes. *Nor did they expect that, as a result, the total time spent in setup might well be almost the same before and after successfully reducing setup times.*[10]

LOT SIZING WITH MULTIPLE PRODUCTS

In the previous section, although there was a work center making multiple products, our analysis approach aggregated these products into one representative "average" product and we considered the effect of the average lot size on the lead time performance. It is useful to refine the analysis in order to look more closely at distinct lot size policies for different products. This will bring out additional problems with traditional approaches to lot sizing, and also give you insight as to what policies to implement.

To keep the graphs simple, I'll consider only two distinct products. As before, these products will compete for one work center, and also, I'll assume that the total demand for each product is fixed. How do lot sizing policies affect lead times in this case? I will not develop the equations this time, but simply present the results. Note that you now have two management decisions to make, namely, values of L1 (lot size of product 1) and L2 (lot size of product 2). Let us fix L2 for now, then Figure 7-11 shows the impact of varying L1 on the lead time for both products. As expected, the graph of lead time for product 1, as a function of its own lot size, maintains the U-shape for the single product case. However, the value of L1 now has a significant impact on the lead time for product 2 as well. Why? Simply, when the value of L1 is very

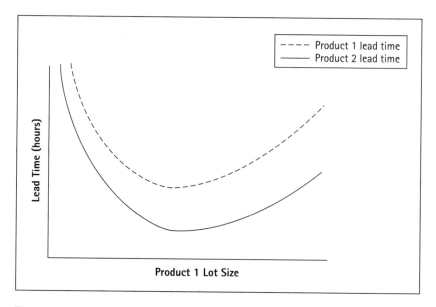

Figure 7-11. Impact of Product 1 Lot Size on Lead Time for Both Products

large, an order for product 2 often gets stuck behind a large run of product 1. On the other hand, when the value of L1 is very small, the utilization of the work center is very high. Both situations result in long lead times. Somewhere between these extremes lies the best situation when product 2 experiences the shortest lead time. Thus product 2 also has a U-shaped lead time graph with respect to the first product's lot size.

There is actually another, more subtle effect taking place as well. If the lot sizes of the two products are significantly different, the variability in the job times becomes significant as well. For example, if 80 percent of the orders are for product 1, which typically takes 2 hours (for setup and run), and 20 percent are for product 2, which takes 17 hours, then the value of VRJ for this work center becomes 1.2. Since this value is squared in the lead time formula, to become 1.44, its impact is even greater. Thus a rule of thumb for multiple product lot sizing is as follows.

....

QRM Nugget: To reduce variability at a work center, a good rule of thumb for multiple products is to keep their lot sizes such that total values of setup plus run (TSU + L × TJ1 for each product) are close to each other.

....

Now let us make an observation that applies to both traditional efficiency measures and EOQ models. *Neither of these approaches accounts for the impact of one product's lot size on the delivery performance of other products.* For instance, an EOQ calculation for product 1 might result in a value of L1 on the far right of Figure 7-11. This has a negative impact on the responsiveness of the company with regard to product 2 orders, with ripple effects on costs of expediting, reduced customer satisfaction, lower market share, and so on. Nowhere are these effects or costs figured into the EOQ calculation, which decides on the L1 value without ever considering any other products at all.

So how does one decide what lot sizes to run in a factory with multiple products? In simple situations, a manager can ask an analyst to create some "back of the envelope" calculations similar to those above. However, in larger factories with dozens of work centers and hundreds of products, this can become a daunting task. Even after creating cells you might find that analysis of a single cell requires modeling 12 machines and 20 products. Added complexity arises if you want to model the impact of machine failures, labor-machine interactions, rework or scrap, and other practical issues. In such cases, managers can turn to commercial software packages based on more complex implementations of the queuing theory techniques used here. Since these packages provide very fast answers (just seconds on a PC) relative to other capacity planning systems, the method is also called Rapid Modeling Technology or RMT.[11] The RMT approach is much simpler and faster than other full-blown simulation methods, and it has been used by companies for both targeting setup and lot size reductions, as well as determining good operating policies for capacity and lot sizes.[12]

THE HIDDEN ERRORS IN EOQ

Since the EOQ formula is still widely used (or else companies have lot sizes that were set by this formula a long time ago and they are reluctant to tinker with those quantities), it is important to underscore the ways in which use of EOQ is incompatible with a QRM strategy. The EOQ formula trades off setup cost with WIP holding cost, but it does so for a single product, without estimating the dynamics and interactions of the shop floor (e.g., how queue times change with lot sizes, or how large lot sizes of one product affect the queue time of other products). As a result, it fails to consider several effects and costs when it calculates the order quantity for a part. These include:

- *Costs of poor quality*: In spite of the best inspection procedures, defects may only be detected when a downstream fabrication or assembly operation is attempted. When you make parts in big lots, this can result in a large quantity of parts being scrapped or reworked, with a correspondingly high cost. Worse is when the consumer discovers the defect, in which case warranty costs are incurred. None of these costs are incorporated in the EOQ calculation.
- *Costs of obsolescence or engineering changes*: Similarly, when you make parts to stock in large quantities, whole batches of parts can become obsolete by a design change. You will need to scrap or rework all the parts. When the large batch size is set by the EOQ calculation it fails to take into account the cost of this possibility.

- *Costs of long lead times*: As explained, large batches result in long lead times. The EOQ formula does not take into account that it can lead to a factory full of large lots that jam up all the traffic, leading to long delivery times, late deliveries, and unhappy customers. The company incurs costs of expediting, high WIP, or even costs of lost sales, all of which are not in the EOQ calculation.
- *Market value of responsiveness*: Ingersoll Cutting Tool Company cut one month out of its lead times for a line of products, and found its sales more than doubled.[13] The EOQ formula fails to factor in the profit that may result from short lead times.
- *Costs of a growing Response Time Spiral*: Large lot sizes result in long lead times, which in turn exacerbate the Response Time Spiral. As described in Chapter 3, the Response Time Spiral has a host of costs associated with it. None of these are factored into the EOQ formula.

In summary, there is *nothing* that can be salvaged from the EOQ formula for use with QRM. In the short term, analysis tools such as RMT can help set good lot sizes. In the long term, the only approach consistent with QRM is to constantly ask: "How can we reduce the lot size for this product even further?" and then to make the improvements necessary for this reduction. Again, tools based on RMT can assist in zeroing in on the key improvements.[14]

PRODUCTS REQUIRING MULTIPLE OPERATIONS

The analysis above focused on one work center. In reality, most products require a series of operations. In implementing QRM you will typically have assigned all these operations to a cell containing the needed resources. To ensure that the cell has short lead times, you also need to understand the impact of some additional variables. These are discussed here.

Transfer Batching or Overlapping Operations

Transfer batching is a strategy that you can effectively use once cells are in place. Under the right circumstances, it can have a dramatic effect on lead time. This technique is also known as *overlapping operations*. The idea is simple: Instead of waiting for all the pieces in a batch to be completed at the first operation, why not move the pieces from the first machine to the next one as soon as they are done? This way the next machine can begin operations even while the batch is being worked on at the first machine; hence the name "overlapping operations." You can use the same argument to move the pieces from the second machine to the third, and so on. It is a good idea, but not necessary, to transfer the pieces one by one—you might move them in cassettes that hold 10, say, while the lot size for the order might be 100. In general, the batch size moved from machine to machine is referred to as a *transfer batch*, and thus this strategy is also called *transfer batching*. A simple example shows the potential of this strategy (see Figure 7-12).

Cell with Four Machining Operations

Consider a cell with four machining operations, and for a given order suppose each of these has a 30-minute setup and a 10-minute cycle time. For simplicity let's first assume that when this order gets to a machine, that machine is available; that is, it is not busy with another job. If the lot size of the order is 18 pieces, and the whole lot is completed at each machine before the pieces move to the next machine, as in Figure 7-12(a), then it will take 14 hours for the order to go through the cell.[15]

In contrast, a transfer batch size of one gives a total time of five hours and 30 minutes, less than half the earlier time.[16] You can achieve an even greater reduction if the whole cell is set up at the beginning; then the first piece does not wait for setups on the second and subsequent machines. In this case the total time is only four hours. However, you can only use this last strategy if the whole cell works on only one order at a time. The way these last two strategies work is summarized in the timelines in Figures 7-12(b) and (c).

Although transfer batching seems highly effective from the above examples, there are some prerequisites for its successful use. One is using reasonably balanced times through the cell. For instance, if the second operation above had a one-hour cycle time, then it would starve the downstream operations—those machines would be set up and then in each hour they would spend 10 minutes working on a part and 50 minutes waiting for the next part. It is not necessary to finely "balance"

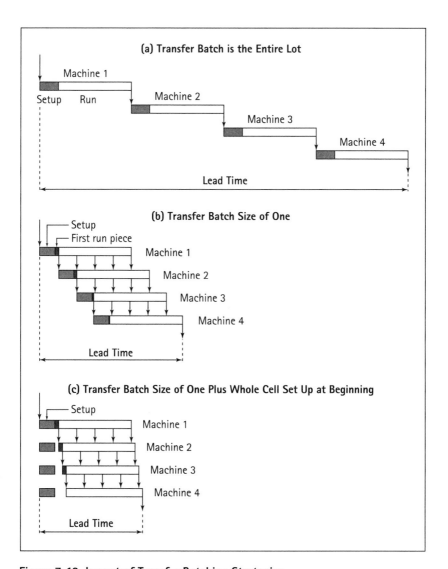

Figure 7-12. Impact of Transfer Batching Strategies

the operation times in a cell as one would in an assembly line, since for QRM the cell will be making a variety of different products in varying quantities. However, the point is to ensure that operation times are not widely different either, or transfer batching will not be very effective.

Another precursor is to make sure that the cell is already running small lot sizes overall. If it is not, then you should reduce setup times to enable the small lot sizes to be run—only then should you use transfer batching. The following example illustrates how the overall lot size can have a far greater impact than the transfer batch.

Cell with Four Products and Eight Operations

With multiple products requiring different setup and cycle times, the calculation of lead times for four products and eight operations becomes complex, so I'll show the results obtained by use of the MPX modeling software. The situation model is based on an actual cell in an aerospace company, whose task is to machine hubs for engines.[17] Figure 7-13 shows the minimal impact of transfer batching when the lot sizes are large (ranging from 30 to 44 for the four products). Next you see that setup reduction enabling lot size reduction can cut the lead time much more than even a transfer batch of one did in the first case. The lot sizes range from 8 to 16 for this case. Lastly, after these smaller lot sizes are in place, we see that transfer batching has a substantial impact, and the lead times are cut by more than 80 percent from their original value.

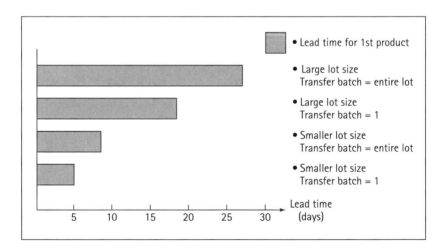

Figure 7-13. Effective Transfer Batching

You may well ask, at this point, "If transfer batching can be so effective, then why haven't we already used it in our operations?" The answer is, in the traditional functional organization, where the next operation might be halfway across the factory, it didn't make sense to move pieces one at a time. It was a better use of material handling resources, such as forklift trucks, to wait and transfer the whole lot of parts together. A second reason is, in the functional operation, where orders queued for days or even weeks in front of each work center, transfer batching would simply result in a "Hurry up and wait!" situation. (Parts would be moved quickly away from the first machine, only to queue up for days at the second one.) Thus transfer batching only makes sense in a cell where (1) machines are close together so pieces can be moved quickly from one machine to the next, as needed, and (2) machine utilization and resource schedules are controlled so that there are no significant queues building up in front of the machines.

Ripple Effects of Variability

Another effect to be considered when multiple operations are performed is the propagation of variability through the system. As you saw in the earlier analysis, variability of job arrivals has an impact on lead time. As far as downstream operations in a cell are concerned, it is the outputs of the upstream operations that produce the arrivals to these downstream operations. So if these outputs have high variability, the lead time of the downstream operations will suffer. But how can you decide if these stations will have acceptable variability of their outputs? Just as queuing theory gives us formulas for the *utilization* and *lead time* performance of a work center, it also gives us a formula for the *output variability* of this work center. This output variability is measured by VRD, the variability ratio for departures. The meaning of VRD is similar to VRA defined earlier; that is, it is the standard deviation of the time between departures, measured relative to the average time between departures. In fact, since the output of one work center becomes the input for other work centers, VRD of that work center becomes the VRA for other work centers. Thus you want your work center operation to be such that VRD is small; otherwise you will have a high propagation of variability through the system.

I will not elaborate on the formula for VRD here,[18] but will give you the key insight from the analysis. In order to keep a given work center's output variability low, that is, to minimize propagation of variability and hence reduce downstream lead times, you want to keep both VRJ and VRA small at that work center. As discussed earlier with the McDonald's example, keeping VRA small means providing work at a uniform rate to the work center. Also, minimizing VRJ means keeping job times similar to each other, and it also requires reducing the occurrence of long down times.

Example of Using Simple Formulas for Analyzing Manufacturing Dynamics

Even these simple formulas can provide a lot of insight into the causes of production problems. Without a well-developed intuition about such manufacturing dynamics, managers often go after the wrong solution, in many cases unnecessarily spending a large amount of money on buying more machines. A good example is provided by Hopp and Spearman for a factory producing circuit boards.[19] One area in this factory consisted of two stations, Resist Apply and Expose. The first station applied photo-resist material to circuit boards, and the second exposed the boards to ultraviolet light to produce a pattern that would be etched onto the boards at a downstream operation. The desired throughput of these two stations was 2.4 jobs per hour. However, this area was frequently unable to meet this target throughput. Analysis of the operation times showed that Expose had the longer average operation time, and the engineers responsible for the area were in favor of installing a second expose machine, at considerable expense.

Instead, analysis using the queuing formulas in this chapter lead to discovery of a better solution. The analysis showed that the desired production rate of 2.4 jobs per hour would result in a queue of 36 jobs in front of the Expose station. However, the clean room area where the Resist Apply and Expose stations were located could hold only 20 jobs. So when this limit was reached, the Resist Apply station would stop production. The factory engineers did not perceive this as a problem because this station was not the bottleneck. Further analysis of the queuing formula for the queue in front of the Expose station showed that a large contribution resulted from the VRA term (variability of arrivals) for this station. This value, in turn, came from the VRD value of the Resist Apply station. Why was the variability of departures so high at Resist Apply? Because it had very long down times—once every 48 hours, on average—it was down for 8 hours. During these failures, the Expose station would run out of jobs (or be starved), and the result would be lost production capacity at the bottleneck.

Hopp and Spearman showed that one solution to this situation was to avoid the failures via a preventive maintenance program at the Resist Apply station, which would require a 5-minute shutdown every 30 minutes. Even though this would mean lost capacity of the same amount as the 8-hour failures every 48 hours—16.7 percent in both cases—it would also mean a dramatic reduction in the VRD value. As a result, the queue at the Expose station would drop to around 8 jobs, well within the space limit. Now the Expose station would never be starved, and it could achieve the 2.4 jobs per hour rate.

The application of these formulas showed the factory that a preventive maintenance program requiring little investment (mostly training) could be used instead of an expensive capital purchase to solve a production problem. Of course, all managers know that reducing down time on a bottleneck is critical, but what is nonintuitive about this example is that the preventive maintenance program had to be applied to the nonbottleneck station. This illustrates the power of these simple formulas for analyzing manufacturing dynamics. This analysis also shows clearly how QRM differs from the theory of constraints (TOC). The simplistic analysis of bottlenecks in TOC would have recommended adding a second Expose machine, an unnecessary capital expense.

USING LITTLE'S LAW

Little's law is a fundamental law of system dynamics and has two important uses in manufacturing management. First, it ensures that managers set consistent targets for their staff, and second, it enables managers to check that they are getting consistent data on performance of their factory. Little's Law is rather simple in appearance. It states that, for any given area of a factory,[20]

WIP = Production Rate $\times$ Lead Time

This relationship is easy to explain intuitively. Suppose a production area has a three-week lead time, and the area completes 20 jobs per week (the production rate). What can you expect in terms of WIP in this area? The answer, from Little's Law, is:

WIP = 20 × 3 = 60 jobs

To understand the answer, consider that each week you introduce 20 jobs into the area, and jobs stay in the area for three weeks. Thus at any time you have three week's worth of jobs in the area, or 3 × 20 jobs of WIP. The law is no more complicated than this!

First Use of Little's Law—Setting Consistent Targets

This simple law can often be of great value. One situation is in setting targets. In many cases I have been asked to help a production manager who is having difficulty achieving the production, WIP, and lead time targets set by management, and have found that these targets are inconsistent—in other words, they are physically impossible, but no one has pointed this out. For instance, if in the above example, the area manager was told to achieve a target of a three-week lead time while producing 20 jobs per week and maintaining no more than 40 jobs of WIP, this manager would never be able to satisfy her boss. If her processes demanded at least three weeks of lead time and she cut her WIP down to the stated target of 40 jobs, she would achieve only 40/3 = 13.3 jobs per week of production. On the other hand, if she put enough jobs in the system to achieve 20 jobs per week, she would find her WIP was 60 jobs. In either case, her boss would be unhappy. This example may seem trivial, but in more complex manufacturing systems with many products and many work centers, it is not so readily apparent that targets set are physically inconsistent. In such cases, careful application of Little's Law can be a good check.

The above discussion also illustrates the importance of using consistent units when applying Little's Law. We measured WIP in terms of jobs. However, if you are going to measure WIP in terms of pieces, then you must measure the production rate in pieces. If you state the lead time in weeks, then the production rate should be per week, and so on. Another caveat in applying this law is that, in the form stated above, it does not concern itself with raw material or other stocked items that are in the system.

A simple rule to use for correct application of Little's Law is as follows. Draw a box around the area where you want to apply the law. Then Production Rate in the formula refers to the total flow through this box, measured for entities that you choose in units of your choice, and WIP and Lead Time refer to the corresponding quantities for those entities in those units. In particular, note that you do not have to include all the entities flowing through this box, only the ones of interest to you. This can be particularly useful, as the following example demonstrates.

Second Use of Little's Law—Performance Reports

I'll use this "box" approach in an example showing a second use of the law, namely, for checking the validity of performance reports. This example is fashioned after a situation where I was asked to assist with a lead time reduction effort. A factory made many types of ovens, over 20 sizes of standard ovens, plus dozens of unique custom ovens. The sales department maintained that in the last few months deliveries of custom ovens had been quite late and many customers were complaining. The production department's response was that the deliveries were late because the sales force was promising delivery in less than the standard lead time. Indeed, a report from the production department showed that the average lead time for custom ovens had been 5 months, which was less than the standard lead time of 6 months. Who was correct?

Additional investigation revealed that over the past year, 36 custom ovens had been shipped, an average of 3 per month. Review of inventory reports for the past 12 months also showed that the average number of custom ovens in process on the shop floor was 24. Both these numbers were considered indisputable because they were based on physical shipments and physical counting of jobs in process at the end of each month. To analyze this situation let us rewrite Little's Law as:

Lead Time = WIP/Production Rate

Application of this formula to the above numbers gives the average lead time for custom ovens as 24/3 = 8 months. In other words, the sales department was correct in this instance; the ovens were taking longer than the quoted six months to be produced. In fact, as it turned out, the production department was not at fault either–they measured their lead time after they had received all major components from suppliers, at which time they could commence final fabrication and assembly. However, key component suppliers were often late by several weeks. Application of Little's Law thus served two purposes. One, it showed that the management reports were not providing accurate information on lead time and sources of delays, and two, it helped to pinpoint the key problem for this company, which was late deliveries from suppliers.

An important property used in the above application of Little's Law is that even though the company made many types of ovens, we were able to apply the law just to the custom ovens. This was possible because all the data being used in the formula were applied to one set of entities in a consistent way. Again, this is a powerful approach because it allows us to gain insight into complex systems without being buried in reams of data.

ADDITIONAL STRATEGIES BASED ON SYSTEM DYNAMICS

Three additional strategies based on principles of system dynamics can be useful in reducing lead time. (These strategies will be described in Chapter 13 where I bring together the main principles of system dynamics that can help to reduce lead time.) These strategies are:

1. Resource pooling
2. Capacity management and input control
3. Creating a flexible organization

You can use these strategies, along with the other principles in this chapter, both on the shop floor and in office operations. For maximum effectiveness, you need to support these strategies with material planning and control methods described in the next chapter. You also need to support these strategies and principles with completely new ways of measuring performance.

Main QRM Lessons

- Industry examples show that managers may not have well-developed intuition about the impact of key decisions on lead time.

- Developing such intuition requires understanding the basics of manufacturing system dynamics.

- Measures of utilization alone are not sufficient to predict lead time performance.

- How to quantify the impact of capacity utilization, lot sizing, variability, and setup reduction on lead time.

- Understanding how the lot size of one product affects the lead time of other products.

- Hidden errors in the economic order quantity (EOQ) formula.

- How transfer batching works in reducing lead time, and how to maximize its effectiveness.

- The ripple effects of variability and how to minimize them.

- Little's Law and how it helps in target setting.

8

Material and Production Planning
in the QRM Enterprise

THE TRADITIONAL ORGANIZATION uses measures of efficiency, utilization, and on-time delivery to track the joint effectiveness of the materials and manufacturing departments. For, in such an organization, the manufacturing managers are measured on how well they can keep machines and workers busy, and how well that work stacks up on an efficiency scale. The materials department, in turn, must deliver the materials to the manufacturing resources in proper time to keep them busy. They must also deliver the materials in the right quantity and develop the production planning sequence so as to enable efficient utilization of resources. For example, combining component requirements for different orders enables all the components to be made with one setup, and scheduling similar products back-to-back helps to reduce the changeover time. This chapter and the next two chapters will show how you need to rethink the measures and resulting material planning procedures when creating the QRM enterprise.

In this chapter I will focus on the *internal* supply of materials, including material planning and production planning. Much of the discussion will focus on the pitfalls of MRP and the need to develop what I call a *higher level MRP* (HL/MRP). In Chapter 9 I will discuss—surprisingly to some readers—the drawbacks of JIT (or flow) and kanban systems and why you need to modify and enhance these systems. I will then outline a new material control and replenishment system to support your QRM enterprise. Finally, in Chapter 10 I will discuss *external* supply of materials, not only back to your suppliers, but also forward to your customers. But before I discuss details of the revised materials planning and control systems, you need to understand the drawbacks of traditional systems in the context of implementing QRM. Only then can you fully appreciate the operation of the systems that must replace them. You may also find that an in-depth under-standing of these points will help you in convincing the materials managers of the necessary changes that need to be made if you are to succeed with QRM.

RETHINKING EFFICIENCY

In developing the QRM principles so far I've covered reorganizing the shopfloor production for QRM as well as discussed the dynamics of manufacturing systems, including the critical issues of capacity and lot sizing. Now it's time to see how you must use your material supply to support this new QRM organization. I'll begin by revisiting the third item in the QRM Quiz:

> 3. In order to reduce our lead times, we have to improve our efficiencies.
> ❏ True   ❏ False

As you know from previous discussions, faith in this principle can be the downfall of a QRM program. But what could possibly be wrong with trying to be efficient? The problem lies not in the *concept* of efficiency, but rather in the fact that most *measures* of efficiency work *counter* to lead time reduction. A typical measure of efficiency looks like this:

$$\text{Department Efficiency} = \frac{\text{Total Standard Hours of Production}}{\text{Total Labor Hours Worked}}$$

In this measure, first you add up the "standard hours" for all the operations completed by a given department or work center. These hours would be based on a standard for each operation, which is usu-ally established by the company's manufacturing or industrial engi-neering departments. You then divide this by the total hours worked by people in the department. For example, let us say a turning depart-ment has two lathes and two machinists who work 35 hours each dur-ing a given week. In the same week, they turn 80 small shafts with a standard operation time of 0.25 hours per shaft, 20 medium shafts with a standard of 0.6 hours, and 10 large shafts with a standard of 1.0 hours. The total of all standard hours for the production in this week is then:

Standard hours produced = 80 $\times$ 0.25 + 20 $\times$ 0.6 + 10 $\times$ 1.0 = 42 hours

The total of hours worked is:

Total hours worked = 2 $\times$ 35 = 70 hours

This means the efficiency of this department would be 42/70 or 60 percent. What were the workers doing during the remaining 28 hours that week? It's not usually the case that they were slacking off—there are several other possibilities:

- Time might be spent in setups or quality meetings.
- A particular job may have been difficult and taken longer per piece than the standard.
- Some material or tooling may not have been available at the right time and the worker may have had to wait in the middle of a job; and so on.

Drawbacks of Efficiency Measures

Although individual companies may have their own versions of efficiency measures,[1] all such measures suffer from the same drawbacks when it comes to implementing QRM; that is, they encourage behavior that is counter to lead time reduction. Here are some specific examples of this.

- *For a typical measure of output for a work center, only good pieces produced contribute to the measure.* The incentive for the supervisors and machine operators is to run large lot sizes, minimizing on setup time and maximizing the number of pieces produced in a given period. In fact, when an operator has achieved a "good setup," so that the machine is producing pieces faster than the standard time, there is every incentive to keep producing the same part since this will have a positive impact on the efficiency measure. However, as I've discussed, large lot sizes result in very long lead times.
- *There is no incentive to help other people in order to keep a job moving.* If an operator leaves his or her machine to assist another operator, it negatively affects the efficiency of the first operator,

since "the clock is still ticking" at the first machine, but no pieces are being produced. Thus the incentive is to "keep your blinders on" and focus on your own work.

- *If workers attend quality or continuous improvement meetings, they are not producing parts, and the efficiency of the department goes down.* Even though it may be in the long-term interest of the company to have people participate in such meetings, the incentive resulting from the performance measure is for them not to hold the meetings or else keep them as short as possible.

- *Efficiency measures go hand-in-hand with maximizing the utilization of machines and people.* We've given numerous examples of the disastrous effect this has on lead times: Chapter 3 showed how this leads to the Response Time Spiral; the anecdote about Metal Products Company in Chapter 5 illustrated how this can reverse any progress achieved with cells; and Chapter 7 gave us quantitative arguments against aiming for overly high utilization.

- *Efficiency measures do not take into account the impact of one product's lot size on the delivery performance for other products.* To maximize efficiency, a supervisor might look ahead three months into the schedule and pool all the requirements for a component, making them in one setup at a press. This might occupy the press for two full days and hold up an order for a different component, which in turn is holding up the completion of a job in final assembly, resulting in late delivery to a customer. This point was also illustrated quantitatively by the graphs in Chapter 7.

- *The focus on efficiency leads the organization to make parts that must later be "written off."* Extending the above example, let us say that in an attempt to run a large ("efficient") batch size on the press, the supervisor looks ahead six months into the schedule and runs the entire quantity for the six months. Next month's requirements in the schedule were based on firm orders, and the five months after that were based on forecasts. Next month there is a shift in the market and this product is no longer selling well. Or worse, suppose that customer complaints result in an engineering change to the component. In either case, the additional five months of parts sits in a warehouse for several years until someone realizes they are obsolete and decides to write them off or sell them as scrap. At this point, the company takes a loss on its balance sheet for these parts. The ignominy of this situation is that in the year these parts were made, management thought the company had done a super (read "efficient") job of making these parts. In fact, if the supervisor had done a diligent job of looking ahead for many products, maybe the press department would have been commended on how good its efficiencies had been that year.

Indeed, an empirical study of five Square D factories over 30 months confirms the inappropriateness of measuring labor efficiency as a proxy for overall factory productivity.[2] Two conclusions resulted from this study: (1) In many cases the month-to-month changes in labor efficiency were in the opposite direction to overall productivity. In other words, it was often the fact that when efficiency went up at a factory, its productivity actually went down. (2) Formal statistical analysis of the data showed little correspondence between the two measures, efficiency and productivity. For three of the factories, the correlation was very weak, in one factory there was no significant correlation, and for one factory there was a statistically significant *negative* correlation between efficiency and overall productivity.

These arguments may be compelling, you say, but if we don't measure efficiency, then what *should* we measure to make sure workers and supervisors are doing a good job? The answer is surprisingly simple, and based on impeccable logic: *If you want to reduce lead time, then measure and reward lead time reduction!* The QRM approach, then, is as follows.

QRM Principle: Measure and reward reduction in total lead time and make this the main performance measure. Scrap all other traditional measures of efficiency and utilization.

This might seem dramatic and perilous to managers. "What, eliminate any measures of utilization and efficiency?" a manager might exclaim. "How will I know that workers aren't slacking off and my operating costs are not going through the roof?" I'll address this concern and drive home the QRM principle via an example of its use at an actual company.

An Experiment with Using Time as the Yardstick at Beloit Corporation[3]

Beloit Corporation is a manufacturer of systems and machinery for the pulp and paper industries. In the fall of 1992, John Amend, manager of manufacturing at the company's Beloit, WI, plant, faced a difficult decision: Should he scrap his long-standing measures of efficiency in the pipe rolls shop?

Background on Pipe Rolls

A pipe roll is a large metal roll that transports the paper in a paper-making machine. Pipe rolls can be several feet in diameter and more than 30 feet long. They are a common replacement part for a paper-making machine, so customers with machines that are being serviced, and thus inoperable, have a strong desire for short delivery times on pipe rolls. Although the equipment needed to make pipe rolls was already organized in cell fashion in one area of the plant, called the pipe roll shop, Beloit's delivery time for pipe rolls in 1992, including engineering, was around 14 to 16 weeks. This left open a large window for the competition to get in. As part of its efforts to improve the performance of the pipe roll shop, Beloit had embarked on a Total Employee Involvement (TEI) program for that area. Although the TEI program had some initial successes, a number of employees were not convinced about it and the TEI effort seemed to be at a plateau. In particular, the impact of the TEI program on lead time had been minimal.

Changing the Yardstick

Around this time, John Amend attended one of my in-house seminars at Beloit and heard about the concept of using lead time as the yardstick. Amend was intrigued by this and saw that it had application to the pipe roll shop. This shop was already organized as a cell, so that was not the issue. Then why did this shop still have long lead times? After examining the operation of the shop, I recommended to Amend that in order to achieve the necessary reduction in lead times, he should eliminate his traditional performance measures of machine utilization and work center efficiency. Instead, I wanted Amend to use the actual lead time of products moving through the shop as the only performance measure. "If you want to reduce lead time, then you have to measure and reward lead time reduction," I insisted.[4]

In addition, the TEI teams were asking for another radical change—elimination of the time standards on job tickets. Amend discussed this with me and I felt that this would only support the focus on lead time. "Trying to work to a given standard for each operation detracts from the goal of reducing overall lead time," I explained. "It makes people optimize each step in an isolated way, and this can be detrimental to keeping a job moving."

However, Amend had a deep-rooted concern: *If you eliminate the other measures, would the workers slack off, making the area inefficient, eventually causing costs to skyrocket and lead times to deteriorate further?*

I attempted to dispel Amend's concerns about soaring costs by citing several examples of companies that had realized both time and cost reductions through such an effort. After further discussions, Amend took the bold steps of:

- eliminating his long-standing performance measures
- removing the time standards from the job tickets
- informing employees in the pipe rolls shop that their performance would be measured primarily by how quickly they could complete a pipe roll—from start to finish

Specifically, for each pipe roll, the clock would start ticking when the raw pipe stock and job ticket were made available to the shop. It would stop ticking when the roll had completed its last operation at the balancing station and had been accepted by the inspector. The elapsed time would be recorded as the performance measure, with the aim being to reduce this time.

The result of these changes in performance measurement was dramatic. The first few pipe rolls for which the lead time was measured took an average of 24 days to complete, with the worst case being 36 days. Three months later, pipe rolls were flowing through the shop in as little as 8 days.

A pipe roll is a very large and complex component requiring many precision operations with very long machining times on specialized machines. Beloit had been trying for a long time to improve the lead times in this shop, but without any major successes. Then why did this breakthrough suddenly happen? Several events contributed to this result.

The New Yardstick Pulled Everyone Together

Since Amend was no longer using time standards, and he was not measuring individual work centers on their efficiencies, employees quickly realized they could reorganize their work any way they desired in order to improve the shop's performance. Before long, the hourly employees and the supervisor were rearranging work, changing long-standing procedures, altering the sequence of routings, and rethinking the task assignments to work centers. More interesting was the fact that, in contrast to the traditional adversarial relationship between hourly workers and the staff of the methods department, the workers now approached the methods staff for ideas to reduce the time it took to do setups and other tasks.

(continued)

The TEI efforts had laid the ground for such teamwork. But these efforts lacked a sharp focus and also lacked the support of the performance measurement and reporting systems. By unifying the mission of the shop to lead time reduction and by making the reporting systems consistent with this mission, Beloit was able to achieve this breakthrough in lead times.

Postscript: QRM as an "Umbrella Strategy"

In today's world, managers and employees at manufacturing corporations face a constant barrage of the latest acronyms, buzzwords, and philosophies, each of them offering ways to solve their problems. JIT, TQC, TQM, and TEI are just a few of the ones we have seen over the last decade. Which one of these ideas should your company emphasize? Using time as the yardstick unifies many of these apparently disparate efforts and provides an umbrella under which these individual philosophies can grow and be nourished, but in a focused way that lines everyone up toward the same goal—reducing lead times.

To summarize the lessons from this experiment, let us revisit the typical manager's concerns: "If I eliminate my measures of efficiency and utilization, how will I know that workers aren't slacking off and that my operating costs are not going through the roof?" The answer to this is, first, why beat around the bush? If your goal is to reduce lead time, then your primary measure has to focus on this, and not on utilization or efficiency. (In Chapter 16 I'll give a more precise measure of lead time reduction and discuss its use in the QRM organization.) Second, as you saw in the case of Beloit Corporation, if you have not hired any more people nor added new machines, you are meeting your production targets, and your lead times are going down, how can your costs be going up? And should you worry about whether people are slacking off if they have taken responsibility and the results are better than before?

This isn't to say you should not track utilization at all. For capacity planning purposes, it is important to have a handle on utilization because, as you saw in Chapter 7, if a resource is overloaded then its lead times can become unacceptably long. The distinction is that one should continue to estimate upcoming utilization to enable adequate resource planning, but not past utilization as an incentive for the workers or managers. Thus measurement systems should focus on capturing operation times for statistical purposes to assist with future capacity planning, but not to reward or punish workers.

PITFALLS OF ON-TIME DELIVERY MEASURES

We continue our examination of traditional performance measures by returning to the fourth item in the QRM Quiz.

> 4. We must place great importance on "on-time" delivery performance by each of our departments, and by our suppliers.
> ❏ True   ❏ False

What could be wrong with this principle? Isn't on-time delivery a cornerstone of JIT? Don't all the books on modern manufacturing state the importance of this? But the reason emphasis on on-time delivery goes against QRM principles is simple. Although on-time performance is a desirable goal as an outcome, its use as a measure is dysfunctional because it actually causes the opposite outcome to occur: On-time performance suffers. This is because of its unintended impact on human behavior, a behavior that ends up feeding our monster—the Response Time Spiral. To understand this, consider the following scenario.

Scene: A *manufacturing manager's office. Bob Sullivan, supervisor of the press department, has been summoned by the manufacturing manager.*

Manufacturing Manager: "Bob, the press department has had a history of being late in its deliveries. In fact, the last quarter your average on-time performance was only 65 percent, and this is hurting our shipments to our customers. I have decided to institute a new incentive system for you, to encourage you to find ways to deliver your jobs on time to the next department. I am going to increase your base salary by 25 percent. However, each month, your actual salary will be this base salary multiplied by your department's on-time delivery percentage. Thus, if you continue with the previous track record, your net salary this month would be 1.25 × 0.65 = 0.81 of your present salary. This may seem unfair, but think of the up-side opportunity I am giving you. Many departments in our company have on-time performance records of 90 percent or more. If you can achieve a 90 percent on-time performance, you will be taking home $1.25 \times 0.9 = 1.13$ times your current salary—a 13 percent raise. And if you can bring your department to the point where almost no jobs are late you will have gotten a 25 percent raise!"

This doesn't seem like such a bad idea, does it? If this approach motivates Bob to improve his department's on-time performance, the company benefits and so does Bob—a classic win-win situation. But before you pass judgment, let's go to the next scene of this drama.

Scene: *A conference room in the materials department. The materials manager is conducting a quarterly planning meeting. As part of this meeting, departments assist the materials manager in setting the planned lead times. We join the scene just as it is Bob Sullivan's turn to discuss the press department.*

Bob Sullivan: "Folks, I have been analyzing my department's schedule and delivery commitments for the past several months. It is clear to me that we are being given more work than we can handle. Because of the backlog of jobs in my department, plus the ongoing heavy load that I can see from the production plan for the next six months, it is unreasonable to expect my department to operate with its planned lead time of two weeks. During the past two months our actual lead times have been more like three to four weeks. If you want to develop a predictable schedule for the downstream departments, then I strongly suggest that you make my planned lead time a full four weeks. That way I can reliably deliver jobs, and the downstream departments will not have their schedules disrupted."

After some discussion of this point, the other supervisors agree that it is in the interest of their own schedules to give the Press Department a planned lead time of four weeks. This revised lead time is implemented in the MRP system.

Scene: *A few months later in the manufacturing manager's office. The manufacturing manager has again summoned Bob Sullivan.*

Manufacturing Manager: "Bob, I've got to hand it to you. The press department has really turned itself around. For the last two months you have been doing better than 95 percent on-time. This also means you've been taking home a bonus of about 20 percent of your salary each month—and you deserve every penny of that, because the whole company has been doing much better in its shipments and our customers are happy for a change! This experiment has been so successful that I'm going to try it in all our departments."

Scene: *A corporate conference room about a year later. sitting around the conference table are the VPs of all the major departments, including marketing, sales, engineering, manufacturing, materials, and finance. At the head of the table is the company president. It is clear from the faces of people around the table that the mood in the room is grim.*

President: "We have reached a new low in our delivery performance. My colleagues in sales tell me that if we don't get our act together very soon, we'll soon have no customers." [*Turning to look at the VP-Manufacturing*] "I see from last year's reports that manufacturing's on-time delivery performance has been dismal. We are promising our customers 8-week delivery, but we usually take 12–14 weeks and sometimes longer. Why can't you get your department under control and run it more efficiently?"

If the last three paragraphs give you a feeling of *deja vu*, you're right! This final scene is the beginning of the act you saw in Chapter 3, where a company is becoming a slave to the Response Time Spiral. How did we get from the success in the press department to this disastrous final scene?

This process began with the new incentive for Bob Sullivan. This led Bob to ask for a much longer planned lead time for his department, to maximize his chances for on-time delivery. After the short-term success with Bob, the same incentive was implemented for all manufacturing departments. Like Bob, every supervisor then had an incentive to pad his or her department's planned lead times as much as possible so its on-time deliveries would look good. (An additional reason for such padding is given in the discussion on MRP systems below.) So instead of every department in the organization trying to reduce its lead times, they are all padding their lead times. Similar remarks apply to suppliers who are measured by on-time delivery.

All this padding results in long overall lead times, but worse still, it feeds the Response Time Spiral. After the initial short-term successes, during which it appears this new reward program is working, the Response Time Spiral, with all its evils, takes over the system. The final result—after a year for the Response Time Spiral to settle in—is a disaster in the organization's deliveries to its customers.

Certainly you want to be on-time to your customers, whether internal or external, but at the same time you should not overemphasize on-time delivery performance. Then what is the correct approach?

QRM Principle: Stick to measuring and rewarding the reduction in total lead time.

Recall that the Response Time Spiral gets its energy from time being *added* to the system. As you *remove* time from the system you shrink the spiral and eventually kill it. Also, as the QRM detectives in each section of the company work on uncovering delays and eliminating their root causes, you further reduce quality problems and other forms of waste. As a result of all these effects, delivery problems eventually disappear and on-time performance will be the result. Note that you get on-time delivery as a *result*, but do not use it as a *motivator*. This is where the QRM approach departs radically from common practice.

I want to point out, though, that for the above QRM principle to work, it has to be supported by many of the other principles already elucidated, such as implementing cells and replacing measures of efficiency. The above principle, implemented on its own in a traditional organization, will not work at all, because the functional organization with its long lead times and tortuous routing of products gives little opportunity to individual departments to change their own lead times. In addition, the materials department has to support the above principle in several ways, which I'll discuss next.

MRP: A COLLECTION OF WORST-CASE SCENARIOS

The concepts I've developed in the first part of this chapter are crucial to understanding the QRM approach to the quiz item number 5:

> 5. Installing a Material Requirements Planning (MRP) System will help in reducing lead times.
>
> ❏ True   ❏ False

Surprisingly, nothing could be farther from the truth. MRP systems do serve an important function of assisting with materials supply; for example, determining when you need to order more stock from vendors. However, an MRP system will not assist in lead time reduction—if anything it fosters an increasing spiral of lead times. I had the misfortune to witness one division of an aerospace corporation attempting to solve its lead time problems with MRP. Instead, over a space of a few years, its lead times grew from 6 months to 24 months. The reason MRP will not solve lead time problems is evident when you combine the basic MRP model with the discussion of the preceding section.

MRP systems work with fixed lead times for each department. This means, to be on time, a department manager needs to quote a lead time that will be consistently achieved, regardless of whether the department is very busy or quite idle. Let's take the example of Bob Sullivan. Even though Bob's department might be able to deliver product in two weeks on average, if Bob wants to be absolutely sure he can deliver, he will give the MRP system his worst case estimate of lead times when his shop is very busy, namely four weeks. As a result of a similar experience with all departments, all the lead times in an MRP system end up being the "worst case" lead times. Take a moment to think about this: Do you really want to run your company using a system that has become a collection of worst-case scenarios?

Since, in the traditional organization, a product routing may take it though a dozen or more departments, and each one has a padded lead time, you end up with a lot of "fat" in the system and a very long planned lead time for each product. These long lead times feed the Response Time Spiral and result in even longer lead times. Companies find they are late. Managers start putting great pressure on "on-time" performance for each department, which only exacerbates the problem.

MRP experts might claim that you can avoid these worst-case scenarios by using the capacity planning function in the MRP system. However, the capacity planning in MRP also leaves much to be desired. This is best illustrated by a detailed example.

Dysfunctional Dynamics of MRP—An Example

Consider a company that makes transfer machines (automated machines used in high-volume production lines). Figure 8-1 shows a schematic of the various components and stages of production for one of their machines. The machine consists of several stations, and one of these is called the mill station. The mill station assembly consists of several subassemblies, and Figure 8-1 shows two of these, the mill head subassembly and the slide subassembly. The mill head subassembly, in turn, is made of several components. A few of them are: head case, rear cover, spindle, and gear. Similarly, a few components are shown for the slide subassembly. The top of the figure shows a timeline in weeks, starting with week 10. Now let us review how the MRP system works. The schedule is developed using the so-called infinite capacity assumption:

> The planned lead time for each department is independent of the load in that department.

This is precisely the fixed lead time assumption I've already discussed. Compare this with Figure 7-5 where you see how dramatically lead times can change with loading of any resource. Clearly the infinite capacity assumption is far from reality.

Continuing with this example, suppose that the machine assembly is to be completed by week 20, and this operation has a 4-week planned lead time. Then the mill station must be completed by the beginning of week 16. Suppose also that the assembly department has a 3-week planned lead time, and so does the subassembly department. Then, working backwards as the MRP system would, we find that the mill head must be completed by week 13, and so the components (head case, rear cover, etc.) must be completed by week 10. These planned completion times of each of the stages of the machine, as you would see in the

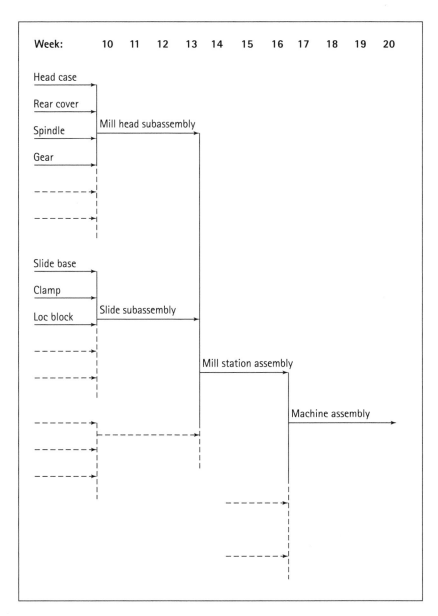

Figure 8-1. Component and Assembly Schedule for a Transfer Machine

204 Part Two: Rethinking Production and Materials Management ....

detailed MRP schedule, are summarized in Table 8-1 and also shown in Figure 8-1 via the vertical lines.

So far everything seems straightforward. But now I'll add a fairly common manufacturing dynamic. Suppose the subassembly department is overloaded in weeks 10 and 11. They are working on other parts and haven't even started on the mill head in week 11. A diligent salesperson, who is checking on the status of his customer's order, notices that the mill head subassembly is a week behind schedule. Through his manager, who talks to a manufacturing manager, pressure is put on the sub-assembly department to expedite the mill head. The subassembly department stops working on other parts (which may now be delayed), and starts work on the mill head subassembly in week 12. Since they started late, even with expediting they can only catch up a little, and they deliver the mill head to the assembly department in week 14— 1 week late. Using overtime, the assembly department manages to make up half the week. Finally, additional overtime is used in the machine assembly stage to prevent late shipping.

What else has been happening in the meantime? The assembly department had been expecting the mill head in week 13, but it never arrived. It didn't have any other jobs at the time, so its workers were idle for the first day of week 13. This department's supervisor, however, has a good relationship with several of the feeder department supervisors, so he asks them to expedite some other parts into his area to keep

Table 8-1. MRP–Generated Completion Times

Item	Completion Date (Beginning of Week Shown)
Machine assembly	20
Mill station assembly	16
Mill head subassembly	13
Head case	10
Rear cover	10
•••	•
•••	•
•••	•

his people busy. By the second day of week 13 these parts start to arrive and he can give his people some work to do.

Well, the company shipped the machine on time, and the assembly department supervisor did the right thing to prevent idle workers, so what's wrong with this picture? As you multiply this scenario by tens of products and dozens of departments, you get a great deal of confusion, disruption, and wasted resources. For example, you get a lot of money going into overtime wages to counter the upstream problems. You get supervisors being pressured to change their schedule by upper management, or by other departments such as the assembly supervisor. These changes further add to the disruption, delayed jobs, and more overtime.

The Problem with Capacity Planning Reports

For the preceding example, why couldn't you anticipate these problems via the more sophisticated reports in MRP II, and thus prevent them? Specifically, wouldn't the capacity requirements planning (CRP) report show us that the subassembly department was going to be overloaded in weeks 10 and 11? There are two issues in regard to this:

1. *Although MRP II's CRP report does flag a capacity overload for a given department in a given week, it is not easy to figure out what to do about it.* Typically, a capacity overload is the result of the sum of loads of a number of jobs being done in that department. One has to first analyze which jobs are contributing to the load, and then decide which one (or more) jobs might be moved to a different week. Moving any job, however, is easier said than done. For the subassembly department, to alleviate its overload in week 11, suppose they decide to reschedule the slide subassembly in Figure 8-1. They can't schedule the job later, since the mill station needs the slide subassembly, so they try to schedule the job earlier. Since week 10 is also an overloaded week, they try to see if they can schedule it for weeks 8 and 9 (assuming they need at least two weeks to get it done). However, this requires that they must complete the components in Figure 8-1, the slide base, clamp, location block, etc., two weeks ahead of schedule, in week 8 instead of week 10. Each component might have its own routing through several departments, or even its own subcomponents, and so they would

have to track the effect of this change on all those departments and subcomponents. This might create an overload in some weeks for other departments, or might require a supplier to bring in a shipment earlier than planned.

As you can see, the ripple effects are starting to mount up. All of these effects have to do with trying to make just *one* change to *one* item in *one* department—imagine the ripple effects of trying to reschedule capacity overloads in multiple departments for multiple weeks. In reality, since this would require tracking hundreds and hundreds of ripple effects, this task doesn't get done at all. You might make some minor attempts to correct this, such as expediting a few items, looking for a few obvious improvements, or adding overtime to combat the overload. More typically, even these attempts are avoided and the CRP reports are just plain ignored, due to another factor which I'll now discuss.

2. *Capacity requirements planning reports are themselves generated using the infinite capacity assumption (i.e., fixed lead times).* The actual arrival of parts to an area may therefore be different from what the MRP system thinks; thus the actual load on the area may be different in a given week from what the capacity planning report shows. Figure 8-2 illustrates this, and I will explain the point further with the transfer machine company example. Consider the assembly department in week 13. Suppose that, due to the overload on several upstream departments, not only is the mill head late, but two other subassemblies are late as well. Let's say that if all the parts had arrived according to plan, it would have overloaded this assembly department. Then the MRP system's capacity planning report will show that in week 13 the assembly department is expected to be overloaded. Suppose that in week 10 the department's supervisor receives a request from two of his workers to take a vacation in week 13. He looks at the capacity planning report, and seeing the overload coming up, persuades the workers to defer their vacation to week 14. But what actually happens? Remember our example—now he has two more idle people in week 13, while in week 14, when the supervisor needs them most to help with the delayed job, they are on vacation. As a result of several such incidents, employees start losing confidence in the capacity planning and other related MRP reports. After a while a company simply stops relying on these reports for decision making.

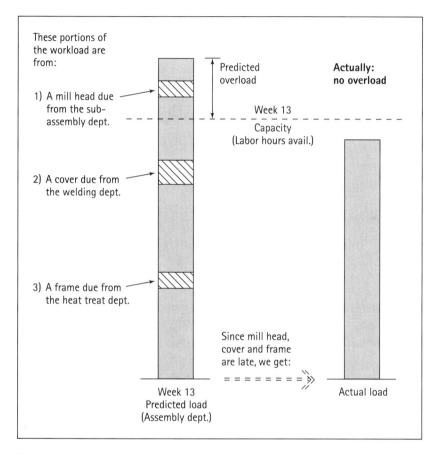

Figure 8-2. Why MRP II's Capacity Planning Reports Are Meaningless

Physical Effects of MRP Dynamics

There are four immediate effects of these "MRP dynamics" on the operation:

1. *Departments inflate lead times.* In an attempt to correct for load variations and possible tardiness, materials analysts and department supervisors collude to use inflated lead times for each department.
2. *WIP increases.* In cases where the load turns out to be manageable, the inflated lead time is far more than what you need, and parts are completed way ahead of time. This results in excessive WIP

and the need for parts storage. It also has a long-term detrimental effect on the organization, discussed in the next subsection.

3. *Last-minute solutions are needed.* In instances where the load is not manageable, the company has to resort to bouts of overtime, expediting, or last-minute subcontracting (which is much more expensive since the subcontractor knows the company is in trouble and there is no time to go out for competitive bids). Or else, if management does not agree to these, then the company simply has late deliveries to its customers.

4. *The system becomes unstable.* A different effect is caused by the conjunction of traditional lot sizing policies and the infinite capacity assumption. In companies where minimum order quantities are in effect, a small change to the final delivery schedule can result in a large change to the published schedule for each department. Thus, the press department supervisor might look ahead to plan the next week's work and labor, but then a small change to shipping dates by the sales manager might produce a whole different schedule for this department by the time the next week rolls around. Such behavior does little to enhance the credibility of published schedules. This phenomenon is called *nervousness*, and has been well documented for MRP systems.[5]

Organizational Effects of MRP Dynamics

As individuals in the organization witness the above phenomena, the following behavior results over time:

- *People ignore MRP-generated schedules.* As explained above, there is a lot of slack in the schedule, and everyone knows this. Plus, people are skeptical about the accuracy of the schedule in the first place.
- *A "hot job mentality" results instead.* Since supervisors know that there is slack in the schedule, and since the schedule is dubious anyway, they don't waste time planning their resources by the MRP schedule. Instead they wait until someone yells for a part, then they work on that part. They find this a far more reliable scheduling method, because then they know they are working on something that is needed.
- *Special orders exacerbate the above situation.* Because the company has such a long lead time, on occasion senior management has to

accept a customer order for delivery in less than the planned MRP lead time, or else it will lose the order altogether. When you load such an order into the system, it shows that the starting departments are already past due.

- *Past due becomes meaningless.* After a while, between the inability of the MRP system to produce doable schedules and the loading of special orders, employees realize that past due has little to do with their own performance.
- *Employees develop apathy and lose any sense of responsibility.* A feeling of helplessness results. Supervisors and workers alike feel they can't do much about being on time because the schedules are impossible to execute.

Table 8-2 summarizes the above points, as well as those from the preceding subsection.

Table 8-2. Long-Term Dysfunctional Effects of MRP

Physical Effects
- Inflated lead times are used in an attempt to correct for inaccuracy of load prediction.
- WIP increases due to the inflated lead times.
- Last-minute solutions are needed due to inaccuracies, including expediting, overtime, and last-minute subcontracting.
- The system becomes unstable, with large changes in schedule resulting from small changes in requirements.

Organizational Effects
- MRP-generated schedules are ignored; everyone knows there is a lot of slack in the schedule.
- A hot-job mentality results, since the only way to know if a part is really needed is if someone is yelling for it.
- Unacceptably long lead times result in management loading in special orders in less than the planned lead time. These orders are past due even before the first department can work on them.
- Past due soon has little to do with a given department's efforts and becomes a meaningless performance measure.
- Employees develop apathy. They feel they can't contribute to improvements because they are given impossible schedules and are judged by meaningless performance measures.

Do You Need to Replace Your MRP System?

These descriptions of the multitudinous problems of MRP might make managers nervous. "Does this mean that to implement QRM we'll have to buy a replacement for our MRP system, and throw away the millions of dollars invested in our existing system?"

This is a timely question. These days the materials management community is a buzz with acronyms for new systems. ERP (Enterprise Resource Planning), APS (Advanced Planning Systems), and MES (Manufacturing Execution Systems) are popular newcomers being examined by manufacturing executives. At the same time FCS (Finite Capacity Scheduling) systems, which have been available for a while but not much used, are suddenly in demand.

History has repeatedly taught us an important lesson about technological solutions, yet how soon we forget. Too often the U.S. manufacturing industry has turned to technology as its savior. The 1970s saw the U.S. automobile industry—most notably General Motors—invest billions of dollars in automation, only to step away from much of that automation 15 years later, instead adopting organizational strategies of total quality, employee involvement, and teamwork, as seen in GM's Saturn and NUMMI initiatives. In the same vein, the answer to the problems with MRP lies not in large investments in new systems capabilities but in aligning the existing MRP system with the strategy of QRM.

ALIGN MRP STRUCTURE WITH QRM STRATEGY

We can summarize the QRM approach to material planning (the principle that must replace quiz item 5) as follows:

> **QRM Principle: Restructure the manufacturing organization into simpler product-oriented cells. Rethink bills of materials for this organization. Use MRP systems to provide high level planning and coordination of materials from external suppliers and across internal cells. Let the cells be responsible for their schedules and provide them with simple planning tools.**

Let us elaborate on the four key points (steps) made in the above principle.

Step One: Restructure the Organization

I covered in detail the first step about restructuring the manufacturing organization into simpler product-oriented cells in Chapter 4 and repeat it here to emphasize that the remaining steps make sense only after you've implemented this first crucial step.

Step Two: Rethink Bills of Materials

Now it's time to rethink the bills of materials for relevant products to complement your shift to the cellular organization. The aims of this step should be twofold: (1) to flatten the bill of materials (BOM), and (2) to eliminate fabrication or assembly operations. Specific ways to achieve both of these aims are:

- *Rethink design decisions.* For example, can you replace a cover consisting of four sides and a top, fastened together in several places, by one injection-molded part? Established methodologies of DFMA (design for manufacture and assembly) can prove useful in conducting this step.[6]
- *Rethink materials choices.* Can a component that needs coating or plating be fabricated from a different material to eliminate this operation? Several examples of this were given in Chapter 6.
- *Reevaluate "make versus buy" decisions, including subcontracting of operations.* The presence of a subcontracting step in the middle of a product's routing can complicate the scheduling of material through a cell. Can you eliminate this operation or bring it in-house? Chapter 6 gave us several new ways of thinking about operations such as heat treatment and plating. Similarly, it might be easier to control the delivery of a part you make in-house. Alternatively, you could have an entire subassembly subcontracted and delivered as one item, rather than getting the parts separately and managing the multitude of parts and their fabrication and assembly.
- *Collapse the bill of materials conceptually.* Even if the BOM physically consists of multiple levels, if you bring all the operations into

one cell you can list the components at one level for MRP. Then you can train the cell operators to assemble the product in the right sequence. This flattens the BOM for the MRP logic.

Ideally you should do such rethinking as part of implementing cells, since there is a lot of synergy between the principles above and the creative rethinking presented in Chapter 6. The principles here extend some of that rethinking by bringing in a materials perspective. I did not present these particular points in Chapter 6 because it was necessary to elaborate on the problems with traditional MRP systems before readers can really appreciate these additional points.

In rethinking all these decisions, don't be caught unawares. Be forearmed with your QRM principles and examples for you will have to battle the cost-based mind-set over and over again. Often, decisions such as make versus buy, and subcontracting, are based on traditional cost-based analyses. Such analyses ignore the large impact of the Response Time Spiral and other costs of complexity and long lead times. It may *seem* as if the simpler operation is more expensive, but the new QRM organization will not incur those other costs, so in the long run it is actually more economical. However, you may encounter disbelief on this point and may have to repeat over and over the many examples in this book to convince your colleagues of the efficacy of the QRM approach over traditional cost-based methods.

Step Three: Use a Higher Level MRP

You should not attempt to use MRP systems for micromanagement of every work center. Instead, you should use the system for high-level planning and coordination of materials. This involves (1) predicting the need for, and ordering materials from external suppliers, and (2) coordinating material delivery across internal cells. I denote this "higher level MRP" by HL/MRP and in the following sections explain how you should structure HL/MRP for internal supply, and why this structure is preferable to the traditional detailed MRP procedures. (In Chapter 9 I'll discuss details of how material is replenished across cells, and in Chapter 10 I'll discuss the way you need to change the external material supply for QRM.)

An Example of Structuring HL/MRP for Internal Supply

Let's consider a company that makes customized power supplies for electronics products. Prior to implementing QRM, a typical power supply product had the bill of materials shown in Table 8-3, along with the manufacturing routings shown in Figure 8-3. There are a total of 16 manufacturing steps for this product. Each step was performed in a different department, such as the punching department and the wave solder department.

After implementing QRM, the product will be made in three cells; two are feeder cells and one is a final assembly cell. Although the QRM ideal is to make a product from start to finish in a cell, in this case the cell would be too large and unwieldy, as discussed in Chapter 4. A logical split that results in manageable cells is to make the sheet metal cases in a case fabrication cell, to make the electronics portion in a PCB assembly cell, and then to assemble the power supply and test it in a final assembly cell. In this way, each cell still makes an identifiable

Table 8-3. Bill of Materials for a Particular Power Supply

Level				Part Number/Description	Qty.
0				PS-140 Power supply	1
	1			MC-140 Metal case	1
		2		Rear panel	1
			3	SS007 Stainless steel blank	1
			3	F6612 Steel lugs	8
			3	F9091 Rivets	6
		2		Left panel	1
			3	[... details of components ...]	
		2		[... similarly for right panel, cover and base ...]	
			3	[... details of components for each ...]	
	1			CB-140 Circuit board assembly	1
		2		MB801 Printed circuit board (mother board)	1
			3	DB801 SMT daughter board	1
				4 SR727 Resistor (SMT)	2
				4 SM525 Microprocessor (SMT)	1
				4 [... several other SMT components ...]	
			3	AC384 Capacitor (axial)	4
			3	[... similarly, several axial components ...]	
			3	[... similarly, several radial components ...]	
			3	[... similarly, several hand insertion components ...]	

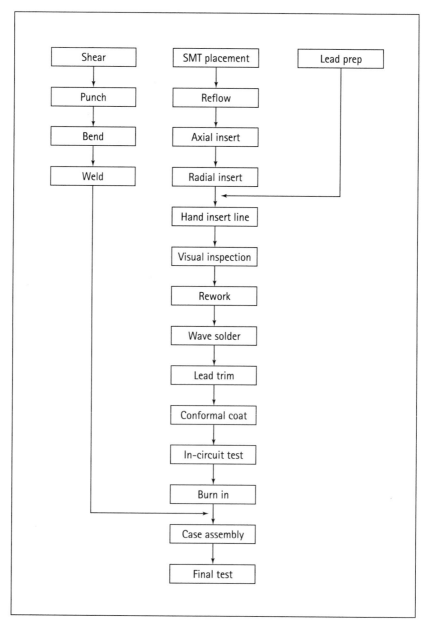

Figure 8-3. Manufacturing Organization Before QRM

"product." (Recall the discussion in Chapter 4.) Another advantage of this organization, in keeping with QRM ideas, is that each of these cells can also feed other cells as needed by various customized products. For example, the PCB assembly cell could feed a different final assembly cell—a situation described in more detail in the next chapter.

Figure 8-4 shows the simplified routings after the company is organized into these cells. To achieve this simplification, several routing steps have been combined or eliminated, and the bill of material has been collapsed using QRM principles (see Table 8-4). Specifically, pre-

Table 8-4. Collapsed Bill of Materials for the Power Supply

Level		Part Number/Description	Qty.
0		PS-140 Power supply	1
1		MC-140 Metal case	1
	2	SS007 Stainless steel blank (for rear panel)	1
	2	Preformed left panel	1
	2	[... similarly preformed right panel, cover and base...]	1
1		CB-140 Circuit board assembly	1
	2	SR727 Resistor (SMT)	2
	2	SM525 Microprocessor (SMT)	1
	2	[... similarly, several other SMT components...]	
	2	AC384 Capacitor (axial)	4
	2	[... similarly, several axial components...]	
	2	[... similarly, several radial components...]	
	2	[... similarly, several hand insertion components...]	

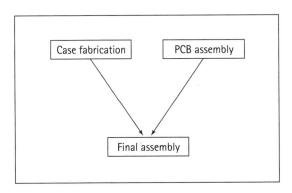

Figure 8-4. Manufacturing Organization After QRM

formed parts are used for some panels that do not need customization, fasteners have been eliminated, and the daughter board has been replaced by mounting components directly on the main board. Also, the product made in each cell is at Level 1, and all components used in the cells are at Level 2—whereas in the previous BOM, there were components up to Level 4.

But none of this will work if you do not have a QRM manufacturing cell. Remember, for a QRM manufacturing cell you organize all the process components necessary to deliver a finished product (or family of related products) into one cell. Recall the definition of a manufacturing cell that I gave in Chapter 4 (p. 90).

So in your new QRM organization with cells and a collapsed bill of materials, rather than looking at the details of replenishing materials and routing steps within each cell, you can structure the HL/MRP routing to consider each cell as one step in the standard MRP logic. For any end-product, such as our Power Supply, each cell will make a given subproduct (such as the Level 1 products in Table 8-4), and all the components needed for this subproduct will be at the next lower level for the purpose of the HL/MRP (such as the Level 2 products in Table 8-4). In this way, there are only three levels (0, 1 and 2) and three steps (the three cells) for the MRP system to schedule. The MRP system will still explode the demand for the final power supply down to the demand for components to be delivered to the cells, but it will only have to explode to Level 2, not to Level 4. (Also, as you will see in Chapter 10, you can remove several of the purchased components from the HL/MRP logic by creating direct links from suppliers to the cells.) It is clear that this results in a simpler system than regular MRP with traditional departments that are organized by function. But today's computers are ultra-powerful and would have no difficulty crunching the numbers for the complex routings, so why does this simpler HL/MRP system work better? There are three reasons.

1. There are only 3 steps, as in Figure 8-4, not 16 as in Figure 8-3, that you need to schedule and control. The ripple effects from 16 steps can quickly become unmanageable, while for 3 steps there is a good chance you can keep them under control.

2. Using the approach in the next section, you'll be giving the cells realistic targets as opposed to unreasonable ones, ones they would soon be ignoring.
3. In the QRM organization, the cell teams are more likely to keep their end of the bargain and deliver product on time.

Step Four: Make Teams Responsible for Running Their Cells

A counterpart to making MRP work at a higher level is to make the cell teams responsible for running all the operations within their cells. This means planning, scheduling, and control of their operations. Keeping our eyes on this goal, which is ever shorter lead times, you must support cell teams in two ways for them to be effective at these tasks.

1. *You must educate them on the dynamics of manufacturing systems.* Specifically, they should understand the main points in Chapter 7. It is not necessary for them to absorb all the equations as long as they understand the key ideas.[7]
2. *You should provide them with simple queuing-based analysis tools to plan and manage their capacity in order to preserve their short lead times.* Traditional capacity planning as in an MRP system or cell loadings based on spreadsheet calculations will not do, since they do not analyze any system dynamics. This means they don't properly trade off capacity utilization, lot sizing, and other effects in terms of how they affect lead time.

In other words, for cell teams to be successful you must provide both education on manufacturing dynamics and tools to predict these dynamics. A case study at an electronics company illustrates the effectiveness of this approach.

Building Customer Satisfaction with Improved Planning at Pensar Corporation[8]

Based in Appleton, WI, Pensar Corporation is a custom assembler of printed circuit boards and other electronic devices. As a supplier for a host of consumer and industrial customers, Pensar's ability to provide timely delivery—in the midst of wild demand swings—is its competitive advantage. Because

(continued)

short lead times and reliable delivery are so critical to Pensar's success, its president, Stan Plzak, has been promoting QRM concepts throughout the company. Danny Gardner, operations vice president, has created worker empowerment teams led by "segment leaders" who engage shopfloor personnel in lead time reduction planning.

What has been Pensar's approach to this challenging task? First, in 1995, they began exposing their company to QRM concepts through training. Next they implemented cells on the shop floor. Then in 1996 they began to use rapid modeling technology, a software implementation of the queuing analysis approaches discussed in Chapter 7. To create "what if" models of lead time reduction scenarios, Plzak and Gardner chose the MPX® software package from Network Dynamics of Burlington, MA. MPX models were built for each of the cells. Then, to get assemblers involved in actual day-to-day planning, they created the empowerment teams to help workers devise and test their own lead time reduction strategies.

The rapid modeling technology embedded in MPX uses queuing analysis to pinpoint capacity and lead time problems and to predict the impact of adjustments to lot sizes, work force, or use of equipment. However, since MPX uses a rough-cut modeling approach, the teams are not bogged down in details of data gathering and analysis. Rather, they can have results of dozens of what-ifs in a matter of minutes.

Plzak is highly enthusiastic about Pensar's newly adopted QRM methods. "We originally trained a team of six, which in turn has got the whole plant involved. MPX has turned out to be a wonderful learning tool. We've got planning down to the application level, where it's become an everyday activity for our assembly personnel." In practice, Pensar segment leaders work with production schedulers to benchmark the master schedule created by the MRP system against MPX models for the cells. Interaction between these two systems results in a schedule and work orders that are precisely matched against the capacity and lead times of the cells.

By August 1996 Gardner had observed up to 30 percent reductions in cycle time after only two months of implementing these planning processes. But just as important, he credited the whole QRM approach for its value as a marketing tool. "A customer came to us recently requesting a sixfold increase in monthly production over previous levels," he said. "Using QRM techniques supported by MPX analysis, we got the relevant work cell teams to run a series of scenarios taking the new loads into account. The upshot of the suggestions from the teams was that we were able to meet scheduling and volume requirements with minimal increases in the work force. In short, we delivered!"

As seen from the Pensar Corporation example, an effective material planning system structure for QRM entails having the HL/MRP system interact with the queuing-based planning tools in each cell (see Figure 8-5). The HL/MRP system is driven by specified demand. This may be firm customer orders, sales forecasts, or a combination of both. Based on this demand, and using the values of cell lead times in its database, the HL/MRP system develops delivery schedules for each cell. (Thus far, this is the standard MRP algorithm, the only difference being that it is operating at a higher level, for each cell, as opposed to doing this for each operation.) These delivery schedules are fed to each cell team. Based on these, each cell team develops its own rough-cut lot sizing and work force policies to achieve the desired schedules.

For example, let us consider the two alternative organizations in Figures 8-3 and 8-4 in the context of the manufacture of 20 PS-140 power supplies for a specific order. For the organization in Figure 8-3, the MRP system would tell each department, such as the axial insertion department, when it should perform the insertions on an MB801 (the part number of the board needed in a PS-140). The system might even look at similar orders in the system and tell the axial department to perform the insertions on 80 MB801 boards instead of only 20. Thus not only would the axial department lose sight of the end product it was creating, it might even lose its connection to a specific customer order.

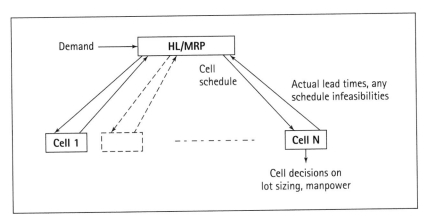

Figure 8-5. Interaction of HL/MRP and Cell Planning Systems

Similarly, the MRP system would schedule work, in appropriate batch sizes and at appropriate times, for all the other departments in Figure 8-3, with the hope that the batch of 20 PS-140 power supplies would be ready at their due date. In contrast, for the organization in Figure 8-4, the HL/MRP system would only specify to the PCB assembly cell that it needed 20 CB-140s (the part number of the assembled board used in a PS-140) by a specified date. Similarly it would specify the delivery requirements for 20 cases to the case fabrication cell, and for 20 PS-140s to the final assembly cell.

If the cell team detects periods when it cannot meet the production targets, or when its lead times differ significantly from the ones being used by the HL/MRP system, it gives this feedback to the HL/MRP system. Based on such feedback, the HL/MRP system develops a modified schedule. The process is then repeated. In the QRM approach, this iterative process is not very complex or long-winded for three reasons:

1. Precisely because HL/MRP is performed at a high level, there are few MRP steps for each product.
2. The BOM is very flat, so the ripple effects are minimal.
3. The HL/MRP-to-cell interactions should be frequent and should look well into the future. This way you resolve problems well ahead of time and there is minimal iteration for near-term schedules.

USING LEAD TIME REDUCTION TO CONTINUOUSLY IMPROVE YOUR PROCESSES

It is worth repeating here the difference in approach between QRM and the Toyota Production System (TPS) (JIT or lean manufacturing). QRM focuses on the relentless reduction of lead time, which eliminates non-value-added waste, improves quality, and reduces cost. Lean manufacturing focuses on eliminating non-value-added waste, which reduces lead times, improves quality, and reduces costs. So the way you continuously improve your processes in a QRM organization is to encourage your cell teams to continuously search for ways to reduce their lead times. You should instill everyone on the team with the mind-set of the QRM detective, asking the "why are they waiting" question again and again. It is through this relentless focus on reducing lead time, with the assistance of the QRM detective's mind-set, that a QRM organization continuously improves its processes and eliminates non-value-added waste.

To achieve this continuous improvement environment you must motivate cell teams to focus on lead time. Your first step to achieving this is to scrap your old cost-based performance measures and use lead time reduction as your primary measure of cell performance. (The specifics for this are discussed in Chapter 16.) The next step is to provide your cell teams with tools to analyze their current operation and search for continuous improvement opportunities. The queuing-based models provide an ideal tool for cell teams. Not only do you want the teams to plan and manage their capacity in order to consistently achieve their short lead times, but you also want them to seek for ways to improve their performance. With their ease of use and speed of what-if analysis, the queuing tools support cells in their continuous improvement efforts. Questions such as: "Which setups should we focus on for a setup reduction program?" "Which machine failures most hurt our performance?" and "Should we dedicate operators to these machines or have them shared across machines?" are swiftly answered in terms of their impact on lead time. This assists teams in focusing their improvement efforts for maximum reduction of lead time.[9]

Having used the queuing tools to target a given setup for reduction, the teams can then turn to established methodologies for (quick changeover) to implement the actual reduction in setup time. Similarly, if a given quality problem, or machine failure, is affecting consistent delivery, then the teams should use tools from TQM, total productive maintenance (TPM), or kaizen to address this issue.[10] In this way many of the known improvement methodologies can complement the QRM approach, driving home the point that QRM is not intended to replace these techniques and methods in which your organization has already invested, but rather to build on their foundation.

Main QRM Lessons

• The numerous drawbacks of traditional efficiency measures.

• The pitfalls of measuring on-time performance.

• Why an MRP system will degenerate to a collection of worst-case scenarios.

• The dysfunctional effects of MRP on the physical and organizational behavior in the manufacturing enterprise.

• Why MRP II's capacity planning reports are meaningless.

• The solution is not in replacing MRP, but in restructuring the system to align it with QRM strategy. This involves rethinking bills of materials, using MRP at a higher level based on the cellular organization and the revised BOMs, providing cells with simple tools to plan and manage their capacity, and motivating continuous improvement for lead time reduction.

9

POLCA–The New Material Control and Replenishment System for QRM

IN CHAPTER 8 YOU LEARNED about material *planning*, which involves creating a schedule for material delivery and identifying this schedule's implications for manufacturing capacity. Now let's turn our attention to material *control*, which is the task of executing the schedule. Specifically, this task involves methods to manage the flow of materials to achieve the desired schedule, or take corrective actions when you detect deviations. Readers less interested in the intricacies of material control may want to move onto the next chapter because the technicalities get quite detailed and are directed to material managers. However, you might want to read the initial sections of this chapter that discuss the differences between JIT and QRM, and the stories of three companies that turned to QRM strategies after deciding JIT did not fit their needs.

REVIEW OF PUSH AND PULL SYSTEMS

Throughout the book I've been discussing the differences between QRM and JIT and why many aspects of JIT just don't "fit" with QRM. To clarify these differences it is useful to begin with a short review of push and pull systems.[1]

Push System

Figure 9-1 shows a typical push system, driven by MRP. The production plan is driven by the due date for a shipment. This date triggers the MRP calculations, which use the planned lead times to determine when material is to be released to the first work center. After release, the material is pushed through this and all the subsequent work centers. Why is this called "push"? Because each work center is supposed to complete work according to the schedule the MRP system gives it. Suppose, in a traditional functional organization, a job needs to be turned and then milled. Let's say the milling department is backed up with a lot of work because one of its milling machines failed unexpectedly last week. But if the schedule at the lathe department dictates that this job should be turned today, then the lathe department will attempt to meet that schedule. After being completed, the job will be "pushed" to the milling department, where it will sit for several days.

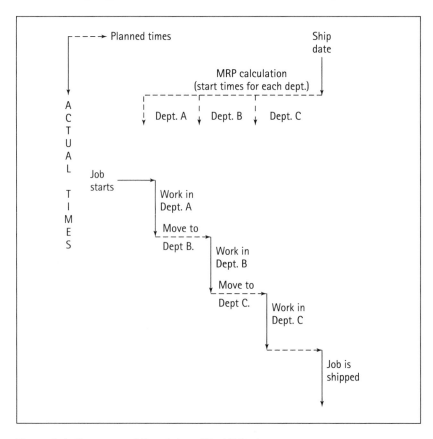

Figure 9-1. Sequence of Events in a "Push" System

In the meantime, the grinding department might be idle, with no jobs to work on. If the turning department had worked on a job that was destined for grinding instead, that department would not be idle. You might wonder why it isn't easy to see the above situation and correct the schedule in the lathe department. The answer lies in the sheer volume of computations needed to keep a typical manufacturing company's schedule on target. This example involved one job and three departments. More likely, a company is dealing with hundreds of jobs, requiring thousands of operations, with different routing patterns snaking through dozens of departments. Now it is no longer easy to determine who is holding up whom, and what job is the best to work on next. In this situation, each department follows its own schedule, "pushing" jobs to the next department, regardless of whether the following department has the capacity to work on this job in the near future.

Pull System

An integral part of implementing JIT is establishing a pull system with some sort of kanban mechanism. (Kanban is the material replenishment method that manages and ensures the success of JIT production via production control cards or containers. To realize the full potential of kanban a company needs to first establish a number of kanban prerequisites like JIT education, TPM program, quick changeover program, zero defects, visual workplace, etc.). In a pull system material movement is triggered when a customer order causes the removal of inventory from finished goods (see Figure 9-2). This signals the preceding workstation to make another piece (or another batch, depending on the number of items that fit in one kanban container). When this workstation depletes its incoming inventory below a preset level, it sends a signal to *its* upstream workstation to make another container of parts, and so on, all up the line. Thus no workstation will ever put more than a preset number of parts in its output buffer, and production is always triggered by demand from the next workstation. Eventually, through the set of pull signals working their way upstream, production at each workstation is triggered indirectly by actual final demand. The level of inventory between each pair of workstations is limited by the number of kanban cards (or containers).

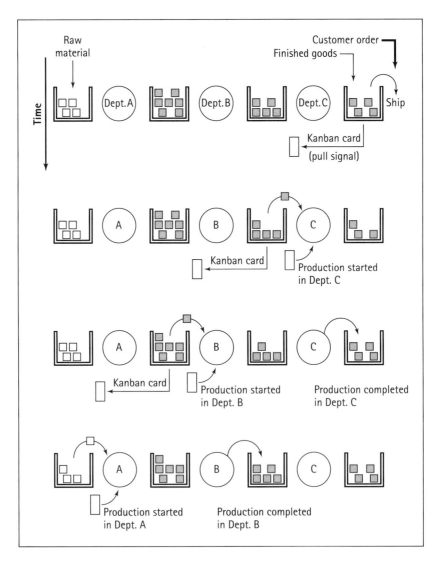

Figure 9-2. Sequence of Events in a "Pull" System

Pull systems have two key advantages over push:

- *Ease of control*: The simplicity of the kanban system's operating rules, combined with the visibility of kanban cards and containers, results in a system that is easy to manage and control.

- *Support for quality and reliability improvement*: The low WIP levels and tight controls in a pull system require high product quality and machine reliability to prevent disruptions. (In a JIT environment you must implement Total Productive Maintenance and quality improvement systems for kanban to be effective.) A pull system also motivates continuous quality and reliability improvements by highlighting the source of disruptions when they do occur.

Researchers have also proved some mathematical properties of pull systems that make them theoretically better than push systems in certain situations.[2]

KEY CONCEPTS OF JIT (LEAN MANUFACTURING) COMPARED WITH QRM

To get at the heart of the differences between QRM and JIT I will refer to concepts that represent the very latest in "lean thinking" from Womack and Jones.[3] Specifically, I will define three key concepts of lean manufacturing and contrast them with the corresponding principles in QRM. These lean concepts are:

1. Elimination of *muda* (Japanese for waste)
2. Implementing flow
3. Implementing pull

Elimination of Muda

The classic JIT literature lists seven types of *muda*, and Womack and Jones add one more to the list. A basic tenet of JIT is the systematic elimination of *muda* through eliminating non-value-added waste, resulting in reduced lead times, improved quality, and reduced costs. In contrast, QRM implies the relentless reduction in lead time, resulting in the elimination of non-value-added waste, improved quality, and reduced costs. Also, once you adopt the QRM approach many additional forms of waste are uncovered that do not immediately surface when applying JIT. These are all the forms of waste caused by long lead times (some of which include the classic JIT waste, but the list is much longer—see Chapter 2).

As an example of this, in 1997 I was part of a team working on implementing QRM at a midwestern metal fabricator (see detailed description in the case example later in this chapter). The plant had already worked with some consultants using a JIT approach, and they had identified what they thought were the key forms of waste that needed to be eliminated. They presented this list throughout the plant to motivate the employees to eliminate waste. Early in 1997, when our QRM team conducted a workshop with a group of 30 managers and employees to identify waste caused by long lead times, we came up with more than a dozen important items that the JIT group had not identified. I was told later that this list was a real eye-opener for many of the key managers in the plant.

Another example that differentiates the view of waste in JIT versus QRM is the way the JIT system requires inventory in many intermediate stages of the materials replenishment system (see the discussion of pull below). But in the QRM approach this kind of inventory is truly "waste," because these are products for which there is, as yet, no end demand. And yet JIT does not recognize this as waste. Quite the contrary, it is institutionalized in the JIT pull system. In contrast, the ultimate aim of QRM is not to introduce any material into the system until there is a firm order for it.

Implementing Flow

Another key concept in lean manufacturing is to make the value-creating steps flow. At the first level this involves breaking down the traditional organization of functional departments with a batch-and-queue mode of operation. This is accomplished by focusing on a given product and laying down all the resources for this product so that an order can proceed continuously without any backflows or stoppages. In doing this, one also rethinks specifics of work practices and machines used. Thus far, this sounds similar to the QRM approach, with one exception: In QRM it is not necessary for cells to have unidirectional flow. (Refer to Figure 4-1c showing how an order might skip two of the operations, but also how it can backtrack in the cell—a "no-no" in JIT cells.) However, the additional JIT aspects of takt time, *heijunka*, and flex fences underline even more the differences between JIT and QRM.

Takt Time

In describing flow, Womack and Jones state, "A key technique in implementing this approach is the concept of takt time, which precisely synchronizes the rate of production to the rate of sales to customers." Specifically, takt time is the time between completion of each piece, if the shipping rate to customers is to be maintained. Womack and Jones again: "The point is always to define takt time precisely at a given point in time in relation to demand and to run the whole production sequence precisely to takt time." Once takt time has been defined, the goal is "for the work team and its technical advisors to determine how to adjust every operation step so that it takes exactly the takt time. This can often be done through careful development of *standard work*, in which every aspect of the task is carefully analyzed, optimized, and then performed in exactly the same way each time in accordance with a work standard."

QRM, on the other hand, is designed to work in situations where products are highly customized and with highly differing specifications, or where there is a large number of possible products with variable demand for each one. In this situation, the organizational structure as a whole needs to be more flexible, allowing more general organization of work within a cell, as well as a more flexible organization across cells, which you will learn more about later in this chapter. In addition, the variability in processing needs and demand for products makes the use of takt time impractical. Although you might define an average time based on a given month's orders, the daily demands on a given machine could be so different that the takt time concept is impossible to implement.

Books on JIT do mention that if there is a change in demand (e.g., sales increase or fall), you need to redefine the takt time and reoptimize tasks using the approach above. However, the detailed nature of the task optimization approach described above makes it clear that this is not an activity that one would undertake on a daily basis.[4] When you have relatively stable demand that shifts a little from week to week or month to month, the flow approach makes sense. But in the QRM context, the variabilities described above can, from one day to the next, lead to huge swings in work content for a given operation. The takt time approach is just too simplistic or unrealistic for QRM then. Instead, the QRM approach tackles this variability in requirements using the many principles in this book, while still managing short lead times. Key among them, and worth repeating here, are:

- Organizational flexibility (resulting from rethinking product design, process design, and organization structure).
- Understanding and exploiting system dynamics that result from this type of interaction and variability.
- Implementation of novel constructs such as time-slicing.
- Use of queueing models to plan and manage capacity and lot sizing to cope with the variability.

Womack and Jones also state that when the specification of the product changes, the right-sized machines can be added or subtracted and adjusted or rearranged so that you always maintain continuous flow. Again, this implies that such changes are not going to be undertaken daily or even weekly, but once in a while. Certainly it is difficult to envision how you would maintain flow in this fashion for customized products that vary daily. The QRM strategy is to create such customized products via the flexible structures that result from the POLCA strategy described in this chapter.

Level Scheduling and Flex Fences

In order to implement flow it is also important to freeze the schedule. Womack and Jones again: "JIT is also helpless unless downstream production steps practice level scheduling (*heijunka* in Toyota-speak) to smooth out the perturbations in day-to-day order flow unrelated to actual customer demand. Otherwise bottlenecks will quickly emerge upstream and buffers (safety stocks) will be introduced everywhere to prevent them." In fact, a major example in the Womack and Jones book, namely the Bumper Works story, shows that one starts with level scheduling and then uses this to set takt times. However, one cannot level the schedule across multiple upstream and downstream steps unless it is frozen within some time horizon. Thus both a frozen schedule and level scheduling are needed. A part of level scheduling in JIT is finding ways to reduce setup times and run smaller batch sizes. Here QRM and JIT agree. But the QRM approach differs in that it recognizes that variabilities (as above) can be ingrained in the nature of a company's business. Indeed, the QRM approach is attractive to these kinds of businesses for this very reason. This is why the QRM organization is designed to cope with these variabilities without negatively affecting lead times. So we might say that in QRM instead of crying "implement JIT-like flow" we advocate "flow with the punches!"

Another issue with implementing the takt time arises when a factory needs components from suppliers with long lead times. If demand increases, even though takt time may be shortened inside the factory, these components may not be available. The typical flow manufacturing approaches attempt to finesse this issue by setting "flex fences," which are ranges of demand increases that a supplier should be able to provide at short notice. Since the flex fences are defined ahead of time, suppliers usually accomplish this by maintaining a buffer sufficient to handle the flex fence. (Womack and Jones also acknowledge that most applications—or misapplications—of JIT in the United States resulted in suppliers maintaining vast inventories of finished goods.) In a QRM company with a wide range of products, this would imply that someone along the supply chain would need to maintain high buffer stocks. More extreme, in a QRM company with custom-designed products, the company often does not know it needs a component until after an order is received and engineered. In either case, the flex fence approach is not practical. As you will see in Chapter 10, the QRM approach changes both the operation of the suppliers as well as the very structure of the interaction between a QRM company and its suppliers.

Implementing Pull

To understand how pull works to support lean manufacturing, I'll summarize the example of a car bumper described by Womack and Jones. When a customer arrives in a Toyota dealer service bay and the mechanic determines that a new bumper is needed, the dealer will "pull" this bumper from the dealer's inventory. (By speeding up replenishment, Toyota is able to ensure that dealers can have a small number of each part available, across a wide range of parts, thus it is likely that the part will be in the dealer's inventory. If not, the part must be found at the next level in the chain, described now.) The sale of a part to a customer triggers a pull signal to the Toyota Parts Distribution Center (PDC), which begins the cycle to replenish that bumper at the dealer. When the PDC ships the bumper to the dealer, that triggers a pull signal to the preceding link in the supply chain, which is the Toyota Parts Redistribution Center (PRC). As this center ships a bumper to the PDC, it sends a pull signal to Bumper Works, the factory that makes the bumpers. Bumper Works, in turn, as it uses up raw material to make the bumpers, sends a pull signal to its supplier of sheet steel.

In addition to the pull signals across organizations, there are pull signals operating within each organization. Taking Bumper Works as an example, a shipment to the PRC does not trigger a pull signal to the steel supplier. Instead the shipment from finished goods triggers a pull signal to final assembly to replenish the parts. As each operation at Bumper Works uses up material to produce a replenishment for the next stage of production, it sends a pull signal to the previous stage. For example, Womack and Jones: "As the [welding] booth used up its reserve of [parts] for Bumper A, the welders would slide the empty parts tub ... to the stamping machine. This provided the [pull] signal to stamp more parts for Bumper A." And so on, upstream all the way to the point where the pull signal leaves Bumper Works and goes to its sheet steel supplier.

Let us understand the implications of this organization. As Womack and Jones aptly put it, the philosophy behind pull is "sell one; buy one" or "ship one; make one." But what "sell one; buy one" implies is that you have one all made and ready in stock to sell; similarly "ship one; make one" means that you have one in finished goods. For the QRM company that makes tens of thousands of different items, this implies that there is inventory of each of these items, or its partly manufactured stock, at *each* stage of the supply chain. Not only does this mean that you have inventory at the end point of each of the organizations in the chain, such as Bumper Works and the PRC, but also it implies inventory between each operation within each manufacturing organization, such as the parts tubs that circulate between the welders and the stamping machine. Multiply these stocks by the number of unique items needed at each stage of the supply chain, and you get a huge requirement for a company that has a large number of end products. Even worse is the case of a company that custom designs and fabricates each product—the pull system fails at the very first step above. There is no product to sell, since the parameters of the product are not known till the order is received, hence "sell one; buy one" is not possible at the final stage of the chain. Similarly, most of the intermediate stages cannot have the required inventory to pull from either, since those intermediate stages whose operations depend on the parameters of the final product cannot start production until the actual order is engineered.[5]

Now that I've discussed some key elements of a lean manufacturing system, I can clear up some misconceptions about its capabilities.

MISCONCEPTIONS REGARDING THE PULL SYSTEM

A common description of JIT is that it enables a company "to give the customers what they want, when they want it." In interpreting this, people then state that lean manufacturing systems can make customized products—but what exactly is meant by customized? It is more accurate to say lean systems are best suited to make products that involve minor customization of a main product, or products whose customization involves choosing from a set of predefined options, as opposed to totally custom-engineered products. To see this, consider this statement from Womack and Jones: "Lean systems can make any product currently in production in any combination, so that shifting demand can be accommodated immediately." Note the use of the phrase *currently in production*, which drives home my point.

Second, there is a misconception about pull. I occasionally hear statements like, "A custom-engineered product can use the ultimate in a pull system where the customer order pulls all the way back to the first operation." This is an abuse of the term pull. (In fact, this is push!) As seen from the detailed example above, which is the core example in one of the most recent books on lean thinking, pull starts with sale of a product already in stock and then works its way back upstream through replenishment of inventories. This scenario simply cannot exist in a custom-engineered environment.

Third, a prerequisite for lean manufacturing methods is a marketplace with relatively stable demand. Indeed, a fundamental basis for lean thinking is the premise that "end-use demand of customers is inherently quite stable and largely for replacement" (Womack and Jones). While this may be true in large segments of the market, it is my hypothesis (verified by many companies that are adopting QRM) that there is a great deal of opportunity in pursuing other market segments such as emerging segments where the pattern of demand is unpredictable and product requirements are changing fast. Or markets where companies need to tailor their products in detail to individual customers. In Chapter 12 I give an example of how Ingersoll Cutting Tool Company hypothesized the existence of such a market segment, went after it, and having found it was able to increase its market share by more than 500 percent. Indeed, the QRM strategy helps companies find such market niches wherever they may exist, and once found, they can blow the niche wide open into a gaping hole—at the same time, they will be the only company that can fill the now large hole with products.

This argument can be carried further. I believe—and this is supported by proponents of other approaches such as agility—that in this era of computer-aided design and computer-aided manufacturing, the marketplace is headed for increasingly more custom-engineered and individually tailored products. This means more and more market segments will be suited to the QRM strategy.

Table 9-1 summarizes the points about QRM versus JIT. The reader can also look at further comparisons in Table 1-1.

EXPANDING BEYOND JIT STRATEGIES—A TALE OF THREE COMPANIES

To show the difference between QRM and JIT, it is also useful to discuss the situation three U.S. firms are facing today regarding the choices in Table 9-1. All three firms have decided that while JIT/lean manufacturing principles have offered a good starting point for improvements, they need to expand their thinking beyond standard JIT strategies because of the nature of their business. After comparing these company situations with the notes on JIT, it will be clear why these companies, and others, are extending their initial JIT efforts through additional manufacturing strategies like QRM.

Table 9-1. Comparison of QRM with JIT

JIT or Flow or Lean Manufacturing	QRM
Systematic elimination of waste leads to continuous improvement.	Relentless reduction of lead time results in continuous improvement and elimination of waste.
Create "flow" by designing production lines so orders can proceed continuously without any backflows or stoppages. One-piece flow is the goal.	Create cells based on families of products with similar operations. However, they need not have linear flow; products can go through cells in various sequences. One-piece flow is not necessary; small batches may be necessary as a consequence of the customized nature of products.
Support flow using *takt* time and level scheduling. Use detailed analysis of tasks and standardization of work to achieve the balanced *takt* times throughout the production facility.	Support the ability to meet demand for widely differing products through organizational flexibility and techniques such as time-slicing, and by exploiting the understanding of system dynamics. Also, use different combinations of cells to create varying end-items.
Suppliers meet flow requirements via pull signals and flex fences.	Suppliers support quick response by changing their operations and via redefining their interactions with the customer.
Pull signals material replenishment: "Sell one; buy one" or "ship one; make one." Implies there is a product ready in stock to sell, or there is a product in finished goods to ship. Hence inventory needs to be kept at each point in the supply chain. Too much inventory when there are a large number of end items. Doesn't work for custom-engineered products.	Tailored to companies that have very high product variety or engineered products. Goal of QRM is not to start a job until there is an order for it. Uses a combination of push for material planning and release, and a modified pull approach to prevent congestion.
Best suited for providing custom combination of predefined options for a baseline product.	Strongest when used for custom-engineered products.
Requires relatively stable demand, and largely for replacement products.	Use to forge new market niches such as emerging segments with unpredictable and rapidly changing demand, or where products must be tailored to individual customers.
Emphasizes on-time delivery as a primary performance measure.	Primary measure is lead time reduction. On-time performance is achieved as a by-product of the strategy.

Company One: HUFCOR

Located in Janesville, WI, HUFCOR is the world's largest manufacturer of operable and portable walls and accordion doors. Its main customers are convention centers, hotels, schools, and churches. Although the company has been in operation since 1900, in the last two decades its market share has grown significantly, from 12 percent to more than 50 percent. HUFCOR is, however, currently facing pressure from competitors who are supplying some products within a shorter lead time. Thus in 1996, Mike Borden, president of HUFCOR, made a commitment to pursue strategies that would achieve significant lead time reduction.

The nature of HUFCOR's products requires that each order be highly customized. I'll describe one line of products, the operable walls. These are partitions or panels that are suspended from steel or aluminum tracks in the ceiling. When in place, they resemble a permanent wall, and their surfaces are finished to match the room decor. Accessories such as chalk and tack boards, and pass-through doors, can also be added. Each customer order is characterized by a number of parameters that make it unique: the width, height, frame type, track type, edge seal material, surface material, and location and types of accessories and pass doors. Customer orders range from one panel to several hundred in quantity.

In view of these highly specific customer requirements, HUFCOR operates in a make-to-order environment. After studies by several different manufacturing experts, HUFCOR management is exploring the use of both lean manufacturing systems as well as other alternatives. There are two challenges to implementing a "pure pull" system at HUFCOR:

1. Because each order is customized, it doesn't make sense to start material into the system until the parameters of the end item are known. Hence, intermediate or end items cannot be pulled from a buffer to trigger replenishment: The end item (or intermediate item) simply does not exist until they know the exact customer requirements. (For a few higher volume components, such as commonly used tracks and rails, pull can be implemented, but this is only a small part of HUFCOR's operation.)

2. The highly customized nature of orders and significant differences in processing rates for different finishes on the products makes it difficult to set takt times for the various operations at HUFCOR. One could compute an average takt time based on forecast demand for three months, but the actual times from day to day might be quite different. The lean manufacturing environment of almost no inventory and smooth flow may be hard to achieve in this situation.

The vice president of manufacturing at HUFCOR, Jim Landherr, is therefore exploring both "standard" pull approaches in some parts of the company, as well as QRM principles in other parts, to see which can be best applied in their environment. Specifically, sections of HUFCOR are adopting the following QRM principles: QRM in order processing via office cells and other office operations principles; queuing models to facilitate capacity planning; QRM to form more flexible cells (again using queuing models to help design cells); and the QRM number to measure performance. All of these practices are currently being implemented.

Company Two: Midwestern Metal Fabricator

This example concerns a large midwestern metal fabricator (I'll call it MMF), that operates a facility employing more than 2,000 people. MMF's customers represent a diverse spectrum of industries, including construction, automotive, aerospace, and electronics. To serve this broad market, MMF manufactures more than 5,000 different products. The factory has more than 100 specialized machines, ranging in size from a large continuous processing line occupying tens of thousands of square feet, to a shear that takes up less than 500 square feet. Products journey through the factory in widely differing routes, employing different sequences, and often looping back to machines they have visited before. The routing for a finished product typically consists of dozens of processing steps.

In recent years, under the guidance of its vice president of manufacturing, MMF has achieved significant improvements in productivity and performance. As part of its ongoing quest for improvement, MMF has now targeted a 50 percent reduction in lead time for its products. To this end, in 1997 the VP-manufacturing formed a team charged with exploring ways to achieve this lead time reduction. The team consists of several MMF employees and a few external "experts," of which I am one. The team recognized that MMF is organized as a large job shop, and it experiences all the symptoms of the Response Time Spiral. The team studied the factory's operations in detail, and also looked at the latest in manufacturing strategies, including JIT/pull/lean manufacturing. After a couple of months of study, the team came to the conclusion that while pull/lean approaches provided a good starting point to motivate improvements, significant extensions would be required to adapt these principles to MMF's operations. There were several reasons to back this conclusion:

(continued)

1. With 5,000 different end-items, in order to implement a kanban-based pull system there would have to be a tremendous amount of inventory both in the finished goods warehouse and at many intermediate stages of production. (See the discussion in the preceding section.) Even with lean manufacturing principles in place and fast response within the system, the level of total inventory required to achieve pull would be so large that its carrying cost would eat into a substantial part of MMF's profit margin.

2. The large variety of products, with variable and unpredictable demand and significant differences in processing rates, makes it difficult to set takt times for different processing centers. MMF could compute an average takt time based on forecast demand for three months, but the actual times from day to day would be so different that no semblance of smooth or balanced production would be attained. In other words, the lean manufacturing environment of almost no inventory and striving for smooth flow is an unachievable ideal in this situation.

3. Many of the machines at MMF are extremely large and expensive—literally tens of millions of dollars each. Further, for most of these expensive machines, the factory has only one of that type. So the lean manufacturing ideal—creating flow by forming families of products and physically dedicating resources to them—cannot be achieved, since these unique resources are required by literally thousands of different products. (For a small number of high-volume products, MMF has been able to create flow for sections of their process. However, this still leaves literally thousands of products that share the remaining machines.) Another possible lean manufacturing approach would be to replace the large machines with much smaller machines that would deal with a few products each, and place these smaller machines together to create flow. There are two obstacles to this approach: one is technological, the other financial. One, the sheer energy and horsepower required for some of the metal processing operations is such that the size of the equipment must be large; smaller alternatives are not available. Two, even if such alternative technology were available, MMF has billions of dollars of assets in existing machines, and could not justify simply writing them off overnight and investing in alternative production methods.

The answers to these problems are still being explored, because all of us "experts" feel that the above factors make MMF's situation nontrivial. Currently the team at MMF is examining how it might apply a combination of several strategies, including pull methods coupled with QRM strategies (time-slicing, POLCA, setup, and lot size reduction), to achieve its target of 50 percent lead time reduction.

Company Three: Ingersoll Cutting Tool Company

Ingersoll Cutting Tool Company (ICTC) is described in detail in Chapter 12. The Ingersoll family of companies is well known for its leadership in the machine tool industry. The following example focuses on one aspect of the company's operations. ICTC makes both stock cutters (standard products) and nonstock cutters (specialized products).

Among the special cutting tools manufactured by ICTC are high production tools used on transfer lines, boring tools, crankshaft and camshaft broaching tools, indexable form tools, saws, slotters, and gear generating cutters. In the highly competitive market for cutting tools, ICTC has found that a reduction in lead time can significantly improve market share. As a result, since 1992 the president of ICTC, Merle Clewett, has made lead time reduction a key company strategy.

The special cutting tools are custom designed. This means that each customer order requires design engineering, process planning, N/C programming, materials planning, and other office activities. Following release to the shop floor, each such order follows a unique routing through various work centers in ICTC's factory. Although you can identify families of similar products, such as custom straight shank end mills, each particular order has differing demands on the various machines it will use. One job may take a lot of time on a five-axis machine for pocketing; another may need a significant amount of machining on a gun drill to create a hole for through-coolant. Orders can be for a single cutting tool or 10 of them, further exaggerating the variability of demand on the machining operations.

The result is that ICTC has already implemented QRM principles of office cells, has one shopfloor cell in place, is exploring time-sliced virtual cells for other parts of the factory, and the QRM number for performance measurement.

SUMMARY OF DISADVANTAGES OF PULL OR FLOW METHODS FOR QRM

Given the vast literature on JIT and its successes it's worthwhile to review some of the less-documented drawbacks of pull systems. It will become even clearer why you must use a different material control system for QRM. Specifically, the drawbacks mentioned below apply to pull systems using the typical kanban control mechanism, as well as to the demand-driven systems described above.

- *Inability to handle custom jobs*: Pull or flow methods are designed for repetitive manufacturing environments. As I've emphasized, an important domain of application for QRM is in one-of-a-kind custom manufacturing situations such as the HUFCOR and ICTC examples stated earlier. Standard pull methods cannot be used for those situations.
- *Proliferation of WIP when product variety is high*: Even if you are not in a one-of-a-kind manufacturing environment, but the number of products made is very high—such as in the midwestern metal fabricator example—a standard kanban system will result in a large quantity of WIP being maintained. This is because in a "pure pull" system the kanban cards have to be part number specific. To see this, consider that the customer purchases a *particular* product, let's call it SR384, from the finished goods buffer. Suppose that a pull system is in place with single piece flow, i.e., each piece used from a buffer triggers a pull (as opposed to a whole container being used). In that case, the customer purchase triggers a pull signal to the previous workstation. If this pull signal were not part specific, then that workstation might replenish finished goods with a part other than SR384. Thus the pull signal to the previous workstation must specify that an SR384 is to be made. Similarly, when this previous workstation manufactures another SR384 for finished goods, it uses up one of the partly made SR384s in its input buffer and triggers a pull signal for the previous operation to replenish it. Again, this pull signal must specify the part number, because this operation, in turn, needs to work on a partly finished SR384—any other part it works on is not relevant to replenishing the SR384 in finished goods when it is next needed. So all up the line the kanban cards must be part specific. (There may be some point back in the line before which parts are generic, e.g., 1″ bar stock; however, up to this point all the cards must be part specific.) Carrying this logic through, you get an absurd situation for a company that makes 1,000 different specifications—it needs to carry stock for 1,000 parts, not only in finished goods, but also in each buffer in front of each workstation in the line. Somehow it doesn't seem right that a method that has as its goal the elimination of waste should end up flooding the manufacturing system with inventory! This is precisely the dilemma that the large midwestern metal fabricator was facing as it contemplated how to implement a pull system with its 5,000 different end products.

- *Late shipments in a growing market*: Even in a repetitive manufacturing situation that is supposedly perfect for pull, if the market is
  growing fast, it will take a while for the pull system to catch up—or
  it may never quite catch up as long as the market keeps growing.
  What this means is that the finished goods buffer will be empty and
  there will always be unfulfilled demand; that is, customers who are
  waiting anxiously to snatch up the next product that comes off the
  line. Why will this happen? Initially the level of inventory at each
  stage of the pull system has been set for the current demand. When
  the demand surges, buffers at each stage of the system are emptied,
  sending frantic pull signals upstream. Since with the pure pull system, signals only go to the next upstream station, it takes a while
  for each workstation to work off its inventory and get to the point
  where it is constantly waiting for the next piece from the upstream
  station. Increasing the rates or capacity at any of these workstations won't help, because they don't have the parts to work on. It
  won't be until the first step in the process is reached and someone
  realizes that this first step cannot keep up with demand that an
  improvement can occur (for instance, the kanban cards can be
  recalculated to reflect the fluctuations in demand). The first station
  can increase its production rates and this improvement can start to
  trickle downstream. By now, however, you've lost quite a bit of
  time and many customers are waiting. Plus, it might be that downstream workstations can't sustain the increased rates, but it will
  take a while to find that out, too. Worse yet, if demand is continuing to grow, then the increase at the first station is too little, too
  late. The number of delayed shipments just continues to mount. The
  only solution is to realize that "pure pull" simply will not work and
  you have to rethink your capacities throughout the line. (In fact, if
  you think through the details, there is no choice but to "push" some
  material into the line in this situation.)
- *Late shipments and excess inventory when product mix is changing*:
  Analogous to the above situation, but with worse effects, is when a
  pull-based line that makes a mix of products faces a changing mix
  in the marketplace. If the demand for one product is surging while
  that for other products is declining, pull will lead to two undesirable outcomes. First, there will be late shipments and unhappy customers for the product with surging demand, for the reasons
  described in the previous item. Second, the pull line will continue
  to make components, assemblies, and intermediate inventory for all

the other products with declining demand, until all the buffers up and down the system are saturated with such WIP. In addition to this WIP being wasteful, since you have more than you need, you may *never* use some of this "extra" WIP.

• *Shifting bottlenecks and complexity of kanban settings when mix is volatile*: Suppose there is no clear up or down trend for products as in the previous example, but rather, customer demand is just plain variable, as is true for many markets. In this case, it is hard to design a pull system that will run effectively. Suppose, in a pull line that includes a mill and a drill, Product A requires more time on the mill and Product B takes more time on the drill. Which machine is the bottleneck will depend on the demand for A and B. If you set kanban levels for each product based on some average demand, then during periods when Product A demand is high, the mill will hold up production up and down the line; that is, there will not be enough kanban cards in the upstream and downstream loops that include the mill. Conversely, if demand for B is high, then the kanbans for the drill will become the problem. You could design the line for worst case, but if there are several products made on the line, that would mean flooding the line with a large amount of inventory at all levels just for this worst case, obviating the advantages of pull. You could also try to dynamically adjust the number of kanbans at each stage of the system, but this would eliminate the simplicity of the kanban system. Since it is not trivial to figure out the root cause of bottlenecks in a typical production line, some analyst would have to continuously monitor the situation and decide if the problem was due to insufficient kanbans in any particular loop. One might even get whiplash and ripple effects from too much tinkering. For instance, there is a short-term problem at a machine, so you mistakenly add too many kanbans for a specific product, and the machine makes too many of this product while ignoring other products. When you discover this you cut back on this product and increase the kanbans for other products. This leads to underproduction of the first product, until you have created your own mini Response Time Spiral within a pull system.

As you look at all these examples, you might say, "Wait a minute! We're not going to operate like brainless automatons—for example, we'll notice when demand is going up and anticipate the need for more production by increasing the number of kanban cards at the first station long before we reach the problem situation above." Aha! As soon as you say you are going to look ahead and make an upstream workstation increase its production *before* it has felt the pull signal, you are actually engaging in a *push* strategy! So it is clear that a pure pull system is not going to work for QRM and that some elements of push may be needed. But what is the ideal way to combine the advantages of both systems without their disadvantages? I'll now show how QRM extends the principles of push and pull into a whole new arena of material control.

MATERIAL CONTROL—DON'T PUSH OR PULL, POLCA

In the context of QRM, where cells are organized to eliminate the Response Time Spiral and focus on reducing lead times, pull and push systems have several drawbacks. As a result you need to develop a different material control method, one that extends these principles into a new hybrid called *Paired-cell Overlapping Loops of Cards with Authorization* (POLCA).

I'll begin by reviewing a suitable QRM material control system. The QRM strategy is particularly effective for companies that make custom designed products in small batches (or even one-of-a-kind). QRM is also effective for companies that, while they don't custom design products, have such a wide variety of options and combinations of specifications that they cannot afford to store inventory for all these options at various stages of their manufacturing system. The way we envision such companies being organized for QRM is for them to create cells focusing on subsets of the production process for like families of parts, and then to process a given customer order through differing cells depending on the needs of that order.

Example—The CFP Corporation

To understand this organization in detail, consider a company called CFP Corporation that makes customized faceplates: rating plates used on products ranging from small electrical cabinets to large earthmoving equipment, nameplates on industrial appliances, and faceplates on calculators, appliances, and control panels. The plates are made from either aluminum or plastic. They range in size from under 1 inch square up to over 18 inches in either dimension and have information printed on them, along with features such as holes, notches, bends, brackets, and fasteners to assist in mounting the plates onto the final product. There are many large companies in this industry that make high-volume products such as automotive displays or faceplates for popular home appliances, but CFP's strength is in making small batches of plates for specialized markets. Specifically, while the high-volume producers typically accept orders for a quantity of 20,000 or more plates of one type, a typical order at CFP may be for just 200 aluminum rating plates that will be used on a particular model of a large backhoe. Another typical order could be for 500 plastic faceplates for a display that will be used in a customized electronic device.

To serve these and other highly varied markets via a QRM strategy, CFP has created several cells (Figure 9-3). First there are two printing cells: P1 focuses on screen printing, P2 on lithographic printing. In keeping with QRM ideas, each cell contains a number of functions and machines, such as equipment to produce the photographic plates or

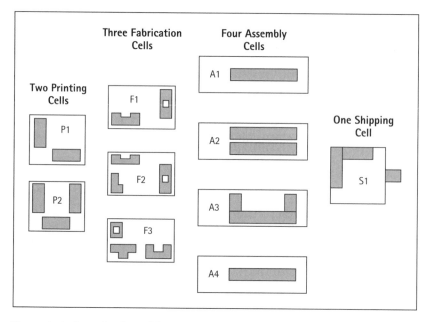

Figure 9-3. Organization of Cells at CFP Corporation

screens, printing machines of various sizes and with abilities ranging from two to six colors, thermal or UV drying ovens, and laminators or coaters to produce a protective layer on the printed plate. The printing cells start with plain sheet stock of aluminum or plastic and complete all the printing operations required for the product, but still in sheet form. Next are three fabrication cells, F1, F2, and F3, which convert the printed sheets into individual plates with the desired features. Operations performed here include punching holes, slots, and notches, cutting the sheets into strips and then cutting the strips into individual plates, and producing angles or bends in the plates. Cell F1 focuses on plastic plates, F2 on light gauge aluminum, and F3 on heavy gauge aluminum. After the fabrication operations, plates go to one of four assembly cells, A1 through A4. Here finishing operations such as cleaning and deburring are performed, along with assembly of fasteners if needed, and then plates are placed in protective packaging. The four assembly cells differ in terms of the size of products handled, the types of fasteners to be attached, and the form of packaging to be used. Finally, all orders go to the shipping cell, S1, where the packaged plates are placed in cartons and shipping containers and then loaded onto trucks.

Each customer order to CFP is served by using the appropriate combination of cells needed to print, fabricate, and assemble that order. Orders can have different demands within the cells too: The punching requirements for an order for large plates with lots of holes may use a lot of time on a CNC turret punch in F3 and not much time on the shear, while another order for small plates may have very little punching time but take a lot of time for shearing. In addition, some large plates may need to be sheared first before they can fit on some of the punches, while small plates may need to be punched first and then sheared.

So in order to serve its market niche for customized plates, CFP has three key requirements for its materials management system: (1) the ability to route products through different combinations of cells as needed by a given order; (2) within a cell, the ability for products to use machines in different sequences; and (3) a good deal of flexibility in terms of capacity requirements for each operation in a cell. The original intent of MRP systems was to enable a company such as CFP to achieve these three requirements, but for all the reasons stated in this book, that intent was not fulfilled successfully. Also, by comparing CFP's requirements with the discussion in the preceding sections, you can see that a pull system will not work for this organization.

How POLCA Works for the QRM Organization

The material control and replenishment system that I've devised for use with a QRM organization such as that at CFP Corporation is Paired-cell Overlapping Loops of Cards with Authorization (POLCA). This is a material control system that operates in the context of (1) a high-level material requirements planning system (HL/MRP), (2) cellular organization, and (3) flat BOMs. POLCA has four key features incorporating aspects of MRP and kanban that enable a company to customize their products while giving it control over congestion and excessive WIP:

1. Release authorizations are created via HL/MRP.
2. Card-based material control methods are used to communicate and control the material movement *between* cells. For material control between workstations *within* a cell, you can use various other procedures, including kanban—examples of several such procedures will be given below.
3. Production control cards, which I call POLCA cards, instead of being specific to the product, as in a pull system are assigned to certain pairs of cells, chosen as follows. Within the planning horizon, if the routing for any order goes from any cell (say *A*) to another cell (say *B*), then the pair of cells *A* and *B* is assigned a number of POLCA cards, all of them called *A/B* cards. The choice of the actual number of such cards is discussed later. Since the intent of QRM is to respond quickly to customized orders, the planning horizon (such as a month) may be longer than the lead time (say, one week), in which case you should use forecasts of sales by aggregate families to decide which routings might be used.
4. The POLCA cards for each pair of cells stay with a job during its journey through both cells in the pair before they loop back to the first cell in the pair. For example, an *A/B* card above would be attached to a job as the job entered cell *A*. It would stay with this job through cell *A* and as it goes to cell *B*, continue to stay with the job until cell *B* has completed it, and while the job moves on to its next cell (say *C*), this *A/B* card would be returned to cell *A*.

Since most cells belong to more than one pair, there will be multiple loops of cards that overlap in each cell. (This will be made clear via diagrams below.)

Applying POLCA to the CFP Corporation

To understand more about how and why the POLCA system works, and to help you visualize its operation, I will go over the detailed operation for the CFP Corporation example. With that in place I'll cover additional details of the system and its advantages for the QRM enterprise.

Figure 9-4 shows the POLCA card flows for a particular order at CFP Corporation. This order's routing takes it from P1 to F2, then to A4 for assembly, and finally to S1 to be shipped. This order will therefore proceed through the POLCA card loops P1/F2, F2/A4 and A4/S1.

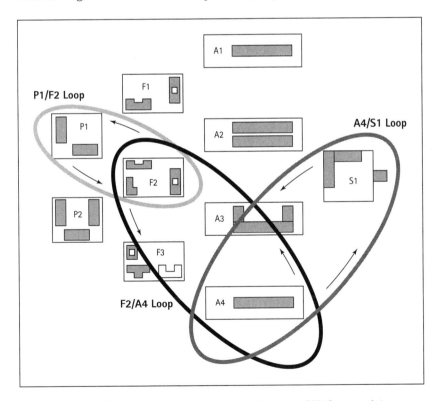

Figure 9-4. POLCA Card Flows for a Particular Order at CFP Corporation

The actual POLCA system includes a procedure of authorizations initiated by HL/MRP, which is omitted at this stage of explanation for simplicity. I will focus here on the material and POLCA card flows. The HL/MRP system prints a routing sheet (also called "ticket") for each job that it launches into the system. This sheet accompanies the job and contains typical information found on routing sheets from MRP systems, namely, the sequence of operations to be done, components to be added (if any) at a given operation, and any other special instructions. However, in addition, this routing sheet states the sequence of cells that the job must visit. In this case it would list P1, F2, A4, S1. (If the company is using a "paperless factory" with all job information on computer screens rather than on routing sheets, this information is made available on the computer screen related to the job.)

Now let's follow the order through the factory. The order's routing begins at P1. If raw material is available, and a P1/F2 card is also present at the beginning of this cell, then the job is launched into cell P1. See Figure 9-5.

As you will see from the details below, by pairing cells together, the POLCA cards ensure that a given cell only works on jobs for which the destination cell also has capacity, i.e., is not backed up. Recall that one of the problems of MRP is that departments would continue to push

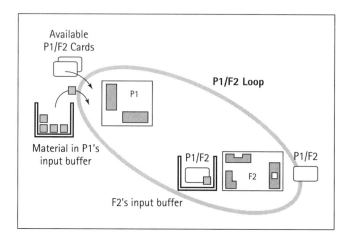

Figure 9-5. Job Launch into Cell P1

jobs to other departments even if they were backed up, adding to the proliferation of unnecessary WIP in the system. After the job is launched into the P1 cell, we have the *first departure from a kanban system*. Although the POLCA card stays with the job through the cell, the flow of material between the workstations in the cell is *not* controlled by the POLCA card, a point that I will elaborate later.

The information on a POLCA card is quite simple (see Figure 9-6). Most important is the acronym for the paired cells for which this card will be used—P1/F2 in this case. This is in large letters, since it should be easily visible to the shopfloor workers. Next are two lines describing the pair of cells, in case a worker is unsure of which cell(s) the acronym refers to. Finally, there is a serial number, which is not used by shopfloor workers. This number is used only by the material planners who control the number of POLCA cards of a given type and modify this number during the periodic planning exercises. The planners keep excess POLCA cards in their office (off the shop floor) and the serial number can help track how many POLCA cards of a given type are outstanding (on the shop floor).

After the printing cell has completed its operations, the job and the P1/F2 POLCA card go to the input buffer for fabrication cell F2; see Figure 9-7. This move, from cell P1 to the input buffer of cell F2, can be the responsibility of either the cell team in P1 or a material-handling operator. I prefer the cell team to deliver the job, because it means the P1 cell team has complete ownership of the job until it is delivered to

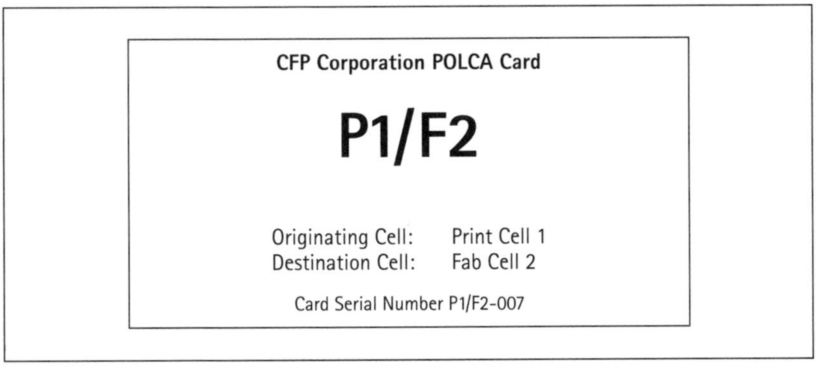

CFP Corporation POLCA Card

P1/F2

Originating Cell: Print Cell 1
Destination Cell: Fab Cell 2

Card Serial Number P1/F2-007

Figure 9-6. A Typical POLCA Card

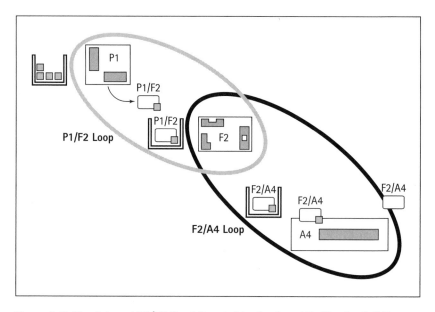

Figure 9-7. The Job and P1/F2 Card Proceed to the Input Buffer for Cell F2

its customer (cell F2). However, if it is necessary to use different staff for material handling, you do not need any more cards or other signals. (The kanban system uses a separate "move card.") All the P1 team needs to do is place the completed job in its output buffer area with the POLCA card attached in a visible manner. Material handling operators doing the rounds of the factory will see that there is a job in a designated output buffer area—this alone is a signal that a move is needed—and then they will see from the P1/F2 marking on the POLCA card that the destination of this job is the input buffer of cell F2. Thus the POLCA card already contains the material-handling instructions.

After the job arrives at fabrication cell F2, since it is destined for assembly cell A4, the F2 cell team needs to have a free F2/A4 POLCA card before it can start our job. However, as Figure 9-7 shows, all three F2/A4 cards are elsewhere: one is with a job waiting in A4's input buffer, one is with a job being processed in A4, and one is sitting at the exit to A4, waiting to be moved back to F2. (Below I'll explain how POLCA cards are returned.) When one of the F2/A4 cards does arrive back at F2, the job is launched into F2; see Figure 9-8.

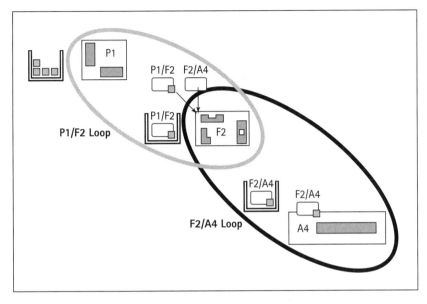

Figure 9-8. An F2/A4 Card Is Available; Job and Cards Are Launched into Cell F2

Now we have the *second departure from kanban*. In a kanban system, the first card (P1/F2) would be removed from the job and sent back to the first cell. However, in POLCA, this card stays with the job through the second cell as well. This means jobs in the second cell will carry two POLCA cards with them as part of two card loops. For instance, each job in cell F2 will carry two POLCA cards with it: one card that came with the job and for which F2 is the second cell in the pair (the P1/F2 card for our example job), and one card that allows the job to be launched into F2 and for which F2 is the first cell in the next pair (F2/A4 for our example job). Hence the use of the term *overlapping loops* in the POLCA acronym. In general, all cells, except those that are at the start or end of a routing, will have two POLCA cards attached to jobs that are in progress in the cell. Clearly, it would be simpler to use only one card at a time. For our example, we could release the P1/F2 card (sending it back to the beginning of cell P1) when the job was launched into cell F2. Then the job would only carry the F2/A4 card into cell F2. However, I'll explain later why using the longer and

252 Part Two: Rethinking Production and Materials Management ....

overlapping loops, with jobs carrying two cards in the cell, will contribute to the QRM need for a high degree of flexibility. When the job is
completed by cell F2, two things occur (see Figure 9-9).

1. The P1/F2 card is detached and returned to the beginning of the
   printing cell, P1. Again, either the F2 cell team or a material-handling person can return this POLCA card. I prefer the F2 cell team
   to return the POLCA card to cell P1, because this should be considered part of "completing their job." However, if a material-handling
   person is needed (e.g., if distances are large), the P1/F2 POLCA card
   itself can be placed in the output buffer area for cell F2. Again, this
   is a signal to a material-handling operator that this card needs to
   be returned to cell P1. (Since the POLCA card is lying in the output
   area for cell F2, and it says "P1/F2" on it, there is no ambiguity
   about where the card should go.)
2. The job is delivered to the input buffer of assembly cell A4 with
   the F2/A4 card still attached. (As previously discussed you can use
   either cell operators or material-handling staff to deliver the job.)

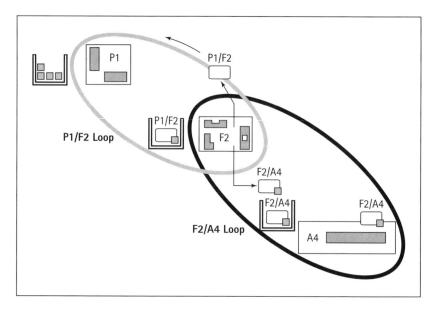

**Figure 9–9. Upon Completion at F2, Job and F2/A4 Card Proceed to A4;
P1/F2 Card Returns to P1**

Now the process repeats at assembly cell A4. To begin, you need an A4/S1 card available. Then the job is launched into the assembly cell, with both the F2/A4 and A4/S1 cards attached to it. When the job is completed, it goes to the input buffer of the S1 cell, while the F2/A4 card is sent back to the beginning of the F2 cell. Since S1 is the last cell in this job's routing, there is no additional POLCA card to wait for, and the job can be launched whenever the cell is ready to start another job. (However, as you will see below, there may be a wait for an "authorization" from HL/MRP.) When the job is completed at S1, it is shipped, and the A4/S1 card returns to the beginning of the A4 cell. This completes the entire POLCA story for the journey of this example job and the various POLCA cards that accompanied it.

Release Authorization—With Some Push

For a viable QRM strategy you need some element of a push system. This is how it works. When the company receives a customer order, the HL/MRP system uses the planned lead times for each cell to determine times when each cell in the product routing *may* begin processing that job. This part of the HL/MRP logic is no different from standard MRP, except that the HL/MRP is based on simpler BOMs and uses lead times for entire cells instead of individual work centers within a cell. (The HL/MRP system treats the cell as a "black box" and just sets a lead time for the cell as a whole, not for the steps within the cell.) But there is one major difference with QRM—the times that are output from the system do not refer to when a cell will start production, or even when it *should* start, but only when it *may* start. Instead of being called *release times* (or release dates), as in standard MRP, these are called *HL/MRP authorization times*. These HL/MRP authorization times are used to modify the POLCA procedures as follows: A job waiting in the input buffer for a cell cannot be started until both the right POLCA card is available *and* its authorization time has passed. To deal with the more typical situations when there are multiple jobs in the buffer that are eligible for starting, as well as multiple POLCA cards, you need to apply the following rules for job processing:

- When the HL/MRP authorization time for a given job at a given cell has passed, we say the job is *authorized* for that cell. Otherwise the job is *premature.*
- Jobs that arrive at the input buffer to a cell are ordered by authorization time, with the earliest authorization time first. If jobs are small, this ordering can be done by physical placement of the jobs. This is the best case, since from JIT experiences, visual management of jobs is known to be an effective scheme. However, if jobs are large, there are two alternatives to the physical ordering: (1) If a computer system is being used instead of paper tickets, the ordering can be listed in the production control computer; (2) if paper tickets are being used, each job should be launched with multiple copies of its ticket. The master copy always stays with the physical material. When the job arrives at a cell, an extra copy of the ticket is detached from the job and brought to the desk or bulletin board where the cell team manages jobs. Here the paper tickets for all jobs in that cell can be easily maintained in order of authorization time. In other words, instead of shuffling physical jobs on the shop floor, you can rearrange the paper tickets. Note that at any point in time, there is a conceptual line that divides all jobs waiting at the input to this cell into two categories: those whose authorization time was prior to the current time, or the *authorized jobs*, and the remaining ones, which are *premature jobs.*
- POLCA cards that are returned to this cell from downstream cells are placed near the input buffer and organized in a way that makes it easy to see if a given card is available (e.g., they can be organized in alphabetical order).
- When the cell team is ready to start on the next job, it picks the first authorized job (i.e., the authorized job with the earliest authorization time). Then it looks to see if there is an available POLCA card that fits this job's destination as printed on the job's ticket. (For example, if the team is at cell F2, and the job is a P1-F2-A4-S1 job, then the team looks for an F2/A4 card.) If there is no available POLCA card for this job, the team picks the next job and tries to find a POLCA card for it. The team continues this procedure of trying to match an authorized job with a POLCA card, until it finds a match, or it runs out of authorized jobs. If it finds a match, this job is launched into the cell. If it runs out of authorized jobs, no job is launched. Note that regardless of which job is launched, the ordering of remaining jobs should not be changed. (In other words, just

because a job had no card does not mean it gets sent to the back of the line. It retains its position.)

- In the case where no job can be initiated the cell team must wait to start another job until one of three events occurs: a new job arrives, another POLCA card arrives, or a premature job changes to an authorized job. If any of these events occurs, the team must go through the above logic to see if they can start a job.
- For a company that prefers to use electronic signals and computer terminals in the cells instead of physical POLCA cards, the logic of matching authorized jobs with POLCA cards is easily programmed into the computer system. Thus the computer system can automatically indicate to the cell team which job (if any) should be launched next.

Planning the POLCA Cards

The procedure used to set the numbers of POLCA cards in each loop is as follows:

1. The HL/MRP system is run for the planning horizon (e.g., three months). This will include firm orders in hand as well as forecasts of products by aggregate families. A caveat is in order here, since we seem to be using forecasts—the antithesis of QRM, and the root of the Response Time Spiral! But there is a big difference between the QRM use of forecasts and the traditional use in an MRP system. A QRM enterprise uses the forecasts only for capacity planning and deciding on the numbers of POLCA cards. *The key difference here is that no material is launched into the system based on forecasts.* The forecasts (by product families) are used only to predict the routings that the products will follow along with rough operation times for the anticipated customer needs. In other words, they are used only for capacity and POLCA card planning. (In contrast, in traditional MRP, material is launched based on forecasts, using up capacity, causing congestion, and the other Response Time Spiral–related problems.)

2. The HL/MRP system then feeds the predicted loads for each cell to the cell teams, who use their cell-level planning tools to respond with their estimated average lead times, as explained earlier (see the discussion surrounding Figure 8-5).

3. The number of POLCA cards for each loop can now be set using a simple application of Little's Law (see Chapter 7). Let LT(*A*) and LT(*B*) denote the estimated average lead time (in days) of cells *A* and *B* over the planning horizon. Also, let NUM(*A,B*) equal the total number of jobs that go from cell *A* to cell *B* during the planning horizon. Finally, let the number of working days in the planning horizon be D. Then the number of POLCA cards for the *A/B* loop (i.e., the loop going from cell *A* to cell *B*) is given by:

No. of A/B cards = [LT(A) + LT(B)] $\times$ NUM(A,B)/D

with any fractional numbers being rounded up to the next integer value.

You should also use this planning procedure as an opportunity to fine-tune the *quantum* of the POLCA cards, that is, the amount of material that should accompany each card. Ideally, for QRM, each POLCA card should correspond to one customer order. However, sometimes customer orders can be very large, resulting in long lead times for all jobs. As explained in the next chapter, one of the steps to QRM is getting your customers to order in smaller, more frequent quantities. However, before this goal is attained, you may need to run jobs in smaller quantities, each with its own POLCA card. To determine this quantum (limit on the size of jobs associated with a single card), in Step 2 above the cell teams should also use their planning tools to try out different values of the quantum, and relay back to the HL/MRP planners what their ideal quantum is. The planners working with the HL/MRP system will then use this information to set the quantum for the POLCA cards for a given planning horizon.

Advantages of POLCA Over Both MRP and Pull Systems

Now that I've described all the elements of the POLCA system, I can detail its advantages and how in the context of QRM it overcomes the drawbacks of both MRP and pull systems. First, the use of POLCA cards assures that each cell only works on jobs that are destined for downstream cells that will *also* be able to work on these jobs in the near future. In other words, if a POLCA card from a downstream cell is not available, it means that cell is backlogged with work (or cells downstream from it are backlogged). Working on a job destined for that cell will only increase inventory in the system since somewhere downstream there is a lack of capacity to work on this job. It is more expedient to hold off putting organizational resources into such a job—those resources would be better used in other ways. This is similar to the logic used in a typical pull system.

Second, the use of HL/MRP authorization times also prevents build-up of unnecessary inventory. As I showed in earlier examples, pull systems have the disadvantage of filling intermediate stages with inventory in the case where there are a large number of products with infrequent demand, or in the case of a high-volume product whose demand is fading. By coupling the authorization procedure using HL/MRP with the POLCA scheme you ensure that you don't make products just because they have a pull signal—you make products only when there is demand for them.

Third, unlike a kanban system where workstations are tightly coupled within a cell and between cells via the kanban cards, the POLCA cards flow in longer loops. There is coupling of cells, but it is more flexible. Remember that a kanban system is highly tuned to produce at a given rate. In fact, in designing a pull system, a good deal of effort is spent in determining the takt time—the cycle time within which each task must be completed in order to meet this demand rate—and then trying to balance work allocation so that every operation ideally takes no more or no less than the takt time. Indeed, the purpose of the tight coupling in a pull system is to find and eliminate the obstacles to achieving the consistent takt times. These obstacles include quality problems, machine failures, parts shortages, and other forms of waste, and hence their elimination leads to greater productivity. On the other hand, for QRM you need to satisfy varying demand for multiple (even one-of-a-kind) products. Recall the example of CFP Corporation where some products need more shearing and others need more punching. Hence you can set some average capacities via the planning procedure just described but the actual rates and bottlenecks will vary from day to day.

This is why it is essential for a QRM enterprise to have a greater degree of flexibility than traditional companies. It is also the main reason for having the overlapping loops in POLCA—by making the card loops longer the additional jobs in the loop act as a buffer to absorb variations in demand and product mix. This allows each cell to balance its capacity as best as it can for the current mix—something you cannot do with kanban cards because of the tight takt time calculations and the tight coupling through the kanban cards (balanced carefully with the takt time calculations).

Detailed Example Showing Advantages of Longer Card Loops

Consider the products that go from F2 to A3 and from F2 to A4. An alternative to this system would be to have cards that are labeled simply F2, A3, and A4. These cards would work as follows. At cell F2, a job would need an F2 card available before it could be launched into F2. This card would stay with a job through F2. When the job was completed and left the output buffer of F2 for the next cell, the F2 card would be returned to the beginning of the cell to be available to launch another job. And similarly for the A3 and A4 cards for cells A3 and A4. Let's say that macro capacity planning dictates the need for four F2 cards, two A3 cards, and two A4 cards. On a given day, suppose cells F2 and A3 are operating slightly faster than planned, but one automated assembly machine in cell A4 has become a bottleneck; not just because of higher than normal assembly times for the jobs, but also because of high variation in the assembly requirements of jobs being processed that day. The two A4 cards may not be sufficient to keep the cell operating at reasonable capacity. For example, when a job with high assembly times is being worked on in A4, a second job that has entered the cell will end up waiting at this bottleneck machine. Thus both A4 cards will be at the assembly station in A4. When the long assembly operation is over, the first job will leave and go on to other assembly and packaging operations in the cell. However, the second job has a short assembly time so it gets done very quickly and also leaves for other operations in the cell. Suppose both jobs are still being worked on at these other operations. Since both cards are with the two jobs, no new jobs can be launched into the cell. At this point we have an undesirable situation: The automatic assembly station, which is the bottleneck today, is now idle because no jobs can come to it! Since we have lost capacity at the bottleneck due to this idle time, we have lost capacity for the whole cell A4.

Now let's consider what would happen with the overlapping loops. Suppose with the macro planning, four cards are allocated to the F2/A3 loop and four are allocated to the F2/A4 loop. (This example is designed to be "fair" in the sense that the maximum number of jobs that can be actively worked on is eight in both cases.) Since the F2 cell is working faster than planned, jobs will not be held up there. Thus, a good deal of the time all four cards (all the cards in the F2/A4 loop) can be active in cell A4. This means a greater likelihood that the bottleneck station has a job waiting in front of it, instead of being frequently starved as in the previous case.

Pull system philosophy would advocate trying to rebalance the operations to alleviate the bottleneck. However, this cannot be done reasonably in the QRM environment because of the high variability between jobs—any attempt to redesign and balance the operations for one set of jobs would be immediately undermined when the next set of jobs arrived with very different requirements. In the QRM environment, you attempt to predict average rates and capacities using the planning systems, but you cannot predict the variabilities between individual jobs. The POLCA system provides you with greater flexibility to "roll with the punches" and alleviate bottlenecks as they arise. The longer card loops provide more cards that can migrate toward the bottleneck, wherever that may happen to be in a given day or week. Also, the additional QRM principles of maintaining some idle capacity and creating a flexible organization will help to cope with the variable requirements of jobs. (In Chapter 13 I'll give a detailed discussion of how to create organizational flexibility.)

Taking the argument for POLCA further, one might envision an extreme case of an even longer card loop: a card loop for each unique routing, from start to finish. For instance, in the above example, there could be a card loop for all jobs that had the routing P1/F2/A4/S1. One could argue that this would give even more flexibility for cards to migrate toward the bottleneck. However, the problem with this idea is that there would be a very large combination of such routes, and a resulting proliferation of control cards. Planning and setting counts for so many cards would be very difficult. A second problem would be that the interactions between cells would be varied and possibly remote. In the case of a P1/F2/A4/S1 job, if no cards were available at P1 to start a job, the bottleneck might be anywhere in three other cells (F2 or A4 or S1). It would take some work to find where the bottleneck was. This would begin to feel like the current complex MRP-based system. By keeping the POLCA loops to two cells and no more, we have deliberately chosen a compromise between greater flexibility and greater complexity—longer card loops would increase both. Particularly attractive about the choice of paired cells are two factors: (1) Each destination cell in a pair is a direct customer of the first cell in the pair, and thus an immediate "supplier-customer" relationship can be established between such cells; and (2) each cell has a limited number of other cells that it interacts with, making it possible to manage the fewer interactions more effectively for situations such as quality problems, rework, delayed jobs, and so on.

The advantages of longer card loops have been discussed by the authors of a strategy called CONWIP.[6] However, the CONWIP strategy has been developed for systems in which all products follow the same linear route. It is not clear how to extend CONWIP to systems with large numbers of varied product routings such as the examples in this section. The only obvious way to do this would be to let each unique routing (such as P1/F2/A4/S1) have its own production card, but this would lead to a huge proliferation of routing cards, with its own complexities, as just explained. Because it uses longer loops and a cellular system that is both more manageable and more flexible, POLCA can be viewed as a way of obtaining some of the advantages of CONWIP without having the proliferation of card types.

Extending POLCA for Multiple Components and Suppliers

The POLCA system is easily extended for the case where an assembly has multiple components arriving from different cells. In the example for CFP Corporation suppose that the products assembled in cell A4 require both light gauge aluminum printed parts from cell F2 and heavy gauge mountings from cell F3. In that case, the POLCA loops for this section of the operation will appear as in Figure 9-10, and will work as follows. The light gauge parts will arrive at A4 along with their F2/A4 POLCA cards as before. In addition, the heavy gauge parts will be produced in F3 and arrive at A4 along with their F3/A4 cards. Also, the routing tickets for this job will indicate that components from both F2 and F3 are needed in the A4 cell's operations. In this situation, when this job at A4 receives HL/MRP authorization, the cell team will know (from the job ticket) that it must look for both sets of components. The team will bring into the cell both sets of the components, now considered as one job (since they will be assembled together and leave as one job). At the same time, just as in the previous case the F2/A4 POLCA card stayed with the job through the A4 cell, now both POLCA cards (F2/A4 and F3/A4) will stay with the job until A4 has completed it. (We want the F3/A4 POLCA card to stay through A4 for the same reason as we wanted the F2/A4 card to do this.) Thus the job will actually carry three cards with it through the cell: these two cards plus the A4/S1 card that it needs in order to enter A4. When the job is completed by A4, it is put in the output buffer, and the two POLCA cards (F2/A3 and F2/A4) are returned to their respective originating cells F2 and F3. As you see, the case of multiple components is just the common-sense extension of the original case.

The extension to suppliers, if desired, is also straightforward. Suppose L1 is a supplier of aluminum sheets to cell P1. Then you can

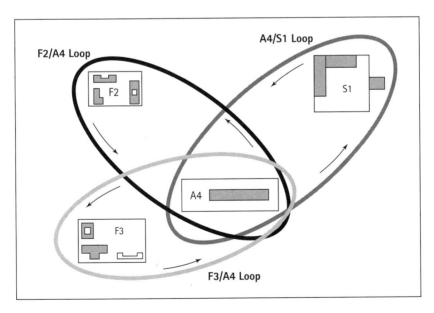

Figure 9-10. Components for Two Cells Are Needed for Assembly in A4

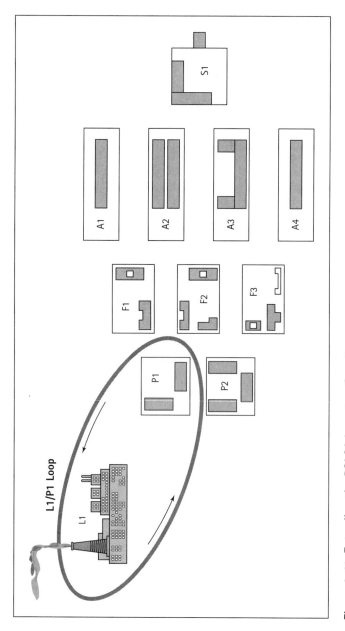

Figure 9-11. Extending the POLCA Loop to a Supplier

just add another POLCA card loop from L1 to P1 (see Figure 9-11). In this case, the supplier must also be governed by the same rules, that is, L1 must receive both an HL/MRP authorization and an L1/P1 POLCA card before it fills the order for resupply.

THREE D'S FOR SUCCESS OF POLCA: DESIGN, DISCIPLINE, AND DECENTRALIZATION

Since no manufacturing system is perfect, even with the POLCA system there will be times when a downstream cell is unexpectedly starved—even though it appears an upstream cell could work on a job for it, but for the lack of a card. In such a situation, it is tempting for the downstream cell team to plead with the upstream cell to run a job for them anyway. However, for the POLCA system to work effectively, it is necessary that the sequencing rules for matching a POLCA card with an authorized job be followed by everybody, all the time. If these rules are compromised, it is only a matter of time before the whole system performance will degrade; favoritism will dictate which jobs are run, there will be no stability of the schedule, and eventually no belief in it either. This will further enhance the hot job mentality, as downstream cells vie with each other to convince their upstream suppliers to work on their jobs.

Then what about the wasted capacity in the downstream cell, in the situation I just described? In the same way as is prescribed for a kanban system, such events (i.e., the cell being starved for work) should be used as an opportunity to learn why the situation has occurred, attempt to eliminate the root causes of this situation, and use this time for preventive maintenance, quality circle meetings, and other continuous improvement activities. This may also be a good time to have your QRM detectives take a look around to eliminate non-value-adding activities.

The converse to this wasted capacity is the need for extra capacity. Since QRM deals with multiple products with varying demand, you need such capacity to accommodate demand surges, or to deal with product mixes that burden one particular resource more than others. Readers with intimate knowledge of JIT systems will not be surprised by this, because Japanese manufacturing systems maintain a certain amount of spare capacity as a backup. As pointed out in many descriptions of Japanese manufacturing techniques, instead of using WIP buffers, these systems use capacity buffers.[7] In a similar fashion, a QRM system needs to maintain some spare capacity to allow the POLCA system to operate effectively. This is another way of stating my earlier principle: You should design a QRM system to operate with a degree of idle capacity.

The final piece of the three Ds for a successful POLCA system is how you control the operation within cells—the issue of decentralization. Remember from the discussion in Chapter 4, the team of workers in your cells:

- Are multiskilled, performing various operations.
- Have ownership of the cell's performance.
- Are dedicated to producing a set of products, which means that its resources are *not* diverted to making anything outside that family.

It is essential for the decentralized operation of QRM that you give cells freedom in choosing their own control strategies. These control strategies will vary considerably depending on the structure of each cell and the preferences of the team running it. Here are a few examples:

- A cell that makes high volume components for other cells might adopt a paced assembly line with no need for additional material control methods.
- Another cell that has machines such as mills and lathes that require setups, but with all products having the same linear routing, could use small batch production with kanban-like bins between each machine and a pure pull strategy within the cell.
- A cell with several machines, but products that go to the machines in different sequences, might just use queuing models to plan capacity usage, using cross-trained workers to move around and alleviate bottlenecks, controlling total WIP in the cell via the use of POLCA cards, but not trying to control the interactions between machines because they are so variable.
- A cell with diverse products and varied routings might use simple finite capacity scheduling software to decide how to run its jobs each day.
- A cell might decide to use the rabbit chase method of the Toyota Sewing System, where production workers migrate up or down the line.[8]

The point is that the POLCA environment gives each cell considerable flexibility in choosing its internal mode of operation and material replenishment. This is where the POLCA system departs significantly from kanban and other flow systems, yet this is essential to preserve the philosophy of QRM. At the same time, the POLCA cards ensure a level of interaction and communication between cells that keep up the pressure to produce the desired products and keep material moving, yet without excessive inventory. The number of POLCA cards in any loop is always a cap on the WIP in that loop.

In the last two chapters I've described in full the systems you need to use for material planning and control within the QRM enterprise. In the next chapter I will discuss how these systems extend outside the enterprise, to suppliers and customers.

Main QRM Lessons

- The concepts of waste, flow, and pull, as defined in the very latest in lean thinking from Womack and Jones, further explain the differences between JIT and QRM.

- JIT doesn't always fit the needs of companies—the tale of three companies that chose the QRM strategy.

- Why JIT/flow strategies and their pull/kanban controls have several disadvantages when it comes to implementing QRM.

- Material control for QRM involves neither push nor pull methods, but a new strategy called POLCA. This springs off the restructured MRP, using a combination of release authorizations and overlapping intercell card loops, to overcome the drawbacks of both push and pull methods.

10

Customer and Supplier Relations

AS YOU PROGRESS ALONG THE ROAD to QRM, with your responsiveness continually improving, it becomes increasingly important to streamline the flow of materials. Not only is it critical to improve the *internal* flow of materials, as discussed in the preceding chapter, but it is equally important to focus on the *external* flows as well. As you will see, managing these external flows—the delivery of material to your company from its suppliers, and the material supply from your company to its customers—takes on a different character under QRM. In the traditional organization, when a company's machined product had a 12-week lead time, the fact that the raw castings took 4 weeks to arrive from a supplier was not a major issue, and the cost of those castings often dominated the choice of supplier. On the other hand, as this company strives to reduce its lead times to 3 weeks, the castings' lead time becomes a serious obstacle. In the case where the company makes custom-designed products, such as housings, in small batches, it is not practical to maintain a stock of commonly used castings. Then how does such a company reduce lead times to 3 weeks? These and related issues will be the subject of this chapter.

There is a duality in this chapter that you should exploit as much as possible. Whenever I discuss a situation concerning a supplier, with you as the customer, and suggest some modifications in the relationship, stop and put yourself in your customer's shoes—now your company is the supplier—and think about how this suggestion would affect your company's behavior for its customers. In the same vein, when I discuss your customers and their behavior, turn that discussion around and examine your own behavior with your suppliers. This will give you additional insight into areas for improvement in your own operation, plus double mileage for reading this chapter.

We'll begin the discussion of the external supply of materials by looking at supplier relations. There is a good reason for starting this way. In dealing with suppliers, you (as the customer) typically have more control, and you can better envision changes you wish to impose on those suppliers. In contrast, it is much harder to change your customers' behavior, and you may be less convinced about the QRM approach if I began with that aspect. However, after reading the discussion on suppliers you will have a better understanding of the motivation behind the QRM methods and hopefully, you'll have the will to approach your customers with the strategies detailed in this chapter. I'll conclude the chapter by looking at methods that affect the entire supply chain. It is often said that you shouldn't concern yourself only with your suppliers and customers; instead you should be concerned about your suppliers' suppliers and your customers' customers. In fact, even this might not be enough for an effective quick response strategy. We will go further by looking at the entire supply chain from primary materials to the end consumer.

SUPPLIER RELATIONS—ANOTHER RESPONSE TIME SPIRAL AT WORK

Two aspects of supplier relations have been discussed in detail in books on quality and JIT, and they are essential to QRM as well. These are:

1. *The importance of quality in supplier selection.* Instead of selecting suppliers based solely on price, as was traditional, modern quality strategy dictates that a supplier's ability to provide consistently good quality products should also be a criterion.
2. *Building partnerships with a few key suppliers.* Similarly, JIT strategy advocates creating long-term partnerships with a small number of suppliers, rather than having a large supplier base that is constantly bidding against each other.

Although I refer to these two aspects in this chapter, I'll not dwell on them, since they are already well-established principles.[1] Rather, I'll focus attention on lead time and delivery issues that determine the need for additional principles unique to QRM strategy. This leads us back to item number 6 from the QRM quiz:

> 6. Since long lead time items need to be ordered in large quantities we should negotiate quantity discounts with our suppliers.
>
> ☐ True   ☐ False

It does seem reasonable, if you're going to order a large quantity, to negotiate with your supplier for a price discount. Listen to Marge Crockett, a materials analyst at a company that makes electromechanical products, as she explains her strategy to her manager:

> I noticed in this week's inventory report that we are running low on those specialty motors that we use in one of our products. The last time I ordered a batch of those motors, it took the supplier three months to deliver them. The shipment came in just three days before we ran out, and I was sweating bullets, because I thought I would go down in history as the first analyst that shut down our final assembly. Since our customer demand has been known to shoot up suddenly in some months, this time I am going to order five months' worth of motors so that we don't run out of them. At our average usage of 2 a day, this means about 200 motors. I know this looks like a big increase in parts inventory, but I am going to make it worth our while—I think I can get the supplier to give us a 10 percent price break for such a large order. I've done the math, and the price break more than justifies the holding cost.

Assuming that the supplier agrees to the price break, this seems like a good way to go. Not only will it give Marge's company a hedge against sales increases, but it will save them money as well. Then why does belief in quiz item 6 go against QRM strategy?

The answer lies in uncovering another Response Time Spiral at work against QRM (see Figure 10-1). To see how this spiral works, consider that your supplier has a factory that works just like yours. If you purchase items in large batches, it encourages the supplier to make them in large batches. Indeed, if the supplier's factory is being managed by traditional cost-based philosophy, they will be delighted to run large batches at (apparently) lower cost. The supplier's other customers, also perceiving long lead times and thinking along the lines of Marge Crockett, will also order large batches. As explained in Chapter 7, all these large batches of orders scheduled through the supplier's factory result in very long lead times from receipt of order to when the supplier can ship the parts. The long lead times, in turn, motivate you and the other customers to order parts in even larger batches. The spiral has begun to unfurl.

The situation is exacerbated by traditional reward systems. Performance of staff in the purchasing department is typically measured by the savings they have been able to negotiate in purchased items. Thus the purchasing agents are encouraged by management to negotiate price breaks and quantity discounts from suppliers—another built-in incentive for your company to order in large quantities.

Now additional consequences of the Response Time Spiral start to kick in. As large batches start running through the supplier's factory their

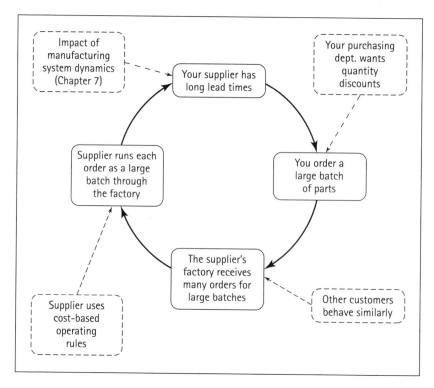

Figure 10-1. Response Time Spiral for Purchasing from Suppliers

quality will suffer, and their costs will rise. In reaction to this, their management may put in "efficiency programs," which will only serve to increase the growth rate of the spiral. At the same time, their lead times will continue to lengthen.

The Hidden Costs to Your Company

The supplier's internal problems also work to the detriment of your own operation. Delays from the supplier may cause late shipments of your product. Poor quality components may shut down your assembly line, or worse yet, create warranty problems for products that were already shipped. To add insult to injury, the supplier's management may share their internal accounting data with you, to show you their spiraling costs and convince you that they deserve an *increase* in price.

There are some less obvious negative consequences for your company as well. One hidden cost to your company is obsolete inventory. Every time you order a large batch of parts, you risk incurring this cost. The market may shift, or technology may change, and so you may never need those parts again. This risk doesn't show up in the typical math done to justify the large order. Even if parts don't become obsolete, they may deteriorate in storage. Again, these potential losses are not usually put into the justification equation.

Another nonobvious consequence is the long feedback loop between your production line and that of the supplier. Suppose you have a batch of parts sitting in storage for several months, and then when you finally pull them out for assembly, you discover that they are all bad. At this point, even if the supplier remakes the parts at no (apparent) cost to you, there will indeed be significant losses to you. Your schedule has been messed up so your analysts need to put effort into rescheduling other products for final assembly and expediting parts for those products. At the same time, the supplier may not be able to track the cause of those defects, since so much time has passed since the original production. As a result, the same defect may recur in the future.

A final hidden cost is that the long lead times from your supplier require you to forecast your production farther and farther into the future. This causes your company's own Response Time Spiral to unfurl, with all its resulting costs.

THE QRM APPROACH TO SUPPLIER RELATIONS

In order to support the QRM strategy, we need to replace belief in quiz item number 6 by the following.

> **QRM Principle: Educate your suppliers on QRM and motivate them to implement this strategy. Through QRM methods, they will be able to produce smaller lots at lower cost, with better quality, and with a shorter lead time.**

If your suppliers are successful in their efforts to implement QRM, you stand to get the best of all worlds. You will obtain the cost reductions you wanted while maintaining low in-house stocks and having responsive and high-quality suppliers. Achieving this, however, requires working with your suppliers over the long haul, since you know from preceding chapters that you cannot implement a QRM strategy overnight and the improvements you are seeking take some time to display themselves. This reinforces the need for long-term partnerships with your suppliers, a point I will return to.

These new policies on order quantities are not just for your suppliers; they need to be supported internally in your company as well. You should not underestimate the difficulty of retraining your purchasing staff. They have been pressured for years to look for price break opportunities, and to pit suppliers against each other. You are changing established rules and policies on them, and they may find it hard to believe this new approach will work. This is why you need to internally support the preceding QRM principle with this one:

> **QRM Principle: Educate your purchasing department on the basics of QRM, and in particular on the pitfalls of ordering large quantities. Rethink performance measures for the purchasing staff.**

In order to avoid organizational discord, you must ensure that the education includes not just analysts, but also purchasing managers and all levels of higher management that oversee the purchasing function. The new behavior of purchasing staff needs to be motivated by appropriate performance measures that replace the traditional cost-savings mentality. These new measures are discussed briefly here (see Chapter 16 for a more detailed discussion).

Supplier Partnerships and Supplier Evaluation

As I mentioned earlier, QRM supports the quality/JIT concept of building partnerships with suppliers. Specifically, instead of having a large number of suppliers competing with each other for each contract, QRM advocates development of a few key suppliers. So far, this sounds similar to quality/JIT strategy. However, a key difference is in the partnership's focus. Whereas quality/JIT strategies focus on quality improvements and/or expected cost reductions, the QRM strategy focuses on working with the suppliers to reduce their lead times; quality improvements and cost reductions become an outcome rather than the primary focus. This has implications for supplier evaluation and performance measurement.

Traditional purchasing policies, as practiced by U.S. companies through the 1970s, gave primary importance to cost. Then, with the advent of JIT, quality and on-time delivery became important as well. So at present the three key criteria for supplier evaluation are what I call the *QCD set*:

Q: Quality (typically measured by percent of defective parts delivered)
C: Cost (unit cost of a part)
D: Delivery (typically measured by percent of on-time deliveries).

For QRM you need to rethink these measures. Recall the discussion in Chapter 8 regarding the fourth QRM quiz item:

> 4. We must place great importance on "on-time" delivery performance by each of our departments and by our suppliers.
>
> ❏ True   ❏ False

Whatever could be wrong with wanting on-time deliveries? The problem, as I explained, is that it leads to padding of lead times by suppliers, so that their on-time performance is good. For instance, if a supplier has a typical lead time of 8 weeks, but occasionally experiences delays in meeting this target, then it is better for the supplier to quote a 10-week lead time and be on time almost all the time. The problem then shifts to the customer who has to take responsibility for deciding their needs 10 weeks into the future. This exacerbates the Response Time Spiral in the customer's operation and the supplier's operation. Both organizations then suffer the consequences. The solution to this vicious cycle lies in the following.

> **QRM Principle: Make lead time reduction the number one evaluation criterion for your suppliers, keeping the QCD trio as a set of secondary measures.**

It is clear how to measure lead time, but how exactly does one measure "lead time reduction"? There are some subtle issues involved, which I'll defer to Chapter 16. For now, I'll illustrate its use with a numerical example below. Another question that arises is, couldn't the supplier still pad lead times initially, in order to perform well in (apparently) reducing their lead times? The answer to this is twofold.

1. From your past relations with this supplier, plus your knowledge of industry standards for those components, you already have a rough idea of what lead times are currently achievable. Thus the supplier cannot quote a ridiculously long lead time. The supplier may, however, quote a slightly padded lead time, such as 10 weeks instead of 8, as in the preceding example.
2. With your relentless focus on lead time reduction, the padding will soon be gone, and this supplier will be forced to make *real* reductions in lead time to continue scoring high on your evaluations.

I'll drive this point home with the concrete example of the padded lead time quote of 10 weeks as above. Suppose the purchasing staff decides on a performance standard of 10 percent lead time reduction every quarter. In order to score well on this rating, the supplier will need to reduce its quoted lead time to 9 weeks by the end of the first quarter. During the next three quarters, the lead time will need to drop below 8.1 weeks, 7.2 weeks, and 6.5 weeks, respectively. Hence, within a year, you will have forced the supplier to find ways to reduce its lead time well below its "true" value of 8 weeks. So the supplier will have to begin implementing QRM methods, and you can expect real gains from the supplier in terms of cost and quality improvements as well.

Additional strategies you need to implement as part of the partnerships with suppliers are as follows:

- *Invest in the suppliers' improvement.* Instead of the traditional arm's length—and even adversarial—relationship with your suppliers, work with them to help them improve their operation. Companies such as Chrysler, GE, and John Deere actually send teams of their own people, at their own cost, to look over their suppliers' operations and help them implement improvements.[2] Note that the QRM approach differs a little in its focus, however. Rather than targeting quality and cost improvements, as companies have typically done in the past, I suggest putting effort into assisting your supplier in implementing QRM methods by using the expertise you've developed through your own experiences with QRM. By focusing on lead time reduction at the supplier, you will get the benefits of quality improvement and cost reduction as well as short lead times for parts. The following case study illustrates one plant's success with this approach and drives home three points made in this chapter: (1) Educate your suppliers on QRM strategy; (2) make lead time reduction the primary focus of supplier improvement efforts; and (3) invest in your suppliers' development.
- *Create systems that enable suppliers to learn how to support your needs.* Ford Motor Company's Taurus achieved record profits when it was first introduced. Although many factors contributed to this success, one important factor was a new relationship with suppliers. Ford committed itself to one high-quality supplier for each purchased component. Then Ford actually took a prototype car to each such supplier so they could see where their part fit, and how the part was integral to the car's performance. In addition, feedback was solicited from the employees of these suppliers. Motorola used a similar approach when it introduced its new pager named Bandit. The company shared proprietary information with suppliers and had them participate as partners in the project.[3] What we are advocating here is that, instead of this being a one-time situation as with the Taurus or Bandit projects, you should put systems in place that enable suppliers to learn about your needs on a regular basis.

John Deere Horicon Works' Success with
Investing in Supplier Improvement

Deere and Company, established in 1837, is the world's leading producer of agricultural equipment and one of the leaders in construction and forestry equipment, diesel engines, and other powertrain components for the off-highway industry. In 1963, Deere began production of ride-on lawn mowers at its Horicon, Wisconsin facility. Today, the John Deere Worldwide Commercial and Consumer Equipment division is the largest producer of premium mowing equipment in the world. Over 3.5 million ride-on lawn mowers have been manufactured at the Horicon factory.

The Horicon Works has developed a reputation as a progressive manufacturer by positioning itself at the forefront of world-class practices. For instance, over an 18-month period beginning in 1982, Horicon reduced raw and work-in-process inventory by 62 percent through implementation of JIT manufacturing. Since that time, Horicon has continued to experiment with and implement cutting-edge strategies such as focused factories, robotics, and most recently, QRM.

Horicon's latest efforts on supplier relationships have their origins in the plant's redefining its business strategies during the 1980s. As a result of this redefinition, the factory was focused on sheet steel stamping, welding, assembly, and paint as core manufacturing processes. With this strategy, purchased part costs began to represent an increasing percentage of the Horicon manufactured costs. This laid the first cornerstone in Horicon's reexamination of supplier relations.

The second cornerstone fell in place when, because of capacity constraints, Horicon's sheet steel stamping department was unable to fill the factory's total stamping requirements. This led to development of external stamping sources. Now the third cornerstone was laid: Discussions began to arise as to whether the internal stamping department should be treated the same as external stamping "suppliers," with the implication that the internal department should compete for business and receive the same level of support as any outside source.

Typically, John Deere's suppliers are small and medium-sized manufacturers. Increasingly, such companies have been under industrywide competitive pressure to reduce overhead and trim costs. Many of them have reduced their employees to the minimum necessary to run daily operations. Planning and implementing new manufacturing strategies is beyond the capabilities of many of these companies, because of lack of experts on their staff. This realization led to the fourth and final cornerstone. A vigorous debate began on the management of internal and external suppliers at Horicon, namely: "Why don't strategic outside sources (i.e., suppliers) receive the level of support provided to Horicon's internal sources (i.e., departments)?" In other words, if an internal department requests assistance with manufacturing improvement analysis, planning, and implementation, then you provide it with technical and personnel resources. Why shouldn't you give similar support to strategic external suppliers? The four cornerstones were now in place for Horicon to construct its answer.

In the fall of 1995 Paul Ericksen, manager of strategic supply management at Deere's Horicon Works, initiated a pilot Supplier Development Program. Ericksen's aim was to resolve the debate via a pilot experiment to support 16 suppliers. An agreement was forged with the pilot suppliers that would entitle Deere to share in any savings obtained from the resulting improvements. Over the next 18 months, the same Deere corporate process engineers who normally worked on Deere manufacturing improvement projects were sent out to work with the suppliers who participated in the pilot. The results of the pilot program were unambiguous. They showed that the price reductions that resulted for Deere enabled it to more than recoup the investment it made in the 16 suppliers.

Based on these results, in 1997 the Horicon Works formed a dedicated supplier development group focused on providing resources to assist strategic suppliers in implementing manufacturing improvements. Services are provided at no cost. If there are no cost savings to the supplier as a result of a project, no price reductions from the supplier are expected. If costs are reduced, the resultant savings are used equitably to both increase the supplier's profit margin and reduce prices to John Deere. If a capital investment is required by the supplier to realize the savings, the supplier is first allowed to recoup that investment from the savings prior to sharing them with Deere.

In keeping with QRM strategy, recent improvement efforts have targeted lead time reduction in the suppliers factories. In addition to providing personnel to work at the suppliers' facilities, Deere has provided QRM education for the suppliers' staff. As a result of these efforts, Horicon has seen reductions of more than 90 percent in lead time at some suppliers and resulting price reductions to Horicon (after providing for the supplier's share as above) have been as much as 15 percent. The program has clearly yielded mutual benefits to Horicon and its supplier base, and is creating a competitive advantage for John Deere in the marketplace.

As a result, suppliers will also do a better job of serving those needs.

- *Include supplier proximity as a factor in supplier evaluation.* Many companies have ignored supplier distance, as long as the quoted cost was sufficiently low. In the 1980s, a U.S. bicycle company outsourced a number of components from Taiwan because of lower costs relative to local suppliers. This resulted in both short- and long-term problems. In the short term, the company experienced many quality and delivery problems, and due to the distance and communication problems involved, was not able to resolve them easily. In the long term, the supplier used the knowledge gained from making these components to forge relationships with other local companies and eventually develop competing bicycles for the Asian markets. Given the proximity of the supplier to the Asian customers, it was easy for their cartel to develop those markets. On the other hand, if the bicycle company had used local suppliers and worked with them to reduce lead times and improve quality, it would have eventually achieved the lower costs it desired. The shorter lead times from local suppliers also would have reduced the occurrence of scheduling problems, as well as the investment in inventory. The proximity of the suppliers would have been key to the company's personnel being able to visit them to help them implement improvements, to cement their partnership through mutual contact and understanding, and to engage in joint development projects. The company might also have had a better chance to retain control of its Asian markets.

The next case study illustrates how one might combat the pitfalls of traditional cost-based sourcing, by looking more carefully at the issue of supplier location.[4]

An Unconventional Analysis of Sourcing Decisions

In 1997 I was asked to advise a study team charged with evaluating an existing overseas supplier relative to a potential U.S. supplier. The company that used the parts was located in the U.S. midwest, as was the alternate supplier. At first blush, it seemed that the overseas supplier was much more economical, with a price of $38 per piece versus the local supplier's quote of $44. Both these amounts also included freight costs. Since the company consumed thousands of these parts each year, it seemed clear that continuing to use the overseas supplier would save the company a large sum of money.

However, I felt that the substantially longer lead times of the overseas supplier might account for some ripple effects that were not being measured by the initial analysis. I therefore suggested that the team attempt to quantify the following additional costs for both suppliers:

- *Freight costs for rush shipments.* Although the overseas supplier's quoted price was based on sea freight, often due to unexpected orders or engineering changes, the customer company paid for air freight instead.
- *Inventory carrying costs.* Due to long lead times, they had to keep more inventory on hand for the distant supplier, to hedge against demand changes. For a proper analysis, the cost of such inventory needed to include interest, rent or other space costs, insurance, obsolescence and all related handling costs.
- *Costs of the procurement process.* Analysis of actual procurement records provided a revealing set of statistics. On average, it took only two faxes and one phone call to complete the placement of an order for a local supplier and they could accomplish this in less than 24 hours. In contrast, it took 14 faxes and two to three weeks to complete the same transaction with the overseas supplier. The time-consuming and wasteful transactions involved in this process included requests for order status; clarification of previously supplied information; delivery date adjustments; confusion regarding sea or air freight choice; changes in pricing or specifications; and applying pressure for faster delivery. These were all the result of long lead times and distance.
- *Costs of poor quality.* As a result of poor communication, language barriers, and just plain distance, the overseas supplier's quality was not as good as other local suppliers used by this company. Of course, the company never paid for parts that were bad. So, to quantify the costs of poor quality, we had to look further afield. I suggested that the team look at the amount of time spent in inspection and sorting, the time invested in assembly before the faulty part was discovered, rework time, and freight costs for rush shipments required to replace defective parts.
- *Costs of variability.* The long lead times, plus the uncertainty inherent in shipping, combined with schedule changes, all resulted in a great degree of variability in the supply of parts and the final assembly schedule. We attempted to quantify the time spent by various people in dealing with this variability.

Several of these costs were difficult to quantify accurately, since precise data were not available (e.g., for the last item above). However, the team focused

(continued)

on making rough estimates, because our goal was not so much complete accuracy as the creation of awareness in top management that the issue merited further study.

When the team completed its study, the numbers showed that the "total cost" of the local supplier was $49, while for the overseas supplier this total cost jumped to $53. More interesting, the team found evidence that through relationships with other local suppliers, the company had been able to improve quality and reduce costs of those purchased parts within a year. By putting in typical numbers for these improvements, the team estimated that the total cost for the local supplier could drop to $45. Compared with $53 for the overseas supplier, this difference was now quite significant. Management had to acknowledge that even with the rough estimates used for the "total cost" calculation, they could not ignore this difference. The controller at the company also looked over the cost estimates and agreed with the general conclusion, so the company proceeded with development of a local supplier for the part.

Additional Supplier Strategies for QRM

The following additional ideas for dealing with suppliers will support your company's QRM strategy.

Challenge Conventional Choices of Make Versus Buy, Materials, and Product Designs

When purchased parts have long lead times and their suppliers are unable to shorten these times, it is worth rethinking existing choices. In Chapter 6 I gave several examples of such rethinking. For instance, the case study from Beloit Corporation illustrated replacing outsourced castings by in-house fabrications, which cut lead time for aftermarket parts by several weeks. Another example in that section described how a plating operation done at a subcontractor could be eliminated by using a different material for the part. Similar approaches were mentioned in Chapter 8 in the section on rethinking bills of materials. As also mentioned there, in most cases you will have to battle the cost-based mind-set: Past decisions on make versus buy, or materials, were most likely made using traditional cost trade-offs, which ignored the impact of long lead times on total system costs. You will need to arm yourself with examples and points from this book to shore up your suggestions. The following case study may assist you with this.

Cost-Based Outsourcing Leads to Delays at Equipment Manufacturer

The materials manager of a large European company in the food equipment industry provided the following example of the pitfalls of cost-based outsourcing. They had to make the cover for one of their large machines in three sections, to allow access for maintenance of the machine. The sections themselves were large and complex sheet metal fabrications. As part of their make-versus-buy decision making, the company put all three sections out for bid to potential suppliers. Based on the bids, plus their own analysis of in-house costs, the company decided to outsource two of the sections and make the third one in-house. Although the in-house section was made on time, the outsourced sections were a few days late, holding up the final assembly of the machine. But that was only the beginning of the problem. When the assembly workers placed the three sections together on the machine and attempted to bolt them together, they discovered that bolt holes on the two outsourced sections didn't line up with those on the one made in-house. Worse, the only in-house machine capable of redrilling and tapping the holes on the large sections was busy with another delayed job and could not be used for several days. So the outsourcing decision ended up costing this company in both rework and a two-week delay in shipping the final product. "With 20-20 hindsight," the materials manager said, "it was clear that we made an incorrect assessment of systemwide impact when we decided to outsource those components."

Make Suppliers Responsible for Maintaining Point-of-use (POU) Inventory

When you are using the principles of POU for purchased parts, it may simplify your planning to let the supplier be directly responsible for ensuring that inventory is available. To do this effectively, you will need to give your suppliers information about your upcoming needs, a point I'll discuss later in this chapter. Where companies have given suppliers responsibility for POU inventory, they have found that not only did it eliminate their internal headaches for managing and ordering myriads of small parts, but also, the suppliers preferred this system. The better visibility of both the customer's needs and the current inventory status enabled the suppliers to support the customer better, at the same time giving the suppliers more control over their own production instead of being "whipped around" by last-minute requests from the customer.

There are, however, some trade-offs you need to consider when implementing POU inventory. These involve considerations of the amount of stock, its location or locations, and ease of supply to the cells. Ideally, each cell would like to have its own bins for POU parts. However, this could lead to a high level of inventory if a number of cells use the same part. After reviewing several POU implementations, I suggest the following guidelines for POU parts.

1. For parts that are unique to one cell, the solution is simple: Bins can be placed right in the cell that uses each particular part.
2. For parts that are shared between cells, further analysis is needed. High-volume parts justify placing bins in each cell. You can place medium-volume parts in a common location that is reasonably accessible to all the cells. Through further analysis, you should choose a location nearest the higher usage cell. If you cannot easily resolve the placement issue, for example, if two high-use cells are far apart, then you can consider multiple common locations. In all of these cases, the company and the supplier will also need to negotiate whether the supplier will be responsible for multiple bins, or whether the supplier will stock one bin and the cell workers will be responsible for restocking the distributed bins from this one location.
3. For parts that are both low volume, and used sporadically by a number of different cells, I do not recommend POU. These parts are best stocked in a traditional centralized stockroom.

Reserve Supplier Capacity Instead of Booking Specific Part Orders

Long lead times at suppliers are often caused by a backlog of orders at the supplier rather than actual factory lead time. This is particularly true for smaller suppliers who need to maintain a solid backlog to feel secure about their future. If you are trying to implement quick response for custom jobs, you can't place orders ahead of time at the supplier (since you wouldn't know what to order). Thus, for suppliers whose lead time is due primarily to backlog, an effective strategy is to carve out an agreement where you buy capacity at prespecified intervals, without needing to place orders for actual parts. Then you can specify the parts to be made at a time closer to that capacity interval.

To drive this home, let us consider a company called CMP Inc. that makes complex machined parts, a few of which require a specialized gun-drilling operation. Since this operation is only required for a small annual volume of parts, CMP cannot justify purchase of a gun drill for its machine shop. This operation is therefore outsourced to a local company that owns such a machine. However, the company that does the gun drilling has a typical backlog of two to three weeks just for the gun drill, and this adds a major portion of lead time to the machined parts made by CMP.

In this situation, CMP could negotiate an alternate contract with the supplier based on an estimate of its needs. Suppose CMP needs, on average, eight hours of gun drilling each week. With its two-shift operation, if CMP owned a gun drill, this would imply a 10 percent utilization of that machine. Even the best QRM arguments might be unable to support purchase of this expensive machine. On the other hand, CMP could buy from the supplier eight hours of gun drill time each week for the next three months. To make this situation concrete for CMP's planners, the contract could even specify the times: let's say, every Tuesday and Thursday morning for four hours. This contract is an extension of the time-sharing strategy to suppliers. In this way, CMP could send parts out every Monday, for return by Tuesday afternoon, and every Wednesday for return by Thursday afternoon. It would thus have two opportunities each week, with a one-day turnaround, for the gun-drilling operation. Of course, the risk is that there may be some weeks when CMP has no parts that need gun-drilling, but has to pay for the time anyway. But this is the price CMP has to pay to maintain its short lead times.

And this price may well be acceptable. After all, what are the alternatives? One is to go back to the traditional contracting method and suffer long delays, incurring expediting costs and possibly losing customers. The other is to buy a gun drill that will need to spread its cost over a very small volume of parts. Examining these alternatives carefully, you may find that the occasional week you pay the supplier even when there are no parts to be done may cost you less than the other two strategies.[5]

Take Advantage of Information Technologies

Technologies such as computer-aided design (CAD), barcoding, and electronic data interchange (EDI) offer substantial opportunities for lead time reduction with suppliers. Sending CAD data directly to suppliers via EDI enables them to work directly with the complete computerized data rather than just a print. This reduces the chance of errors in the parts that are made. You can use EDI in other effective ways too. You can eliminate purchase orders, faxes, and other paper documents, along with the time it takes to send them. In addition, if the EDI data can be directly transferred between the customer's and supplier's files, you can eliminate the chance of clerical errors as well. Again, this reduces the lead time that gets used up in rework. Finally, use of barcoding by both the supplier and the customer enables virtually instant verification of part identity as it is received or as it is about to be used on the assembly floor.

Share Your Forecasts and Planning Processes with Your Suppliers

This strategy has implications for the entire supply chain, so we will discuss it at length in a separate section below.

Alleviate Suppliers' Concerns About These Nontraditional Strategies

While implementing all the strategies discussed in this section, you will need to work on alleviating your suppliers' concerns. One unfortunate effect of JIT implementation during the last decade has been the concern of suppliers that they will have to carry the inventory instead. This concern is not without cause because that is exactly what happened with many of the smaller suppliers to large American companies (e.g., in the automobile industry). However, since you can apply QRM to low-volume customized products, its supplier strategies are largely different from JIT. Through education about QRM and the strategies in this chapter, you will need to convince your suppliers that "inventory shifting" is not your goal.

Strategies for When Suppliers Cannot Reduce Their Lead Time

What if, even after attempting all these strategies, you are still dependent on a few long lead time items whose suppliers cannot reduce their lead time, and you have no other alternative because these are the only suppliers that make those parts? There are still two other strategies you can use in this case.

1. *Use standardization through platform designs along with delayed differentiation.* Both platform strategies and delayed differentiation will be explained in Chapter 14. These approaches enable standardization of parts. The idea is to use these design strategies to convert those long lead time items into standard commodities that are always in the pipeline for delivery but without sacrificing flexibility of product offerings.
2. *Leverage your case through industry associations.* Many small companies feel helpless when a large manufacturer is a sole supplier and simply refuses to be responsive to its smaller customers. In this case, industry associations consisting of large numbers of companies can bring pressure to bear on the big companies by giving visibility to the problems they are causing for their customers, and through making the big companies aware of the large customer base represented by the industry as a whole.

CUSTOMER RELATIONS—ANOTHER RESPONSE TIME SPIRAL AT WORK

It is time to change hats. So far, you were the customer, and we discussed dealing with your suppliers. In this section I'll consider the situation where you are the supplier and discuss how to best manage the relationship with your customer in order to support your QRM strategy. I'll begin by revisiting a quiz item number 7.

> 7. We should encourage our customers to buy our products in large quantities by offering price breaks and quantity discounts.
>
> ❏ True   ❏ False

This is just the reverse of the situation I described with your suppliers. In this case, your own lead time, cost, and quality performance will suffer, with resulting negative effect on your customer. In fact, there is another Response Time Spiral at work between your customer and yourself (see Figure 10-2). This spiral is similar to the one in Figure 10-1, but it is useful to see it in a different context with your company as the supplier.

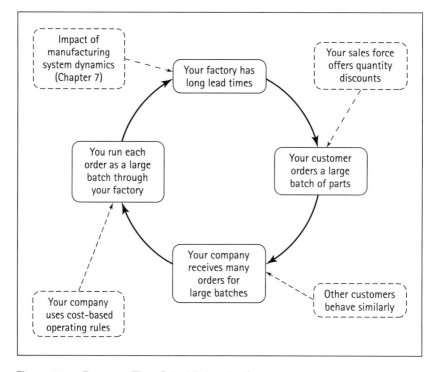

Figure 10-2. Response Time Spiral Driven by Customers

QRM Principle: Educate your customers on your QRM program. Form a partnership with them, which includes a schedule of moving to smaller lot sizes at reasonable prices, while
you implement your QRM program.

Explain to your customers how, in the long run, they will get the benefit
of your producing smaller lots at lower cost. However, they need to be
patient while you embark on QRM, and not continue to order larger lots.

You can still offer the customer quantity discounts, using the concept
of a "blanket order," where the customer guarantees a certain quantity
over the year, but orders the parts in smaller batches. The quantity discount can be based on the size of the blanket order, rather than the
individual batch size.

In the worst case, if a customer insists on ordering a large batch, you should produce the order in several small batches. This may seem most unreasonable. If the customer wants to place a large order, why not economize on setups and make a higher profit on that job? The answer combines many of the arguments given in previous chapters. An important reason is that a large order can get in the way of your other customers' orders, blocking a critical resource from use by those other orders, delaying them and creating dissatisfaction in the rest of your customer base. Breaking up the large order allows other jobs to get through critical resources. Then there are all the other costs of large lots listed earlier, such as increased scrap and rework, and expediting (see "The Hidden Errors in EOQ" in Chapter 7). The basis of QRM strategy is realizing that, in the long run, these exceed the costs of multiple setups.

Also, ask yourself, why does your customer insist on ordering large batches? You know that the customer isn't going to use all these parts on their assembly line immediately. The answer probably lies in the fact that, in the past, your company had long lead times and was often late in its deliveries. The buyer at the customer's company was not willing to risk exposure, and thus wanted to have enough parts on hand. This resulted in the buyer ordering a large batch of parts (recall the story of Marge Crockett earlier in this chapter). However, as you progress down the QRM path and your lead times get shorter and more reliable, your customers will feel more secure about ordering small batches. In this way you will eventually obtain the customer behavior that you desire.

Your QRM strategy can go beyond passively satisfying your customer's needs. In fact, it can become an active element in securing new customers in more ways than you might first imagine.

YOUR QRM PROGRAM CAN SUPPORT YOUR MARKETING EFFORTS

Once your customer realizes that your lead times are indeed shorter and more reliable, there are many benefits to the customer for ordering smaller batches, such as lower inventories and better responsiveness to *their* customers. Thus, you can expect the customer to be receptive to your strategy.

Keeping your existing customers happy is not the end of the story, however. You can capitalize on your QRM implementation to create novel marketing strategies and secure new customers.

How Does a Company's QRM Strategy Benefit Its Customers?

In the fall of 1996, Bruce Backer (vice president of sales and marketing) and Kris Plamann (marketing specialist) of Pensar Corporation had an intriguing idea. Pensar, a custom assembler of printed circuit boards based in Appleton, WI, had already reaped substantial benefits of implementing QRM within its organization (see the case study in Chapter 8). Bruce and Kris felt that it ought to be possible to leverage their company's QRM strategy into a marketing strategy as well.

The key idea would be to identify the benefits of Pensar's QRM strategy to its potential customers. Of course there would be some obvious benefits from the ability to respond quickly to customers' needs. However, Bruce and Kris wanted to uncover the nonobvious benefits and then use them in an effective marketing strategy.

They collaborated with the Center for Quick Response Manufacturing to conduct a workshop that would uncover the customer benefits of a company's QRM strategy. In the spring of 1997, representatives from a number of companies that had implemented QRM worked with the Center's personnel to identify and categorize these benefits, based on their own experiences. Table 10-1 summarizes the result of these efforts. For a company that has implemented QRM, Table 10-1 provides a powerful addition to its marketing arsenal.

Table 10-1. The QRM Company's Marketing Pitch

Ten Good Reasons to Buy from a QRM Supplier

1. Gain new accounts that were not manageable in the past
2. Spend less effort on forecasting
3. See your planning cycle become smoother
4. Reduce your costs of expediting and rescheduling
5. Speed up your new product development
6. Take early advantage of the latest component developments
7. Engage in rapid joint development of new products
8. Experience lower costs due to fewer failures during assembly and post-sales
9. Plan your downtime more accurately
10. Benefit from lower materials inventory

Additional Customer–Related Strategies Supporting QRM

Some additional changes to your operating strategies can go a long way toward serving your customers better.

Change Your Truck Loading Criteria to Include Responsiveness

In an effort to minimize transportation costs, shipping departments have traditionally loaded trucks and other shipping containers to maximize space. This might result in a given order having to wait until a full load is created for a truck going in the vicinity of that customer. Although the shipping cost is, indeed, reduced in this way, the resulting longer lead times can create other costs for the company in other ways that you now understand. The transportation department is, of course, not trying to add delays, but is attempting to meet its goals of cost. To change this behavior, top management needs to use new performance measures. For example, management could make the transportation department's lead time (time from the order arriving to that department to the time when it has reached the customer's receiving dock) the main performance measure and provide an appropriate increase in budget to allow for lower utilization of truck space. The shipping department should also consider using smaller trucks or containers. Again, this may mean somewhat higher shipping costs, but reduced lead times. Such change does not come easily. Transportation managers have had many years of training on how to minimize shipping costs, and they may not be willing to accept the new approach.

Outsource the Entire Shipping Responsibility

A key component of modern competitive strategy is to focus on one's core business. Thus, yet another approach to reducing the shipping lead time is to outsource the entire responsibility for shipping to a company that specializes in that operation. Even a large company such as BASF Corporation has taken to shipping more than 95 percent of its products via outside trucking companies. "I'm a chemical producer, not a transportation manager," stated Bernd Flickinger, vice president of purchasing and logistics at BASF.[6] In response to this trend, many firms that started by offering transportation services have expanded their offerings to include warehousing and complete transportation management, including working with manufacturing companies to reduce their over-all transportation costs. Some logistics companies go so far as to place their service representatives at the manufacturer's plant. For example, Roadway Express has a representative placed on-site at the Framingham, MA, plant of Bose Corporation, which makes stereo speakers and other audio equipment. If your company wishes to utilize such approaches to support its QRM strategy, the key to choosing an outside logistics company is to use short lead time as the primary performance measure, with cost and other factors as secondary measures.

Take Advantage of Available Technologies for Electronic Commerce

With today's Internet-based technologies, it is possible for your company to let a customer examine your product offering, configure or even customize the product they want, obtain a cost estimate, and place a purchase order, all in a short interactive session on their computer. This approach eliminates a number of steps and delays, gets you a firm order, and lets the customer do much of the thinking and design work. Electronic commerce technologies thus offer a tremendous tool to support quick response to your customer. However, if your products are customized, such an approach is not trivial to implement, a point I'll discuss at length in the next chapter.

Work with the Customer's Designers and Manufacturing Engineers

To increase part commonality you need to work with the customer's designers and manufacturing engineers. In the absence of specific feedback, engineers tend to create new parts and components with each new design. However, if you can show your customer how, by creating commonality of components across their product line, they will make it possible for you to greatly improve your responsiveness, they may be willing to make compromises in their product offerings and functionality. Sometimes, no real compromise is needed, just an awareness that will allow different sections of the customer's company to use the same component. Also, it is not necessary to go for exact commonality; even similarity of parts will help. Consider a company that supplies shafts to a customer making large custom generators. The supplier starts with bar stock and then machines the shafts to create various customized features. Since the customer makes a variety of generators using shafts of widely varying diameters and material, the supplier cannot stock all the types of bar stock needed. The result is, for certain types of shafts, the supplier must quote a long lead time that includes the time to get the bar stock. However, if the customer can rationalize the different types of shafts used, the supplier could stock just a few types of bar stock and always have stock on hand, since the investment required would now be reasonable. The supplier might even be able to premachine some of the common features and have the shafts 50 percent ready prior to receiving the order specifying the detailed features. In this way, the customer's efforts in thinking through the product features and options can assist the supplier to shorten its lead time.

Finally, you need to ask your customers to (electronically) share their schedules and forecasts with you, which I'll discuss in detail in the next section.

APPLY QRM STRATEGIES TO THE ENTIRE SUPPLY CHAIN

Products delivered to an end consumer go through a whole chain of suppliers, starting from the mining of minerals, to the manufacture of stock materials, the fabrication of components, assembly of the product, and eventually, delivery to the end customer (possibly through distributors and retailers). This set of links, from raw materials to the customer, is known as the supply chain. In order for a QRM strategy to be most effective, you need to communicate all the strategies in the preceding sections (relating to both suppliers and customers) up and down the supply chain, leading to a partnership of manufacturers that is extremely responsive to the consumer. To motivate the need for this type of partnership and information-sharing, let us examine what occurs in a traditional (adversarial) supply chain. Before proceeding with this, however, I should clarify my intent.

Recently there has been much publicity on the issue of supply chain management[7] and there have been a number of success stories in the literature.[8] Although the topic of supply chain management is vast, my treatment here is limited to the specific context of QRM. For example, I will not discuss supply chain optimization models, which look at optimizing the location of inventory throughout the supply chain, but focus instead on structural changes that need to be implemented both *within* organizations and *across* companies in the supply chain. These changes are in keeping with QRM strategy, in that they fundamentally transform the behavior and dynamics of the supply chain to inhibit the growth of Response Time Spirals. One might say that while supply chain optimization models attempt to find the best solution with a given behavior of the supply chain, the QRM approach is to change the nature of this behavior. In fact, it turns out that a QRM strategy is essential to support effective supply chain management. Despite the high degree of hype about the latest information systems that enable supply chain management, without a solid foundation in QRM strategy these information systems will not be very effective. Witness a quote from a well-known manufacturing consultant:

> Cutting customer lead times by speeding up information flows will not significantly reduce the structural depth of supply chain inventories. On the contrary, much evidence suggests that cutting lead times without reducing manufacturing and transit lot sizes causes sharp increases in supply chain inventories.[9]

So my focus in this section will be (1) to understand the need for supply chain management, and (2) to explain the specific techniques that will provide maximum synergy between a QRM strategy and effective supply chain management. I'll begin by reviewing the workings of a typical supply chain.

The Whiplash Phenomenon in Supply Chains

Figure 10-3 shows a typical supply chain. Probably the worst effect of a traditional, adversarial supply chain is a phenomenon known variously as *whiplash*, the *bullwhip effect*, or *volatility*. Small changes in end-consumer demand lead to increasingly large changes being propagated to successive levels of the supply chain. A number of studies documented this effect, beginning with a landmark paper by Forrester more than 30 years ago.[10] A typical scenario goes like this: Consumer demand increases by 5 percent at the product manufacturer; this leads to a 10 percent increase in demand at the first tier suppliers; the second tier experiences a 20 percent increase; and the third tier is hit by a 40 per-

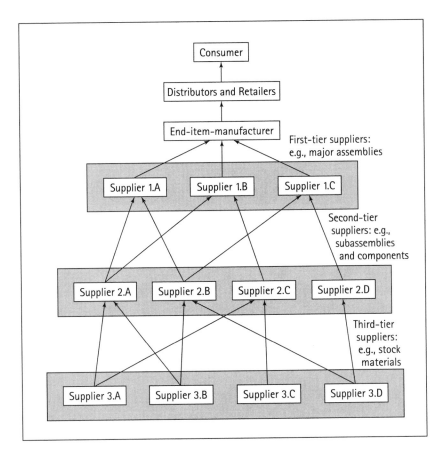

Figure 10-3. A Typical Supply Chain

cent increase. When consumer demand goes down by a small amount, the reverse occurs. Thus, small changes in consumer demand result in large changes for the suppliers. Like the end of a flag waving in the breeze, the supply chain finds itself being whipped around by small gusts (see Figure 10-4).

This whiplash effect is not just an irritant to the suppliers, it causes true inefficiencies in the system, leading to a loss of profits for the entire industry. One study shows that eliminating this whiplash effect could increase profits by up to 44 percent.[11] What causes this amplification of volatility? There are several cumulating factors:

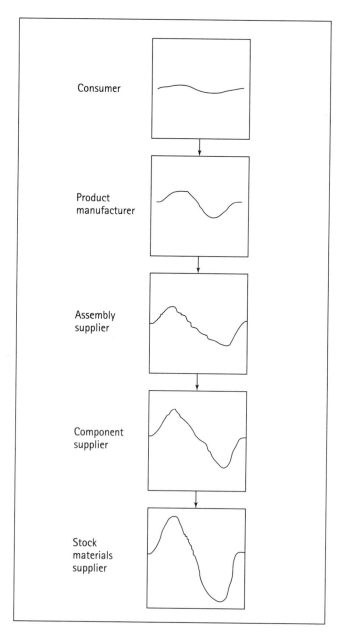

Figure 10-4. Amplification of Volatility in the Supply Chain

- Batching of orders leading to increasing batch sizes for lower levels of the supply chain.
- Using demand data from the immediate customer to update forecasts, rather than using demand data from the end-customer.
- Overreaction by each supply tier to a demand shock.
- Placing large orders with suppliers when there is a "boom" cycle, creating backlogs that require placement of even larger orders by everyone, then canceling of the orders when there is a "bust," leading to a huge downturn in the orders.
- Forward buying practices for seasonal items that amplify the seasonality down the supply chain.
- Inability of the humans in the system to anticipate the behavior of the complex feedback loops caused by all these factors.

I'll give two simple examples from two different industries to highlight this whiplash effect or volatility.

Example One: Retail Chain

Consider desk lamps sold by a retail chain. One particular store sells 10 lamps per week, on average. However, the lamps are packed in cases of 40, so the store places its order for 40 every four weeks or so. The regional manager for the chain receives several such orders each week. On a given week, suppose she gets orders for a total of 160 lamps. However, she needs to place her orders in terms of pallet loads, and a pallet load is 300 lamps. While shipping the order to her, the transportation manager for the factory needs to fill a truck, so he convinces the regional manager to take two pallet loads. So in a given week, a particular demand for 10 lamps at one store can trigger a whole sequence of events leading to an order for 600 lamps at the factory. Since this order will take a while to be consumed, the factory will experience no demand for a long period of time, and then a large order will appear again. The regular consumption at the stores has turned into "lumpy" demand at the factory.

Example Two: Tool Manufacturer

Consider a machine tool manufacturer in a capital equipment industry. Here you'll find a leveraging effect of another nature. Suppose its customers replace 10 percent of their equipment annually because of obsolescence. Now if these customers expect a 5 percent increase in demand in a given year, they will order an additional 5 percent of new equipment in addition to their normal 10 percent. This results in a total demand that is 50 percent higher than the regular annual demand. Thus a 5 percent increase in demand at the consumer level has led to a 50 percent increase at the machine tool manufacturer.[12]

These and many other examples have been documented for several industries, and they illustrate the difficulties faced by the traditional supply chain. Now we discuss how you can attenuate these demand distortions.

Multi-Tier Concurrent Planning

You need to adopt several strategies to reduce the whiplash effect, but prime among them is for multiple tiers in the supply chain to plan together instead of each supplier doing its own forecasting and planning. This may seem difficult or even impossible to you if your industry is characterized by adversarial customer-supplier relationships as well as suppliers bidding fiercely against each other. Yet segments of the U.S. apparel industry have achieved just such concurrent planning—an industry where historically the norm was highly adversarial customer-supplier relationships combined with cutthroat bidding wars between the suppliers at each tier. Now, instead of each tier planning separately, there are some success stories where retailers, apparel manufacturers, and textile suppliers plan together, replacing the traditional adversarial relationships with cooperative partnerships.[13]

I'll call such multi-tier cooperative planning *concurrent planning*. How is such planning accomplished? First, the final manufacturers must be willing to share their forecasts, not just with their first-tier suppliers, but with the whole supply chain. And second, groups of planners (and marketing/sales representatives) from the entire supply chain need to convene periodically to develop shared plans and forecasts. This is to ensure that the opinion of one planner or company is tempered with the opinions of others, and the discussions benefit the industry as a whole.

It should be noted from this that supply chain management software alone will not provide a full solution to the whiplash problem. Such software can certainly help companies in implementing some of the solutions discussed here, but key to concurrent planning is the group process involving participants from multiple tiers. Without this group process, any software solution will be based on the planning outlook of a single company and will not have sufficient impact on eliminating the whiplash effect.

Another key element of concurrent planning is to collect sales data, preferably at the point of sale by electronic means such as barcoding, and to share such data as soon as possible with the supply chain. Doing this enables everyone at the concurrent planning meetings to base their discussions on actual facts.

Some companies have formalized such shared forecasts into their MRP and reordering systems: An MRP module actually gives a supplier a plan for several weeks, of which (say) the first two weeks can be considered a firm purchase order. Here both the customer and the supplier win: The manufacturer saves on purchase order time and obtains shorter lead times, while the supplier has improved visibility into the long-term forecast and can reduce safety stocks and better plan its own production.

The benefit of sharing forecasts with one's suppliers is not always obvious to people in the customer's firm, however, as a different anecdote illustrates.

Shared Forecasts: Benefit or Burden?

In 1997 I was working with an equipment manufacturer to help reduce their supplier lead times. One of the strategies we discussed was giving the suppliers better visibility into the company's production plans. Surprisingly, I found substantial reluctance from the company's planners to share their schedules with their suppliers. "We have a lot of 'garbage' in our schedules," one of the planners remarked. When I asked what she meant by this, the planner clarified: "Our schedules contain many speculative orders, based on estimates from our sales department on when they hope to close specific sales leads. These speculative orders are then exploded through the MRP system to generate due dates for in-house production and for components from our suppliers. But many of those machine sales are realized later than expected, or not at all. If we gave those demand figures to the suppliers, we would have a lot of complaints about misleading them and making them produce parts that we never ended up ordering!"

(continued)

> To address this concern, I'll provide a contrasting story, one that goes to the other extreme. Bose Corporation, a manufacturer of audio speakers based in Framingham, MA, not only gave its key suppliers full access to its computer schedules, but also asked them to place representatives at its plant, giving them the authority to place their own orders as needed to keep the Bose factories running. The director of purchasing explained this strategy in an interview:
>
>> You're trying to give your supplier as much insight as you can into your business. But as a matter of fact, that person has his nose on the windowpane on the outside of your company, trying to look and see what you're doing. Common sense tells you that if you bring that vendor inside your company, give him access to your MRP system, empower him with your purchase orders, remove the planner, remove the buyer, then you have this experienced person who brings all the scar tissue, and all the gray hair from putting up with your surprises over the past several years. This person knows his business, and this person knows your business.[14]
>
> The experience at Bose Corporation has also shown that the information acquired by the supplier's in-plant representative—whether through formal meetings or informally over coffee—can help keep the supplier's factory operating more smoothly.
>
> Returning to the concern of the planner at the equipment manufacturer, we see that such a concern arises because of the traditional adversarial view of suppliers. She saw the suppliers being given "arms length" numbers, without bringing them into the planning process and without allowing them to understand the background for the numbers. On the other hand, the Bose Corporation experience shows that if we view suppliers as partners helping us run our business, and train them to understand our business and planning processes, then they should be able to plan demand for their items at least as well as our own planners. Further, knowing their own factory's constraints and their own business uncertainties, they can probably do a better job of serving that demand. They will also serve the demand in the way they know best for their own business. Thus both our business and their business end up being winners.

Another strategy you can use to improve the performance of the whole supply chain is to implement a single point of quality control for each item. It is not unusual for a retail item to be inspected at a supplier, again at the main manufacturer, then at retail receiving facilities, and finally once more by the store clerk that restocks the shelf. Implementing quality programs that include supplier certification can eliminate duplication of effort all along the supply chain.

It is no easy task to transform an entire industry to enable such partnerships, particularly an industry that has a long history of adversarial relationships. However, such transformation has already been accomplished in some segments of traditional industries such as the apparel and automotive industries, so we know it can be done with sufficient will and effort. The reward for this effort is that, in the end, everyone will be a winner, as inventories are eliminated, quality improves, costs for the entire supply chain go down, and the consumer's orders are satisfied rapidly at a competitive price. Such success will result in further growth of this industry segment, with rewards to all the companies in the supply chain partnership.

Main QRM Lessons

- Understanding another Response Time Spiral at work between customers and suppliers.

- Make supplier lead time reduction your #1 performance measure for supplier selection; then educate your suppliers on QRM and assist them in implementing this strategy.

- Don't forget the importance of educating your purchasing staff on QRM; also rethink their performance measures.

- Educate your customers on your QRM strategy and work on a joint schedule of moving to smaller lot sizes.

- How your QRM program can support your marketing efforts.

- Motivate your entire supply chain to implement concurrent planning.

Part Three

Rethinking Office Operations

11

Principles of Quick Response for Office Operations

I EMBARKED ON THE DESCRIPTION of QRM principles by discussing their application to the shop floor and materials flow. This was deliberate, since the factory is considered to be the core of a manufacturing enterprise; this is where products take on their physical shape and characteristics. However, a good deal of the lead time to serve a customer, and a substantial amount of the cost of operating a manufacturing enterprise, take place outside the shop floor. In fact, office operations represent a significant portion of both cost and lead time, and yet companies have largely neglected the office as a source of productivity improvement. This chapter begins Part Three of this book, which focuses on the application of QRM principles in this oft-neglected area of office operations.

In manufacturing companies, office activities that affect lead time include three key ones: request for quotations (RFQ) processing (all activities relating to capturing an order); order processing (activities following receipt of an order up to its release for production); and new product design and development. This chapter and the next lay down a set of QRM principles related to office operations in general and apply them specifically to the tasks of RFQ processing and order processing. These key tasks affect lead time for existing products. I also discuss the situation where products need some amount of engineering or other customization, but I will not dwell on the issue of introducing a new line of products since I focus on the entire new product development process in Chapter 14.

OFFICE OPERATIONS: A NEGLECTED OPPORTUNITY

In my dealings with dozens of manufacturing concerns, I have found office operations to be a neglected opportunity for developing competitive advantage. Most managers don't realize how significant this opportunity can be—and that they can do something about it. Let us begin by understanding why reducing lead time in your office is so significant.

- *Office operations can account for more than half your lead time.* As an example, consider statistics from lead time reduction projects conducted by the Center for Quick Response Manufacturing with companies that delivered customized products involving a small degree of engineering. It was found that more than 50 percent of the total lead time after receipt of the order occurred in office operations that took place prior to releasing any jobs to manufacturing.[1] The statistics are surprising even for companies whose products do not require customization. Orders for standard, oft-made products sometimes take one or two weeks to be processed through the office, even though all the steps involved ought to be routine. An extreme example is provided by the automotive industry. One would expect that car orders from dealers to the factory would be relatively straightforward to handle; after all, the options available on a car are clearly demarcated. And yet, at a division of a U.S. car manufacturer, orders from dealers take *three* weeks to be processed through various office steps before they are released to the manufacturing scheduling department. Because of the need for schedule balancing and a typical backlog of orders, it then takes another five weeks before the specified car is actually made by the factory and shipped to the dealer. So of the eight weeks it takes a dealer to get a specific car, about 40 percent is spent in office processing.
- *Office operations can account for more than 25 percent of your cost.* In Chapter 2 I discussed the breakdown of the cost of goods sold (COGS) for a product. At a typical U.S. company, this breakdown is: materials and purchased parts—50 percent; direct labor—10 percent; overhead—40 percent. Of the overhead costs, some are accounted for by depreciation of assets, marketing campaigns, and research and development costs. However, a significant remaining amount represents office workers and their support expenses.
- *Office operations play a significant role in your ability to capture orders.* In Chapter 2 I also discussed how implementing QRM in quote generation can have a redoubled effect on your order capture

rate: It can significantly increase the number of quotes that convert into firm orders, while at the same time reducing the amount of resources required to generate the quotes.[2] The latter means you can send out more quotes at less cost than before. Couple this with the increase in acceptance rate, and you end up with a large increase in firm orders.

Despite the significance of the above points, if you look at the last two decades of manufacturing improvement trends and fads you'll find them focused on the shop floor and materials flow.* So it is worth asking, if office operations are indeed such a significant contributor to competitiveness, why has U.S. industry neglected them? Studying the answer to this question is useful not just for historical reasons, but because it may also give you some insight into shortcomings in your own operations. Several factors have contributed to the neglect of office operations in manufacturing companies (also see Table 11-1):

- *A traditional focus on shopfloor operations.* At the turn of the century, entrepreneurs like Ford, along with Gilbreth, Taylor, and other proponents of scientific management methods, achieved tremendous productivity gains by minute examination of shopfloor operations. Ever since then, manufacturing companies have assumed that improving methods and efficiencies on the shop floor is the key to profitability. As a result, a focus on the shop floor lies deep in the marrow of the U.S. manufacturing corporation.

As an indication of how little attention companies devote to understanding office operations, consider the following three facts. First, a study of 18 companies by researchers at Harvard Business School showed top managers typically had a very simple, truncated, and inaccurate description of the flow of jobs through the office. At the same time, people deep within the organization saw only their own individual details. In each department, people thought that someone in another department understood the entire flow, while no one in the organization really did; everyone could give only a partial description.[3] Second, my own experience with examining office operations at over two dozen

---

*One exception is the recent emergence of Business Process Reengineering (BPR), which has put office operations under the spotlight. There are, however, some important areas where QRM goes beyond BPR; these will be discussed later.

Table 11-1. The Neglected Opportunity in Office Operations

Why Office Operations Are Significant
• They account for more than half the lead time in many companies.
• They can account for more than 25 percent of the cost of goods sold.
• They can have a substantial impact on order capture rate.
Why They Are Neglected
• A traditional focus on shopfloor operations, stemming from the success of scientific management methods.
• Costing based on direct labor.
• Absence of lead time measurement for office activities.
• Lack of appreciation for the impact and benefits of lead time reduction in the office.
• Cost-based mind-set and misconceptions about QRM methods.

manufacturing companies in the Midwest has been the same: There is no established knowledge of the complete set of office procedures required to service a given type of customer order. A specific case study will illustrate this in the next chapter. Third, a study of several U.S. banks conducted by the Wharton School of Business revealed that managers had little knowledge of the sequence of steps and processing procedures involved in completing particular transactions.[4] Contrast these three examples with manufacturing operations: In most organizations, not only is there a standard routing for each product's journey through the shop floor, but they even have setup and run times for each step. This reinforces the observation that companies have focused their measurements on manufacturing, while neglecting what goes on in their office operations.

- *Costing methods based on direct labor.* The scientific management movement established accurate data for the direct labor content of each manufacturing operation. An indirect consequence of this was that management designed costing systems that were primarily based on this direct labor data. Rather than trying to measure the remaining organizational costs involved with making a given product, they allocated this "overhead" as a multiplier on the direct labor. Office operations became a part of this fixed multiplier, and the resulting total cost of a part thus depended—for cost estimation purposes—on only two controllable quantities: material and direct labor. Managers and engineers therefore focused their efforts on reducing these two, reinforcing the traditional focus on the shop floor, and further neglecting office operations.

- *Absence of lead time measurement for office operations.* Despite the fact that office operations account for a significant portion of lead time, and despite the fact that most companies state their desire to reduce lead time, few companies accurately track the components of lead time spent in the office. There may be rough data available about the time spent by orders prior to the release to manufacturing, but seldom is there any data on which office operations were done when, and if jobs were waiting, what was the cause. In contrast, shopfloor systems diligently record the start and finish of every manufacturing operation, and workers need to fill out comments on delays and other problems. This is one more indication of the focus on the shop floor and the neglect of office performance. The result is the oft-stated dictum: You get what you measure. Since management does not accurately measure lead time for office operations, there is little incentive for improvement. The following quote from the manufacturing manager at a tooling company echoes the sentiments of manufacturing managers around the world:

  > Our normal quoted lead time is eight weeks. There is some design work to do for most of our tools, so you allow four weeks for the office operations and four weeks for manufacturing. In reality, though, I get jobs barely one or two weeks before their due date. I have to rush them around the shop floor—through 12–14 work centers—to be on time. Sometimes I don't even see the order folder till after the due date! In either case, guess who gets the blame for the late delivery?

- *Lack of appreciation for the impact and benefits of lead time reduction.* Although most managers understand, superficially, the importance of short lead times, few fully appreciate the ripple effects and indirect benefits of achieving quick response in office operations. For example, in my discussions with managers I have found that the various benefits detailed in Chapter 2 are an eye-opener to many. So much so, that on occasion managers do not really believe these benefits exist, and need to see hard data and hear case studies to be convinced.

Table 11-2 illustrates one situation where the benefits of rapid quoting are typically underestimated. As seen from the calculations in the table, it may be possible for a company to increase its profits tenfold by rethinking its approach to quoting. Yet, in all my work with manufacturing companies around the world, I have never met a senior executive who had already performed an analysis of this nature, and seen the opportunity in quite the same way as Table 11-2 shows.

Table 11-2. The Tremendous Opportunity in Quick Response Quoting

This calculation is based on an actual U.S. equipment manufacturer that spent 20 hours of staff time on each quote, with a closing rate of 10 percent of the quotes (see column "Traditional Quoting"). The lead time for quoting was over 100 hours or 2.5 weeks (since, during 80 percent of this time, a job was not being worked on, but was waiting for some reason). Suppose this company could respond to quotes with just 1 hour of staff time by using Quick Response quoting methods described later in this chapter. We consider two situations. First is the "conservative" case where the company's closing rate remains the same but the quick quotes are less accurate, resulting in lower gross margins per job. The second, "optimistic" case is where the closing rate increases to 30 percent thanks to the speedy response to inquiries, plus quick quotes are no less accurate (which is usually the case, as discussed later in the chapter). We see that even in the conservative case, the company makes a higher profit than by the traditional method, while in the optimistic case the profit is greater by an order of magnitude.

	Traditional Quoting	Quick Response Quoting (conservative)	Quick Response Quoting (optimistic)
Assumptions			
Employee Hours/Quote	20	1	1
Number of Quotes per Year	250	250	250
Average Price of Quoted Jobs	$100,000	$100,000	$100,000
Physical Results			
Lead Time to Quote	12 days	under 1 day	under 1 day
Closing Rate	10%	10%	30%
Average Gross Margin on completed jobs	20%	10%	20%
Financial Results			
Gross Profit	$500,000	$250,000	$1,500,000
Cost of Quoting Efforts[†]	$375,000	$ 18,750	$ 18,750
Net Profit	**$125,000**	**$231,250**	**$1,481,250**

[†] The cost of quoting is based on an average salary of $20/hour for the staff involved in quoting, plus $5/hour for benefits. To this is added 200 percent overhead to cover office equipment, space, utilities, consumables, and remaining organizational costs.

- *Organizational mind-set and pervasive misconceptions.* The cost-based mind-set, discussed in previous chapters, only serves to reinforce the focus away from office work toward shopfloor operations and materials management. Beyond that, however, are misconceptions about what it would take to implement quick response. These serve as additional barriers to lead time reduction in the office.

In summary, then, while office operations represent a significant opportunity for lead time reduction and productivity improvement, manufacturing companies have not devoted enough attention to improving this aspect of their operations.

THE RESPONSE TIME SPIRAL FOR OFFICE OPERATIONS

To fully comprehend the need for the QRM approach, we must first visit an old friend—the Response Time Spiral. There is no reason why this insidious spiral should be confined to the shop floor and materials portion of the company—the office is not immune to its attack. Let us study the evolution of the Response Time Spiral in the office of a traditional organization.

The offices of modern manufacturing companies are large, complex organizations, with many departments such as order entry, process planning, cost estimating, design engineering, manufacturing engineering, scheduling, and so on. Long lead times are needed to ensure that a job gets resources in any one of these departments. For example, if a job is sent to the design engineering department, it would be unreasonable to expect that it would be worked on right away. There would typically be several jobs ahead of it, and it would need to be "scheduled" to be worked on. The same is true for all other departments the job needs to visit.

Department managers thus need to plan ahead for three reasons. First, they need planning to assist in getting jobs done by their due dates. Second, they need to look ahead to see what jobs are due to arrive in their department, so that they can plan their work force needs and other resources. Finally, they need to plan ahead to see what due date commitments they can take on for new jobs.

As you know, as lead times lengthen the accuracy of these plans declines. Changes in customer requirements, unexpected new jobs, or errors and rework all cause ripple effects invalidating the original plans.

Department managers, however, having experienced many such unexpected changes and ripple effects, learn to insert safety margins into the times that they plan. As a result, the total lead time for a job is much longer than it needs to be, and jobs are started into the system much earlier than necessary. The number of jobs in process in the whole organization is therefore unnecessarily large.

Because lead times are so long, the company is sure to receive jobs that need to be done in less than the standard lead time. Perhaps top management is under pressure to accept an order to prevent a major loss to the competition. Or perhaps an important customer changes their mind on one of their options, but still wants the same delivery date. Whatever the reason—and there are many such reasons—a "rush job" is pushed through the departments. Scheduled jobs make way for this job, which gets through in time but leaves a trail of devastation behind. Regular jobs are now taking even longer than expected, despite the safety times applied by the department managers.

Each department then goes through a process similar to this: The department begins to get a reputation for always being late. The department manager tries many different approaches to improve the department's efficiency, but they all fail, for reasons discussed earlier in this book. Finally, after being reprimanded by top management, the manager decides that this department needs to quote a longer lead time in order to meet its commitments.

As all departments go through this cycle, the lead time for office operations grows. With even longer quoted lead times, there are more inaccuracies in the plan and an even greater need for rush jobs. These result in even longer lead times. The spiral has begun to grow.

Roots of the Response Time Spiral in the Office

The spiral I just described evolves from the same forces as the one discussed earlier in the book for the manufacturing arm of the company. Specifically, it is a result of scale-based and cost-based thinking that is in place from past manufacturing strategies. These strategies result in a functionally organized enterprise. This issue was discussed in depth in Chapter 3, but I'll review the key points here.

A keystone of the industrial revolution was discovering how to break a complex job down into simple and specialized tasks. These tasks, such as order entry, could then be performed by people with limited skills, trained only for that job. This way salaries could be kept to a minimum. Each such specialized task was completed by a group of workers managed by a more experienced specialist, and this organization became a department. The goal of the supervisor was to keep the department cost down. This meant she or he had to do two things: (1) each resource had to produce as much work as possible (translate this into maximizing the efficiency of each person), and (2) it was imperative that no one ever ran out of work (translate this into maximizing the utilization of the resources). The former criterion led to work methods such as batching of tasks, which gave precedence to efficiency over keeping jobs moving.[5] The latter led to a backlog of work in each department. Both these outcomes meant that lead times in each department deteriorated. Since jobs typically had to go through numerous departments, the lead time for an order grew to weeks and even months.

Worse yet, quality deteriorated and problems took very long to resolve. Because employees had limited responsibility and narrow training, often a problem was not detected until the job had been processed through several departments. Even after detection, employees had neither the expertise nor the responsibility to deal with the reason for the problem, so it was left for the managers to resolve it. In this hierarchical organization, communication was poor and there was frequent backtracking of jobs, further exacerbating the lead time problem.

The large number of jobs in process made the situation even more frustrating. As employees were implored to do something about the delays and late jobs, they felt increasingly helpless, surrounded by a growing volume of jobs. Morale plummeted, and quality and efficiency deteriorated further.

A few additional reasons for long lead times arise from the fact that office operations are geared toward handling information (an issue I will deal with more thoroughly in Chapter 12). In many cases jobs wait for information, such as a price quote from a component vendor, or a telephone response from a customer clarifying an issue about product options. Other delays arise from the use of computer systems. In this information age, it may come as a surprise that computerization did not necessarily reduce the lead time of office operations. On the contrary, when implemented incorrectly, it led to a proliferation of independent departmental systems that could not share information, resulting in redundancies and non-value-added work throughout the system. Throughout the 1980s, U.S. businesses invested a staggering $1 trillion in information technology, with no signs of any productivity growth as a result.[6] One result of the proliferation of independent systems is the fact that employees may have to use multiple computer systems to complete a task. Also, incompatibility among multiple computer systems can lead to the need for data to be entered repeatedly, as well as the possibility of inconsistencies between the systems which then require problem solving and rework.

Eliminating the Response Time Spiral

To see how to cure the office of the Response Time Spiral, let us revisit the very first question in the QRM quiz:

> 1. Everyone will have to work faster, harder, and longer hours in order to get jobs done in less time.
>
> ❏ True   ❏ False

As I've mentioned before, this is the traditional way of trying to get jobs out fast, and it is usually unproductive and even counterproductive. The detailed reasons for this were provided in Chapter 3, but briefly they are as follows. The traditional approach is unproductive because the best one could expect by pushing people to speed up their work, or by optimizing paperwork operations, might be a 15–20 percent improvement over current work rates. On the other hand, QRM aims for 50–95 percent improvements, so it is clear that such an approach will fall far short of the mark. The traditional approach can even be counterproductive because it results in the functionally organized enterprise and the Response Time Spiral, which then result in longer, rather than shorter, lead times. Instead, the QRM approach is:

QRM Approach: Find whole new ways of completing a job, with the primary focus on lead time minimization.

For shopfloor operations you've seen that this whole new way involves major restructuring of the manufacturing portion of the organization. The same is true in office operations. In fact, you can apply many of the principles used for shopfloor lead time reduction to the office. For example, the ideas of focusing on a product family and creating cells, so successfully applied on the shop floor, can be extended to the office environment. However, since the shop floor processes primarily materials, while the office processes primarily information and paper, the details of implementation are necessarily different. I'll separate this discussion of key principles into three main categories:

1. *Organizational Principles*: These lay down the framework for the whole new way of completing work in the office in order to minimize lead time. They form the office counterpart to Chapter 4 (which focused on the shop floor).

2. *Information Handling Principles*: In Chapters 8 and 9 you learned how you need to change material handling approaches to support the new structure of the manufacturing organization. In a similar way, since the office deals primarily with information, several basic principles of information handling need to be put in place to support the QRM organization in the office.

3. *System Dynamics Principles*: Just as in the context of manufacturing operations, another key element of quick response is understanding and exploiting the dynamics of delivery systems. Restructuring the office will not be completely successful—and may even be a disaster—if you do not follow certain basic system dynamics principles.[7] Chapter 13 will present the key system dynamics principles that complement the company's restructuring efforts and ensure short lead times.

ORGANIZATIONAL PRINCIPLES REQUIRED IN THE OFFICE

A word of caution: After looking over these principles, you should not attempt to reorganize the whole company right away. U.S. managers are notorious for operating on the "follow the latest buzzword" philosophy, which works like this. A CEO or vice president attends a seminar or reads a book, and is enamored with a particular new buzzword, for example, "zero inventories." The manager returns to his company and fires off a volley of memos requiring everyone to adopt this philosophy overnight. In this case it might be: All departments have to cut their inventory by half within the next quarter.[8] Without the painstaking effort of laying the foundations for inventory reduction, such as quality improvement and preventive maintenance efforts, this rash attempt at chasing the new philosophy turns into a nightmare on the shop floor. Although JIT and TQM are by now well accepted productivity improvement techniques, such "jumping on the bandwagon" and overnight management edicts accounted for the failure of many early JIT and TQM efforts. As a more recent example, books on reengineering actually promote such radical and large changes, and that has accounted for the failure of many reengineering projects as well.[9] The QRM approach is to:

- Focus on a market segment.
- Determine the product of office processing for this segment.
- Identify the office processing steps required for this product.
- Look for subsegments amenable to simpler processing steps.
- Determine if this subsegment represents a significant market opportunity.
- Focus your initial QRM effort on the Focused Target Market Subsegment (FTMS).
- Create an office cell to serve the FTMS.
- Redesign processing operations for the FTMS.
- Provide resources and support to ensure fast flow.
- Eliminate traditional approval and control systems.
- Consider integrating office and manufacturing cells.
- Support these efforts with information handling and system dynamics principles.

I'll lay down the details of this approach and its key principles here and in the next chapter. These principles will be described in a logical sequence, so that each can be understood in the context of preceding principles. This is not the way to proceed with actual implementation, however. The specific steps for implementation are somewhat different, and will be described in Chapter 17. Thus the reader should use this chapter to understand the principles, and then use Chapter 17 to help define an implementation plan.

Focus on a Market Segment

Begin your QRM efforts in the office by looking for a segment of your business where there is opportunity for success via a quick response strategy. Ask your marketing people: Is there a noticeable trend toward shorter lead times in any area of your market? Ask your salesforce: Which clients or business segments are complaining about long lead times, and are you losing sales because of this? Ask the customer service people: Is there a market segment where your deliveries are late as a norm, and customers are losing patience and likely to jump ship? Use all this information to narrow in on the most promising opportunity. By a promising opportunity, I mean one where a reduction in your lead times could make a splash—it could dramatically increase your market share, or open up a new market segment, or make your customers stand up and recognize your performance.

Determine the Product of Office Processing for this Segment

By *product* I mean, not the physical item manufactured for a customer, but rather, the outcome of the office processing steps for this segment. Consider again the example from Chapter 4 involving SteelShaft, Inc., a company that makes customized shafts for motors and generators. When an order is received at SteelShaft, it is processed by the following office departments: inside sales, order entry, design engineering, process planning, materials planning, scheduling, and shop release. After this, the order is handed over to the manufacturing arm of the company. What is the product of these office processing steps? It is a completed "shop packet"—a folder containing blueprints, routings, processing instructions and any other information needed to manufacture the shafts for this customer.

Now let us consider a different example. Suppose, when you ask the questions to determine the market segment, your sales manager replies:

> It isn't the product delivery time that is hurting us, but rather, the time to generate quotes for customers. This is a source of great frustration for my salespersons. In many cases, by the time we deliver the quote, the customer has long since received and reviewed several other quotes, and has already issued the order to someone else!

So you need to base the strategy for capturing this market segment on quick response in quote generation. In this case, the product of office processing steps is a quote for the customer.

Identify the Office Processing Steps Required for This Product

This step is easier said than done. As I've mentioned before, even major banking corporations have discovered they do not know their own processing steps. In the next chapter, I'll introduce two tools, process mapping and tagging, that will help you to identify your office processing steps. Note, however, that it is not necessary to completely understand all the details of the current procedures. This is because you will be redesigning the procedures anyway, along the lines described below. You just need enough detail on the current operation to know the main processing steps, and more important, have good insight into the problems, exceptions, and difficulties that you will encounter during these steps.

Look for Subsegments Amenable to Simpler Processing Steps

As you look at the processing steps used for this market segment and observe the exceptions, special cases, and other anomalies that you must deal with, ask yourself if there are some portions of this market that are not cluttered with such complex processes. In other words, might there be classes of jobs that could be better served with simpler processing steps? This is an apt question for a brainstorming session with representatives from sales, customer service, and key office departments that process jobs for this market.

To illustrate the operation of this principle, let us return to the SteelShaft example. The company makes custom shafts of varying diameters, materials, and with custom-designed features, so in general, an order for a new shaft could require a lot of processing through various departments such as design engineering and process planning. However, consider an order for shafts made from 8640 stainless steel (a steel alloy often used for shafts due to its strength and impact resistance), the diameter lying within a common range made by the company, and where the features involve only straightforward slots, keyways, and face milling. Since all these items are routine for SteelShaft, such an order could itself be routine, requiring only a few simple processing steps in the office.

Determine If This Subsegment Represents a Significant Market Opportunity

Having identified a subset of jobs that you can process with simpler steps, discuss with your sales people whether there might be a significant mass of customers interested in this subset alone. For SteelShaft's situation, the question would be: If they limit their initial effort to customers who want 8640 steel shafts between 0.5″ and 3.0″ diameter, and only allow the following features on the shafts (a list of features is placed here), *and if the lead time for such shafts were to be 3 weeks instead of the current 10 weeks*, is this a sufficient market opportunity to pursue? Note in particular the phrase in italics, because that is key to a QRM strategy. This segment may be very small today, but are there a lot of customers who would jump at the chance of having custom shafts in 3 weeks?

If you determine that the opportunity is indeed significant enough to pursue, then proceed to the next step below. On the other hand, if discussions determine that such a subsegment does not represent a significant market opportunity, then brainstorm to find another subsegment, and repeat the analysis with this subsegment. If such subsegments repeatedly fail to live up to the mark, then this entire market segment may not be the best area to start your QRM efforts. You may need to go back to the beginning and start over with a different market segment. In my experience, such starting over is extremely rare. Usually, one can find a suitable subsegment in one to three tries. The reason is simple: The people involved in the various brainstorming activities—both the initial segment determination and the following subsegment determination—have a great deal of experience in dealing with customers and good intuition about the marketplace, hence their initial guesses tend to hit the mark. I call this subsegment a focused target market subsegment, or FTMS.

Focus Your Initial QRM Effort on the FTMS

Clearly demarcate the boundaries of this subsegment. Identify precisely the materials, features, and numerical values of any dimensions that will define the products that lie in this FTMS, so there can be no misunderstanding. Your QRM effort will begin with this subsegment. In fact, you will intentionally leave other parts of the organization unchanged. Contrast this with some reengineering approaches that advocate making large changes in the whole organization. Also contrast this with the "overnight management edict" method described earlier. The QRM approach, instead, creates a success in a limited portion of the enterprise and then builds on this success, expanding to other parts of the enterprise later. This stepwise approach is discussed in depth in Chapter 17.

But if you are not going to change the rest of the organization, then what about products that lie in the overall market segment, but not in this FTMS? Should you turn down the orders for them? Not at all. Just continue to handle them the way you do now—after all, you do have these processes in place and working today, even if they are not ideal—and keep these products in mind for a later QRM project.

Create an Office Cell to Serve the FTMS

As you'll see below, the office cell I describe is based on a team. Teamwork is all the rage nowadays, and many companies are putting in teams to try to reduce lead time. However, implemented incorrectly, these teams can be ineffectual. Therefore I need to dispel another misconception presented in the QRM Quiz:

> 8. We can implement QRM by forming teams in each department.
>
> ❏ True   ❏ False

The problem with this belief is that a team which has all its members in one department may be useful for efforts such as departmental quality and process improvements, but it will do little to reduce overall lead times. Instead, you need the following.

> **QRM Principle: Cut through functional boundaries by forming an office cell containing a multifunctional, cross-trained, collocated, closed-loop team responsible for completing all the office operations for the FTMS.**

This is the white collar version of a factory cell. Since the term *team* is used (and even overused) in many contexts today, I will refer to this particular type of team as a Quick Response Office Cell, or Q-ROC. (This is pronounced "queue-rock." The purpose of the hyphen is only to elicit the correct pronunciation of the term from an uninitiated reader.) Although the ideas for this concept derive from the same principles as cellular manufacturing, the details of implementation in the office are different. I've used several important terms in the above principle so I'll further define and discuss the following: multifunctional, cross-trained, collocated, and closed-loop.

Multifunctional

In contrast with the traditional functional organization, where each major function is performed in a different department, the Q-ROC will perform multiple (hopefully, all) functions needed for this subsegment. But does this mean you need one representative from each department of the company? How does one determine which functions are needed? This is actually an iterative process. Begin by identifying the core steps of processing needed to serve the target segment. This can be accomplished using two procedures. First, review your previous work identifying the set of simpler office steps needed to serve this target market. Second, use the method of value-added analysis, described in the next chapter, to decide if a given processing step deserves to belong in the set of core steps. Be ruthless in the application of this analysis; the more steps you remove from the core steps set, the smoother will be the flow. After this, the tasks that remain in the core steps define the functions that you need.

So we've determined what functions will be performed in the Q-ROC. However, it is not necessary that each function be represented by one person. Through various procedures described below, we are going to find ways for each person in the cell to perform multiple steps of work. In addition, through brainstorming and other process improvements, I am going to further refine the allocation of work in the Q-ROC. This is the iterative process referred to above, and it will become clear as I define more of the principles.

Cross-Trained

For a Q-ROC to be effective, it is imperative that its members be cross-trained to perform multiple functions required for the particular target segment. There are several reasons for doing this.[10] One obvious reason is that it enables the Q-ROC to keep operating even when a cell member is sick or on vacation. Another is that it enriches the office worker's job and reduces the drudgery associated with some repetitive tasks found in a traditional organization. For example, instead of having to sit at a computer and enter data from purchase orders all day, a cell worker may go through a cycle of talking to a customer, entering data into a screen, and doing some initial work on the actual order. Reducing the drudgery will help to keep morale high, thereby reinforcing quality and productivity efforts.

These are good reasons, but there are two more fundamental reasons for engaging in cross-training. One is that it enables a single person to accomplish multiple processing steps. The elimination of handoffs goes a long way to reducing lead time in office operations. Of course, in large companies you can understand if a handoff from one department to another adds days to a job. What is surprising is that such delays are substantial even in small companies. A rule of thumb that I've come up with after numerous studies of office operations is the following.

> QRM Nugget: Each handoff in the office adds at least one day to a job. Hence every handoff that you eliminate saves at least one day of lead time.

Ralf Hieber, a researcher at the Center for Quick Response Manufacturing, recently verified this rule at a small company in northern

Wisconsin. Even in this tiny organization, where all the order processing staff sat at desks in the same room, we found that a handoff from a sales person to an engineer, or from an engineer to an estimator, would add a day or more to order processing lead time. Even more surprising was that a handoff to a clerical worker, who was just expected to send a fax but do no other processing of the job, also added an average of a day in lead time.

The second fundamental advantage of cross-training is that it results in improvements in productivity. This occurs because, as cell workers learn more about the whole set of tasks, they can see major opportunities for improvement that were not obvious when people only worked on a piece of the job at one time. These improvements are surprisingly large and frequent in occurrence.[11] The scope and extent of cross-training for a Q-ROC will be discussed further in a later section.

Collocated

After deciding on the members of the office cell, it is critical that they be located together in an area of the office. Many companies make the mistake of creating teams whose members remain located in their functional departments and report to their traditional functional managers. Such teams have to schedule their meetings in company conference rooms, and while they may have some impact on lead time they do not approach the full potential of a Q-ROC. Indeed, the following is another misconception about teams.

> **Misconception: We can implement cross-functional teams while keeping our organizational structure intact.[12]**

A QRM team must be collocated—which is another reason I prefer the term *office cell* to *team*—and it *will* cut across functional boundaries and change reporting structures. By creating the physical space and relocating people into it, management sends a clear signal to the whole organization that it is moving away from the traditional way of doing business, and there is no going back to the old way. You need to change reporting structures because the cell members cannot reasonably report to their previous functional bosses. But whom should they report to?

This issue involves rethinking performance measurement and the whole reporting structure, which I cover in Chapter 16.

Closed-Loop

For best results, the Q-ROC should be able to complete all the office processing steps needed to serve the subsegment. Ideally, the job should not have to leave the cell. This is the meaning of the term *closed-loop*.[13] The term also implies that the Q-ROC must have authority to make decisions; without this, you have not given them full ownership of the process. The idea of delegating decision making may be difficult for some managers, particularly when large dollar amounts are involved, so let us clarify the meaning of authority and ownership. The QRM approach does not imply unlimited authority for the Q-ROC. Rather, it implies full authority within preset boundaries. For example, management can lay down a policy that the Q-ROC can issue, without outside approval, quotes that do not exceed $50,000. Any quote that ventures above that limit will need management approval. Such a policy is within the QRM framework—but only if, for the quotes below $50,000, the Q-ROC is not in a constant state of having to show its work to senior management for its tacit approval.

What if the number of steps required to achieve the closed loop is so large that the cell would have to contain 20 or 30 people? There are several ways to tackle this. The first is to ask if the subsegment can be narrowed further. Second, consider whether through cross-training, more steps can be done by one person. Third, use the principles discussed in the next section to find ways to eliminate and combine more of the tasks.

If, even after applying all these efforts, the cell is still unwieldy,[14] then you should partition the processing into logical subsets, with a Q-ROC for each subset of tasks. The goals in creating the partition should be:

- Each Q-ROC must be clear what its product is.
- Each Q-ROC must be closed-loop with respect to that product.
- Each Q-ROC should be given full ownership of delivering that product to the other Q-ROC (or Q-ROCs).

Redesign Processing Operations for the FTMS

With the preceding steps in place, you are now in a position to make a significant impact on the processing operations themselves. As mentioned frequently in this book, when operations are redesigned for QRM, the primary focus of the redesign should be minimization of lead time. Thus, in this step, you should focus on whole new ways of accomplishing the processing, not with a view to efficiency or cost, but with a view to completing all the steps as soon as possible.

At this stage it is also critical to "think outside the box" and to explore entirely new paradigms for accomplishing the tasks. I'll illustrate just such a whole new way for the activity of responding to requests for quotes (RFQs).

Rapid Quoting for Custom-Made Products

In the arena of custom-made products, the task of quoting has always been time-consuming. Each request from a customer requires review from skilled designers, manufacturing engineers, tooling experts, cost estimating personnel, and others in the organization, to ensure that you can make the product and to quote a cost that is at the same time profitable yet competitive. So the processing of quotes in such companies consumes a lot of people's time. On the other hand, as shown in Table 11-2, if rapid quoting were possible, it could significantly improve profitability. But how does one take the complex job of quoting custom products and perform it rapidly and accurately? Researchers at the Center for Quick Response Manufacturing have come up with some novel approaches for this.[15] In order to understand these approaches, it is helpful to begin by dispelling some myths about quoting processes.

Myths About Quoting
1. *The more time spent on the quoting process, the more accurate the quote will be.* In fact, you can obtain rapid estimates that have the same degree of accuracy as those obtained through lengthy and detailed cost estimation efforts.
2. *Every order we get is for a completely different part. Hence we cannot exploit any commonality to speed up cost estimation.* On the contrary, using product routing patterns, instead of detailed part shapes, it is indeed possible to speed up the quoting process even for custom-made parts.

(continued)

3. *If we don't consider the details of each manufacturing step during the quote process, there could be unpleasant surprises later, and profitability will suffer.* Regardless of the details considered during the quote process, in a custom manufacturing environment there is a great deal of variability in manufacturing processes when orders are executed. For example, setup times on a given machine can differ by hours for similar products made on two different days. There would have been no way to estimate these two different setup times during the quoting process, because they are not due to the specifics of each product but rather, the result of unexpected events on the shop floor. It is clear, then, that the use of too much detail in the quoting process is unproductive, in the sense that it does not lead to a better estimate of what costs will actually be incurred if the product is made. Thus, there is room for alternative strategies for quoting of custom-made parts.

New Strategies for Rapid Quoting

Although it may not be possible to rapidly quote for each and every customer request, at the Center for Quick Response Manufacturing we have found that up to 80 percent of RFQs at custom manufacturers yield to two main strategies. These are based on classifying the majority of a company's products into two categories as follows.

1. *Modified-Standard Products.* In looking at the spectrum of products requested by customers, one often finds a set that consists of minor modifications to features on standard products made by a company. As a simple example, if a company makes a standard line of cabinets, the modifiable features for customizing the line might include the width, height, and depth of the cabinets. A more complex example of this methodology involves customized cutting tools and is described in the next chapter. The approach to developing rapid quotes for such products involves three main steps. First, a family of modified-standard products is selected and the customizable features are identified, along with the range of allowable modifications for each feature. Second, cost drivers are developed for each modifiable feature. Third, a simple model is constructed on the basis of the features and cost drivers, which enables virtually instantaneous generation of quotes using a spreadsheet or computer program. In some instances the model can be reduced to a set of simple tables. The use of this approach by Ingersoll Cutting Tool Company, and its subsequent success, is illustrated by a detailed case study at the end of Chapter 12.
2. *Routing-Similar Products.* Where custom-made products differ significantly from each other, so that they cannot easily be classified into modified-standard families based on features, you can turn to exploit-

ing similarities in the manufacturing processes. Again, we have found that in spite of apparent differences in external shape and features, a majority of products follow a small set of routing sequences. For example, in a company that makes custom metal rating plates, each one completely different in terms of size, printing, holes, notches, and mounting features, the sequence shear—print—turret punch—cut to size—pack, could be followed by a number of jobs. To capitalize on the routing-similarity approach to quoting, you begin by finding a routing sequence that is used by a significant fraction of products. Then you find the cost drivers for each operation in the routing. This is a more complex task than in the modified-standard approach, because each operation may now have significantly different features being created. However, by performing some data analysis on previously manufac-tured products and then using the latest methods in generative process planning, it is possible to create simple prediction models for the cost of each operation. What is interesting about this approach is that while the prediction models do not go into each operation in great detail, when you compare the total cost estimate with conven-tional approaches, the routing-similarity based models do just as well, if not better. In one trial, for example, cost estimates made by an experienced estimator at the time of quoting were compared against actual costs incurred for jobs and also against the predictive model. It was found that, in general, the model performed better than the expe-rienced cost estimator.

Although both the methods described here require a considerable amount of up-front work before they are usable, they do illustrate the importance of looking for whole new approaches, and questioning traditional beliefs (myths) when redesigning operations for quick response. The results of these approaches, plus the benefits displayed in Table 11-2, also remind us not to underestimate the impact of speed in responding to customers.[16]

Basic Principles for Redesigning Operations for All Situations

The preceding example may give the impression that it is necessary to hit upon a major breakthrough in order to change office processing operations. This is not the case. While it is true that each individual sit-uation benefits from brainstorming about its own characteristics, as in the case of cost estimating, there are a few basic principles you can use in redesigning operations for all situations. These principles yield sig-nificant benefits even without adopting a whole new paradigm.

Principle One—Combining Steps

This principle has already been mentioned—it's to combine two or more steps into one step. For example, in a traditional organization, a person in inside sales might talk to the customer, jot down some information, and pass it on to a clerical worker. The clerical worker would create a folder with appropriate forms and then hand the folder to another person in order entry, who would enter the information into a computer system. In the cellular operation, it would be reasonable for one person to talk to the customer, create the folder, and enter the information into the computer (perhaps even while speaking to the customer).

Principle Two—Eliminating Steps

The next principle is to eliminate steps in view of the new mode of operation. Techniques that support both the above steps are cross-training, quality improvement methods, and application of electronic technologies. For instance, in the example above, you could decide that paper folders are not necessary and that all information will be kept as computer records. This would eliminate the steps of creating and maintaining a current folder. (I'll have more to say about the use of electronic technologies in the next chapter.) For another example, let us return to the possible Q-ROC at SteelShaft. Instead of one person doing the design and another doing the cost estimating, a single person could be cross-trained to do the design and cost-estimating for the FTMS. From this you can also see the benefits of cross-training and combining steps—the designer might be able to see quick ways of modifying the design in order to reduce the cost of the part. In the traditional organization, such modification might need several iterations back and forth between the design engineering and cost estimating departments, or it might never happen at all. As a final example, in companies where office processing is complex and involves many steps, there are a number of checking steps in place to make sure things are on track and mistakes have not been made. Once you have created the simpler processing steps and reorganized into a Q-ROC, you could eliminate such steps altogether. If past data on errors shows that you need such steps, then you should put in place a quality improvement program with the goal of eliminating these checking steps.

Principle Three—Redesigning Steps

Extending principle two, you should ensure that the redesigned process-ing steps aim at implementing quality at the source, rather than through inspection procedures. You can use both task design and tech-nology for this. As an example of task design, if it is found that a par-ticular data item about each customer order is repeatedly being missed by the field sales people, or incorrectly represented, then effort should be put into designing a form for field sales people that requires them to go through logic that will derive the correct data value. As an example of technology, consider the following procedure. An inside sales person takes down an order by hand, writing down the part number for an item previously ordered by a customer, along with the design changes the customer wants for the current order. When this folder arrives at the design engineer's desk, she notices that the description does not make sense in the context of the part number, implying that the part number was incorrectly recorded. This starts a cycle of backtracking of folders between departments, phone tag with the customer, and related delays. In this case, if the inside sales person were to look up the database *while talking with the customer*, he or she would see immediately that something was wrong: Either the customer had the wrong part number or he or she had heard it wrong. In any case the problem would be dis-covered and resolved at the source.

Principle Four—Continuous Flow of Work

A further principle is to ensure a continuous flow of work through the operations. In other words, you should minimize the traditional tendency to batch different jobs for the sake of efficiency. As an example, office workers may hold back jobs with similar requirements, with the intention of working on them all at one time. This can be exacerbated by older technologies: If a given type of transaction requires logging in to a special computer system, and this takes a while to do, then jobs requiring this transaction get stacked up and done once in a while. So in redesigning operations for QRM, it is important to surface all such constraints that invite batching tenden-cies and replace them with technologies or procedures that will result in a smoother flow of work.

Provide Resources and Support to Ensure Fast Flow

A cell designed for QRM may require resource levels that seem high relative to a traditional organization. A case study serves to illustrate this point. (See box on opposite page.)

Clearly, utilization calculations alone will not justify an investment in the printer. It would have to be attacked in terms of response time improvements, using all the weaponry I've presented in this book. For example, what does it *really* cost the company to have these parts shipped late? Are future sales of equipment being impacted? Are customers turning to other sources for replacement parts? Is a lot of time being spent on placating irate customers? And so on.

Another example comes from a company that created a Q-ROC but did not invest in providing on-line access to some of the company databases. While the cell workers had computers that enabled them access to e-mail and certain text-based systems, the company did not want to invest in extending a broadband local area network required for access to other company databases. This required the workers to go to another office when they needed to access 3-D drawings and certain other information. As a result, the workers would batch their work and take it to the other office once in while, quite against the QRM principles I've just espoused.

The moral of these case studies is that if you create a quick response cell, then you must supply it with all the resources necessary to achieve that quick response. If there is any obstacle stopping the flow of work, be prepared to invest in resources and equipment to eliminate the obstacle. Such investment may well appear unusually high compared with the traditional organization, as seems the case for the $15,000 printer that would be used only twice a day. Yet, if you did your homework correctly in picking the market the cell serves, this investment will pay for itself many times over with the usual QRM benefits of increased sales, growing market share, higher productivity, lower cost, and greater profitability.

Eliminate Traditional Approval and Control Systems

The traditional, functionally structured organization requires a management hierarchy to keep it operating on track. This management

Cost-Based Resource Allocation Degrades Customer Service

A Midwest manufacturer of capital equipment had a special department devoted to serving "breakdown orders." These involved jobs where one of this company's machines had broken down at a customer's site, and the customer's production line was stopped as a result. The company's field service people determined that the breakdown was due to defective or broken parts; then this department would fabricate the parts and ship them to the customer.

The company's stated policy was to ship such breakdown parts within 48 hours. However, customers complained that it was often three to five days before parts were shipped. Complaints mounted and eventually both field sales representatives and the customer service department asked the VP-operations to conduct a project aimed at reducing the lead time for breakdown parts.

At this stage the company asked me to review their process for serving breakdown orders. During this review, one policy on resource allocation stood out and serves to illustrate the principle here. The breakdown department consisted of office operations and fabrication operations. In the office portion, the customer problem was recorded, the root cause determined, and then instructions were generated to fabricate any needed parts. This included preparing blueprints for the parts, printing out the routings, and obtaining the relevant NC programs. The fabrication side consisted of cross-trained machinists and several types of NC machines, allowing for a wide variety of parts to be made.

Since the broken parts were ones previously made by the company, the blueprints, routings, and NC programs were already in-house. We discovered, however, that it took more than a day before these were delivered to the shop floor. Fully 50 percent of the promised lead time to the customer was being used before any tools touched any metal. Upon investigation, it turned out that the breakdown department did not own a printer large enough for the blueprints. Instead, these were printed at the company's design engineering department, which was in another building two miles away. The facilities were linked by a computer network, so the breakdown department did have access to all the databases. People processing orders in the breakdown department would therefore have blueprints printed out at the main building, and once a day someone would make a trip over to bring the prints back.

Staff in the breakdown department had complained on many an occasion about the lack of a printer. However, each time they put in a request for one, the answer from top management was the same: "You only process a few jobs each week, so your printer would be utilized hardly 1 percent of the time. You want us to spend $15,000 on a printer that will lie idle 99 percent of the time?"

structure is rife with checkpoints and sign-off requirements. For instance, a design for a critical component may not leave the engineering department until the engineering manager has approved it; a quote cannot be sent to a customer until it has been approved by two levels of management; and so on. When implementing Q-ROC, it is essential that this system of approvals and sign-offs be dismantled. There are two reasons for this: (1) an obvious one connected with additional delays, and (2) a less obvious one connected with the quality of output, as I'll now explain.

Suppose SteelShaft creates a Q-ROC whose job is to generate quotes for certain types of custom shafts. At the same time, suppose that SteelShaft's management is not comfortable with a quote going out to a customer without having it seen by a senior manager. An obvious disadvantage of this policy is that quotes will, at times, sit on the manager's desk waiting for approval. This will work against the cell's goal for quick response. If the manager is backed up with other work, he or she may even barely look over the quote in detail so that the sign-off may be a rubber stamp rather than any real check of the process. However, a second problem with this approval process is more insidious, as I will illustrate with a hypothetical situation.

> Jim, a member of the newly formed Q-ROC at SteelShaft, sat at his desk one afternoon, worried about getting his work done on time. He had to take his son to a hockey game, but he had committed to his cell colleagues that he would finish the last portion of a quote and deliver it to their manager for approval. What remained was for him to verify that a few detailed features on a shaft could indeed be done on SteelShaft's standard machining centers and wouldn't require a more expensive precision machining center. The quote, as it was presently written, was based on using the lower cost machines. He glanced at his watch and realized that if he spent time on this task, his son would be late for the hockey game. Then the solution dawned on him. "Of course!" he exclaimed to himself. "I can just give the quote, the way it is now, to our manager. If he signs off on it, then it won't be my fault if we are wrong, because he will have approved it. On the other hand, if he doesn't sign off on it and wants it reworked, he will send it back to my desk—I'll be gone by then and I can redo it tomorrow. But at least I will have kept my word to my colleagues, because the quote will be on our manager's desk before I leave today."

What we observe here is a reversal of the benefits of cells. I discussed at length in Chapter 4 how ownership of the process results in many quality and productivity improvements. By requiring the sign-off, SteelShaft has taken away ownership of the full process. This removes the responsibility from the cell workers. As a result, the quality of their work drops. Therefore, an essential principle of creating cells is to eliminate management approvals of the cell workers' decisions with respect to their normal processes.[17]

Another management mechanism you can simplify or eliminate is centralized project planning, scheduling, and control. Again, in the traditional organization with many departments, you have to track projects on their journey through the maze of departments—it's the only way to make sure they are on target. Some companies even institutionalize this activity into *yet another* department with a title such as project management, whose sole function is to provide such planning, scheduling, and control for the whole company. In fact, these companies attempt to solve the problem of long lead times and late deliveries via systems technology—another typical misunderstanding about how to cut lead times.

....

Misconception: Installing sophisticated project management and scheduling systems will help to reduce lead times.

....

Why will this approach yield minimal results? Because the root of the problem is, once again, our old friend the Response Time Spiral—this time it's back to haunt us in the office. And as you know, attempts to control this spiral only result in longer lead times. The spiral has to be killed. To appreciate the QRM approach that replaces the above belief, you need to understand the activities involved in such project management systems. Planning involves checking the resource requirements of committed projects against resources available in the departments. Scheduling is the process of releasing jobs to departments and establishing due dates for each department to complete the allocated work. Control is the function of checking the actual progress of jobs against the planned targets, and initiating corrective measures where needed to get jobs back on track.

In the cellular organization you greatly simplify these tasks. In the situation where a job is serviced from start to finish in one cell, the team itself can track the progress of the job and be responsible for planning, scheduling, and control. In this case the need for any central function is eliminated. In the situation where a few cells are involved in serving a customer, the task is still much simpler. Cells can do their own capacity planning, and then you can do the scheduling and control in one of two ways. Either you can have a small central department coordinate the intercell schedules—it will be a small department because now the job involves only coordinating a few of the more complex jobs between a small number of cells, and this is a much simpler and limited task—or the cells can have periodic meetings of their representatives to coordinate their own schedules and decide on corrective actions where needed. In the latter case, the central function is once again eliminated.

Consider Integrating Office and Manufacturing Cells

As I have repeatedly stated, the goal of reorganization for QRM is to put all the steps necessary for serving a market segment, from start to finish, into one cell. The logical conclusion of this goal is a cell that contains both the office processing and the manufacturing portions of the work. Indeed, some companies have attempted this and have been very successful. The office workers are located on the shop floor next to the machines, and this further improves the productivity of the group. In noisy environments the office workers may be enclosed in a glass-walled room. This provides a noise barrier while still integrating them visually with the rest of the workers.

However, these solutions are not necessarily always appropriate. In particular, two situations warrant retaining the separation between the office and manufacturing operations. If the cell obtained by combining office and shopfloor operations would become unwieldy, in terms of the number of people, then a logical split is at the office-manufacturing boundary. Also, if orders going through one office cell are routed to a number of different manufacturing cells, then again there is no logical way to combine the office and manufacturing operations. In this case too, the use of separate office and manufacturing cells may be the best QRM solution possible. (This does not contradict the concept of an

FTMS. It is often the case that a single FTMS for office operations needs parts made in different manufacturing cells.)

In the next chapter I'll continue to discuss the application of QRM to the office, this time focusing on how you deal with information as well as key tools to assist with implementing your Q-ROCs.

Main QRM Lessons

- Office operations can account for more than half your lead time and more than 25 percent of your cost, yet most companies neglect this opportunity for improvement.

- New paradigms for office operations can make a tenfold improvement in your profits.

- The Response Time Spiral attacks the office too, and must be eliminated by finding whole new ways of completing jobs, with a focus on minimizing response time. These whole new ways are derived by applying three types of principles: organizational principles, information-handling principles, and system dynamics principles.

- Do not attempt to reorganize the whole office area in one effort (this is often a reason for failure of reengineering or TQM projects). Key organizational principles include identifying a focused target market subsegment (FTMS) and beginning the QRM effort by creating a Quick Response Office Cell (Q-ROC) to serve only this FTMS. Several detailed principles are necessary to support implementation of the FTMS and Q-ROC concepts.

Tools to Support Q-ROC
Implementation

EARLIER IN THIS BOOK you saw how once you restructured the factory floor for quick response you had to rethink the entire material delivery organization to support this new structure. On the shop floor, the object that you process is material. In the office context, however, the object is information. In this chapter I will continue the discussion of the principles behind implementing quick response office cells (Q-ROCs). Specifically, I will look at how you need to rethink the ways in which you deal with information. I will also present some tools you can use that will assist with the initial analysis, design, and eventual implementation of a Q-ROC. Finally, I will look at some of the common concerns that eventually end up becoming barriers to implementing QRM. A case study of Ingersoll Cutting Tool Company will bring together all of these QRM principles and tools.

Implementing a Q-ROC takes more than the simple application of the key principles provided in this chapter and Chapter 11. In Chapter 17 I'll recommend a specific sequence of steps you need to follow and in Chapter 13 I'll lay down system dynamics principles that will help you in maximizing your chances of successfully implementing Q-ROCs. In addition, it is useful to keep in mind the points made in Chapters 4 and 5 on factors relevant to implementing a manufacturing cell since most of these factors are relevant to Q-ROCs as well.

INFORMATION HANDLING PRINCIPLES

There are several QRM rules when it comes to handling information in the office. They are discussed in this section.

Implement the "At Most Once" Rule

The first QRM rule states that an employee should handle any information item once at most. As motivation, consider this example from my study of an American manufacturer in 1995.

> During a phone call with a customer, a sales "account rep" jots down information on a sheet of paper. This information is transferred to a formal order form. The rep then logs onto a computer system, looks up some customer information in the system, and writes it down on the order form. This form is forwarded to the order entry department, where a clerk enters some of the information from the form into another computer system. Then the clerk enters some of the same information into yet another computer system...

While reading the above, were you counting how many times a piece of information was copied, transferred, and entered? In this age of relational databases and sophisticated query programs, there should never be a need to handle a given information item more than once. In fact, our rule says *at most once*. The reason for this is, often the information is already available on the computer, and provided you have rapid lookup capability for such information, there is no reason to enter it at all.

This rule isn't concerned only with the productivity of employees. Granted, reducing the number of times an employee needs to copy or enter an information item increases their throughput. A more important reason is that multiple handling of the same information item can introduce errors and inconsistencies. The backtracking of jobs and rework by employees resulting from these inconsistencies is a greater contributor to low productivity than just the time for duplicate data entry. So in redesigning processes for quick response, look for opportunities to cut down the number of times an employee handles a data item, or even eliminate the handling entirely if you can replace it by an automated lookup.

Reexamine Whether Information Items Are Really Needed

In the more complex organizational flow, forms were designed to handle a huge variety of different jobs going through the same department.

As a result, these forms required a vast amount of information to be filled in, "just in case." In the cellular organization, with each Q-ROC dealing with a focused segment of the business, you can eliminate altogether those information items that are superfluous. This requires a conscious effort, because cell workers coming from traditional departments may have it ingrained that a given data item is essential to an order. In rethinking their processes, Q-ROCs should be ruthless at eliminating information that does not contribute to getting their jobs done.

Provide Local and Rapid Access to Information

The aim of this rule is to minimize the Q-ROC's dependence on external information, since such dependence causes delays. The first step is to examine carefully what information a cell needs on a regular basis. The next step is to see if you can make all such information available locally to the Q-ROC. You can put it in the form of physical copies, such as catalogs and reference manuals, or in databases that are accessible through the Q-ROC's computers. If this means owning duplicate sets of reference manuals or catalogs, bite the bullet and incur the cost for these, much as we justified the printer in an earlier section.

In the case where such information is necessarily external to the Q-ROC, such as cost data from a supplier, investigate whether you can establish fast information links with this external source, to access this information on an as-needed basis. With the tremendous growth in the Internet, it is likely that you can obtain much of this information from external sources via Internet-based technologies.

Take Full Advantage of Electronic Commerce Technologies

There are today a range of technologies available to transmit information to conduct business electronically. These include e-mail, office networks, shared databases, EDI (electronic data interchange), and the Internet. You should design your new procedures around these capabilities, rather than replicating the existing paper-based methods. For example, can you eliminate paper forms and always use electronic forms whenever there is client contact? Can blueprints be replaced by

3-D CAD files? Taking it a step further, can you completely eliminate the practice of creating folders for orders, instead keeping the entire customer record on the computer, and transferring the job between Q-ROC workers via electronic transmission rather than handing off folders? The age of electronic commerce has arrived. Fundamentally new paradigms are emerging for conducting business electronically. By putting yourself on the leading edge of this trend your Q-ROCs may be able to achieve amazingly fast response times.

Invest in Compatibility of Information Systems

It may come as a surprise that in this day of information networks, companies still struggle with moving data between their computer systems. Witness this quote about a researcher's experience at a Fortune 500 company in 1995: "It was soon discovered that much of the information that was available was stored in various databases in different departments within the foundry. Information was departmentalized, with little, if any, integration of related information across departmental boundaries."[1] For several of the above principles to be applicable you need to have the ability to share data with various information systems. For example, the *at most once* rule requires a unique data item to be accessible across applications, as does the idea of creating an electronic order folder. One step along the way to quick response in the office is to investigate the different information systems currently in place and make them compatible. But with QRM you don't have to make your systems compatible overnight. You can start to put Q-ROCs in place, and then gradually improve their responsiveness with each new capability that you add to your information systems. The cell workers will also help to drive the priorities of what system compatibilities will most improve their performance.

TOOLS TO ASSIST IN Q-ROC IMPLEMENTATION

In this section I will focus on tools that can assist with the initial analysis, design, and eventual implementation of a Q-ROC. To begin with, you can and should use many tools from well-known management methodologies to assist in this process. Examples of such tools are:[2]

- Pareto charts
- Fishbone diagrams
- Affinity diagrams and other tools that assist with group brainstorming activities
- Simple statistical tools such as histograms and stratified analysis
- Tools used in kaizen and TQM efforts
- Simple capacity modeling and simulation tools
- Team-building exercises

I will not give descriptions of these well-known tools, but rather, focus on three tools that are specific to QRM implementation: process mapping, tagging, and value-added charts.

Process Mapping

The technique of process mapping is one way for a business to understand more about its office processes, about which it often has little knowledge. In the context of Q-ROCs, the aim of process mapping should be to understand how all the requirements for a given FTMS flow through the organization until the product for that FTMS is completed. (By *product* I mean not a manufactured item, but rather, the outcome of office processing steps for that specific FTMS—see the discussion in the previous chapter.) You can eliminate much unnecessary time devoted to process mapping activities by always keeping this aim in mind. In many instances, countless hours are spent detailing how jobs flow through an organization, when most of the steps are irrelevant to the FTMS at hand.

The outcome of a process mapping exercise is a flowchart detailing the sequence of activities performed for a given product, with a summary description of each activity, along with all decision points and loopbacks to earlier activities. This chart is developed using two main techniques: interviews with employees who do the work, and group discussions to resolve discrepancies or misunderstandings about the flow. In addition, the technique of tagging can help to unearth activities that you missed by the first two techniques.

I will give an example of an actual process mapping exercise, along with the final flowchart, in a detailed case study at the end of this chapter. As that case study will illustrate, a secondary aim of process mapping is to make managers and workers accept that there is considerable opportunity for improvement, and hence, to prepare them for the fact that there will be major changes to the current way of operation.

Tagging

This technique involves attaching tracking documents to each job flowing through the organization. The aim is to get data on where the jobs actually went, how long they spent in various steps, and other useful information. Figure 12-1 shows a typical form used for tagging—I refer to this form as a *tagging sheet*. Conducting a tagging exercise is not just a matter of copying the form in Figure 12-1 and attaching the copies to incoming job folders. In the years of conducting such exercises I've developed three principles to support a successful tagging exercise.

Principle 1: Get Employees Lined Up Behind the Tagging Effort

The knee-jerk reaction of employees to such an exercise is that management is looking for inefficiencies, so that it can find how and where to reduce the headcount. Another concern is that when the tagging data is analyzed, they will get called on the carpet for mistakes or delays in particular jobs. If companies do not alleviate these concerns up front, the tagging exercise can be unfruitful or even an outright failure. One company that engaged in tagging without addressing this concern found that a large number of the tagging sheets simply disappeared somewhere along the way.

The only way of alleviating this concern, in my experience, is through frank and open discussion of the need for lead time reduction. Managers should hold meetings for all employees who may be involved in the tagging effort. The discussion should begin with facts such as: customer complaints, missed deliveries, customer requirements for shorter lead times, loss of market share to competitors with shorter lead times, or a market opportunity that should be exploited. Then you should introduce employees to the concepts of QRM as a means of tackling these problems or opportunities. You should outline a few key principles of QRM, particularly those that apply to the office. After this, employees should learn that process mapping and tagging are typical tools needed to support implementation of these QRM principles.

Tagging Log No.	18
Customer	Brunner Company
Order Date	Monday, July 1, 1997
Order Number	RS-140-483
Delivery Date	September 20, 1997

Name and Dept.	In		Out		Tasks Performed	Comments
	Date	Time	Date	Time		
John Inside Sales	1-Jul	9:05	1-Jul	4:10	Spoke to cust. and created folder	Went to sales mtg. for 4 hrs.
Heidi Engineering	2-Jul	10:20	3-Jul	3:55	Reviewed cust. info. and created rough design	Interrupted by two rush jobs. Back to John for missing info. on options
John Inside Sales	4-Jul	10:55	5-Jul	1:15	Got info. from field sales rep.	Telephone tag with field sales rep.
Heidi Engineering	8-Jul	9:05	9-Jul	2:05	Reviewed new info. and modified rough design	Other jobs got ahead of this one while it was with John
Steve Estimating	10-Jul	11:30	11-Jul	1:20	Estimated cost	Took 6 hours for a parts supplier to call back w/price
Ramesh Manu-facturing	12-Jul	10:45	15-Jul	9:00	Manufacturing review	End-of-week jobs took all my time on Friday 12 Jul.
Eric Inside Sales	15-Jul	11:15	16-Jul	9:30	Determined markup	Had to check with Carl about policy for this cust.
Beth Clerical	16-Jul	11:45	18-Jul	10:00	Created Quote Document	Many other jobs ahead of this one
Carl Inside Sales	18-Jul	3:30	18-Jul	4:30	Quote approval	

Figure 12-1. Example of a Tagging Sheet Filled Out by Employees

It is important that you emphasize that this will not be another efficiency study, and in particular, there will be no witch-hunt for inefficient employees. Realizing QRM leads to increased sales, many companies have gone so far as to guarantee employees that there will be no headcount reductions. This counters the typical fear that "If we get jobs out faster with higher productivity per employee, won't there be fewer jobs for us?" True, there may be reorganizations, and job descriptions may change, but you've assured them that there will be work for everyone.

As an additional motivator you should share with your employees the results of case studies of tagging efforts at other companies. One instance of such a success story is the Ingersoll Cutting Tool case study at the end of this chapter.

Principle 2: Design the Tagging Exercise

You want to design the tagging exercise to provide results with a reasonable degree of statistical validity. Specifically, this means four things.

1. *Don't tag a selected sample of jobs, such as every fifth job.* Although apparently a statistically sound sample, it may bias employees' behavior: When they see a tagged folder, they might give it preferential treatment in order not to look bad on the tagging sheet. Instead, if all folders are tagged, the employees might initially try to work harder on each one, but soon the organization reverts to its normal behavior because everyone can't work harder on every job all the time—the usual problems of missing information, rush jobs, and other ills of the functional organization will be sure to come in their way. This point about tagging all jobs may seem to contradict our earlier statement that process mapping should be focused on the FTMS, not the whole business. Shouldn't the same be true for tagging? The answer involves a compromise: For departments or employees that are along the main path of jobs for the FTMS, try to tag all jobs coming their way. Otherwise these employees may bias the sample by giving preferential treatment to tagged jobs. However, this approach may result in the tagging of jobs too far removed from the FTMS, and so in some cases you may have to decide not to tag jobs of certain types. By thinking ahead and asking the question, "Will it make a significant difference if we get a bias in this step of the process?" you can decide whether or not to tag all jobs going through that step.

2. *Choose a tagging period that is both representative in terms of incoming demand and long enough to get a significant number*

of tagged jobs out at the end of the process. Don't make the error of compiling data for the first 50 (say) jobs that exit the tagging process. This will significantly bias your sample. Why? Because the fastest or easiest jobs will be overrepresented in the sample, while the problematic and long lead time jobs will still be stuck in the system and not show up. There is a simple solution to this problem. Base the sample of tagging sheets that will be analyzed on a window of time when the jobs *started*, but continue the tagging effort until all these jobs have exited the system. For instance, if the tagging starts on Monday, May 1, and by June 15 all the jobs that arrived between May 1 and Friday, May 12 have exited the system, and this amounts to 64 jobs, you might decide to end the tagging effort. In extreme cases, if some jobs are stuck for very long and the QRM project is being held up, you can call an end to the tagging effort, but record how many jobs were still in the system and in what stage. When the tagging effort is stopped, in addition to the 64 jobs that were in the input window, there will be two other classes of tagging sheets: those that exited the system but had arrived after the input window, and those that are still in the system. These sheets should not be discarded since they're still useful in terms of providing insights and anecdotal evidence. However, you should not use them to derive any statistical data on lead times or rework.

3. *Keep a log book of incoming jobs.* In this log, record the date and time of arrival of each job, assign it a serial number in the log, and put this serial number on the tagging sheet as well (see Figure 12-1). This will help to track which jobs have exited the system and whether any jobs are buried in some remote part of the organization.

4. *Keep a sense of perspective and don't overdesign the data collection effort.* Although you want to avoid the statistical traps, as above, it is not necessary that the data be too detailed or of high fidelity. The reason is, you are going to change the processes anyway, and you are looking for major insights, not areas for fine tuning. This was why I used the qualifier "reasonable" at the beginning of this section.

Principle 3: Review the Completed Tagging Sheets Early

You want to review some of the completed tagging sheets early in the exercise and apply a mid-course correction if needed. Companies have conducted a lengthy tagging exercise, only to find at the conclusion that the information on the tagging sheets was not very helpful. For example, in a case where a folder stays at Jack's desk a long time, his comment might read, "Waiting for information." This is unhelpful. We don't know what information Jack was waiting for, and who was supposed to provide it. Similarly, "Returned to Patty to redo" could be rewritten with more specifics on what needed to be redone. By reviewing a few of the tagging sheets early in the process, the team involved in the QRM effort can provide feedback to employees, enabling them to do a better job in writing their comments. This can be done individually with specific employees, or in a group setting, depending on the nature of the problems with the tagging sheets.

Analyzing the Tagging Data

Once you have obtained a sufficient number of tagging sheets, the next question is, what types of data should you look for? These are the main analyses you need to conduct from the tagging data:

- *What sequence of steps did jobs actually follow?* This helps to verify or add to the process map you have charted thus far.
- *How much time was spent in each step?* Which steps accounted for the largest portions of the lead time? (You can also feed this and the preceding analysis into the value-added charting analysis described in the next section.)
- *What fraction of time were jobs being worked on versus waiting?* More important, what fraction of time was value being added? (A definition of "value added" is provided in the next section and is also discussed in the case study at the end of the chapter.)
- *What were the main reasons why jobs waited?*
- *What were the major sources of problems or errors and their frequencies?*
- *How many times did jobs revisit a given department or step?*
- *Can jobs be divided into subsegments, with different response times or routings for each subsegment?* In particular, are there subsegments with significantly shorter or longer times, or with simpler or overly complex routings? What fraction of jobs flow along these various paths?
- *What insights can be gained from the anecdotal data, regardless of statistical analysis?* The anecdotal data can suggest some obvious

solutions, even if an event does not occur frequently enough to be statistically significant. For instance, if Jim waited for information from the accounting department only four times in a sample of 230 jobs, but it is easy to give Jim access to that data over a computer network, then the solution should be implemented anyway.

Figure 12-2 and Tables 12-1 and 12-2 show the results of such analyses. The data are from actual tagging exercises at three U.S. firms, but I have disguised the products to protect the identity of the companies.

Since filling out the tagging sheets does consume a bit of time, I have occasionally encountered resistance, not so much from the workers but from supervisors, with the complaint that this exercise will reduce the productivity of their department. The two root causes of this

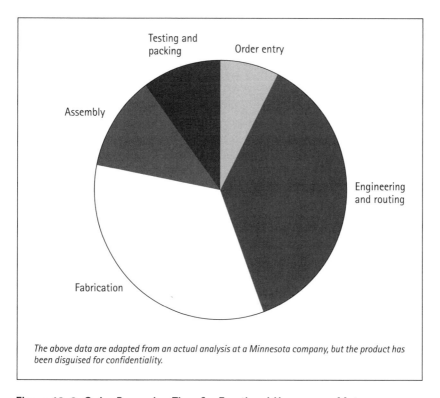

The above data are adapted from an actual analysis at a Minnesota company, but the product has been disguised for confidentiality.

Figure 12-2. Order Processing Time for Fractional Horsepower Motors

Table 12-1. Department Returns of Order Folders for Lightweight Cabinets

Department Name	Average Number of Visits[†]
Inside sales	3.2
Design engineering	1.7
NC programming	1.1
Cost estimating	1.2
Inventory control	1.5
Production control	2.8
Routing	2.2
Purchasing	1.3
Quality	1.9

[†] For each lightweight cabinet order, we counted the number of times the order folder visited each department. These data are the average of those numbers for the orders in the tagging sample. (The data are adapted from an actual analysis at an Illinois company, but the product has been disguised for confidentially.)

resistance are (1) the fear of what may emerge from the study, in terms of the supervisor looking bad, and (2) an overemphasis by management on measuring departmental output, even during the study. You need to tackle the former concern the same way as you tackle the employee concern. For the latter, management has to make a strategic decision to give up on a little output during the study, for the sake of potential

Table 12-2. Non-Value-Added Time in Quoting for Customized Equipment

Department Name	Average Value-Added Time per Job	Average Departmental Lead Time per Job	Non-Value-Added %
Customer service	3 hours	5 days	92.5%
Engineering	12 hours	12 days	87.5%
Cost estimating	2.5 hours	9 days	96.5%
Quote writing	2 hours	9 days	97.2%
Total	**19.5 hours**	**35 days**	**93.0%**

improvements in the future. In fact, the amount of time spent by employees in filling out the sheets is minuscule relative to their other work, but it is still necessary to acknowledge these concerns and deal with them, or the tagging exercise may be killed before it has a chance to produce any useful data.

Value-Added Charts

This approach involves creating a time-phased chart of how value is added to a given job as it moves through its processes.[3] Figure 12-3 shows a typical value-added chart. The value-added chart is usually created as a result of process mapping and tagging exercises. A typical order (or averaged data) is selected from the sample, and a plot is made for this order or data. Along the horizontal axis, you plot elapsed time, from the arrival of the order to its departure from the portion of the system that is being observed. Along the vertical axis you plot the percentage of value that is added at each stage of the process. Keep in mind that you are not going for complete accuracy here but, as before,

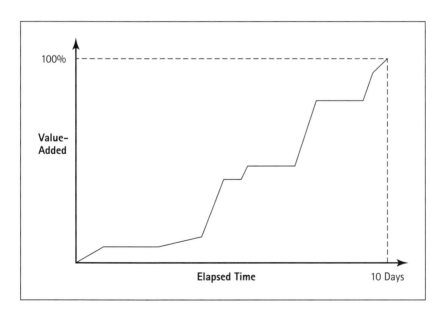

Figure 12-3. Example of a Value-Added Chart

looking for insight. Thus, you can use rough measures of cost of resources at each step as a proxy for value-added. But there is one important rule you must observe: Always ask, "Would the customer pay for this step?" If the answer is no, then there is no value added, regardless of the amount of cost incurred.

Let us consider two examples. A folder is returned to design engineering because an alert employee in NC programming picks up a discrepancy between what the customer asked for and what was designed. The customer is not likely to pay for both the original design work and the rework—"It was your fault," the customer would say, "so you should absorb the cost of the error." Hence, according to our strict definition, there is no value added during the rework. As a second case, an order folder waits in your company's office while a credit check is being done on the customer. Credit checking is done for the benefit of your company, not for the benefit of the customer. By our definition no value is being added.

Continuing in this vein, you obtain an approximate total for value that is being added. This can include time, materials, and departmental overheads, wherever value is indeed being added. Then you break this total into its constituent amounts over the periods when the value is being added, with the result similar to Figure 12-3.

At this point you might wonder, since the data for the charts is coming from the other analyses, what's new here? In fact, there isn't much new in terms of analysis, but what is useful is the way of presenting the information. Essentially, value-added charting is another weapon in your arsenal to aid in highlighting problem areas, and for convincing managers or employees of the need for change. In looking at a chart similar to Figure 12-3, there are large areas where the line is horizontal, or no value is being added at all, plus areas of low slope, where only a little value is being added over a long period of time. These help to quickly flag problem areas, plus the chart conveys the message that much of the time is being spent with little value being added.

Beware of the temptation to suboptimize individual steps based on looking at the value-added chart. Looking at horizontal portions of the chart, one might have an urge to put in some quick fixes. However, if you truly intend to apply the organizational principles, then the solution involves a major reorganization of the whole sequence of tasks;

suboptimization could detract from this goal by taking away time and energy from the larger opportunity that should be pursued. The purpose of the value-added chart is to give insight and convey a message, not to provide detailed data for improvement. Note that my own view here differs from other descriptions of value-added mapping, which see the outcome of the mapping exercise as a tool to enable detailed analysis and to search for specific improvements. That, in fact, is why I use the term value-added charting for this approach, to differentiate it from conventional value-added mapping methodology.

CONCERNS WITH IMPLEMENTING OFFICE CELLS

In spite of presenting to company managers process maps and tagging analyses illustrating large amounts of non-value-added time, they still may not accept the idea of implementing a Q-ROC. Our study of companies attempting to implement Q-ROCs show there are some common concerns that become barriers to implementation.[4] These concerns are discussed in this section, followed by a case study that brings together QRM principles and tools that address these common concerns.

Lack of Familiarity

A degree of anxiety arises simply from a lack of familiarity with the Q-ROC concept. As I noted in Chapters 4 and 5, although cellular manufacturing has been around for decades, there are still concerns and even misconceptions about it. The situation with office cells is more pronounced, since the concept is still young. The way to combat this concern is through education for both managers and employees, combined with case studies of successes at other companies and even visits to other firms that have implemented office cells.

The Use of Human Resources

This concern takes two forms. One is that by dedicating people to a Q-ROC, we will take resources away from some of the functional departments, which may not be able to meet their targets. This issue is further exacerbated by department managers' fear of loss of "territory," as the number of people under their control declines. You must resolve the issue of resources by realizing that non-value-added work will be greatly reduced, so the organization as a whole will still be capable of meeting its targets, only in a different structure than before. In this context, managers must also realize that their roles are going to change as companies move to team-based cellular organizations.

The second form of concern occurs when there is only one expert in a given area—no Q-ROC can do its work without this expert's knowledge, and yet this person cannot be dedicated to a single cell because that would disable the other cells completely. An example helps to make this problem concrete. In 1995 I studied the quote generation process at a manufacturer of fasteners to see how they could reduce the lead time from several weeks to a couple of days. After some analysis, a logical solution appeared to be the formation of three Q-ROCs, each focusing on a different FTMS. Each Q-ROC would have three employees who could tackle the tasks of inside sales, manufacturing analysis, cost estimating and quote writing, but they needed input from a tooling expert to complete the cost estimate for some of the complex jobs. (A good deal of the cost for making the fasteners depended on the cost of the tooling that would be used.) The company had only one such tooling expert. How could it implement the three Q-ROCs?

Employing three experts would be overkill—data showed that the tooling expert would be needed less than 25 percent of the time by each cell. Sending the quote to the tooling expert would destroy the flow in each Q-ROC and degrade lead time. Training the people in each Q-ROC was not feasible in a short period, because this expert's knowledge had been acquired through 10 years of experience.

The answer appeared in an office version of the time-sharing strategy described in Chapter 6. The tooling expert would be given a schedule, with specific times during which he had to be present at each Q-ROC. (The tooling expert also had other responsibilities in the manufacturing arm of the company that used his time when he was not involved in quoting.) How does this solution differ from just sending the jobs to him for his input? The three-part answer is similar to the one I gave in Chapter 6.

1. The key is that ownership of the quote process is not lost by a Q-ROC.

2. A Q-ROC is not at the mercy of the expert's availability; it is guaranteed a certain amount of attention each day, and doesn't have to ask him for special favors to get its quotes done.

3. During the time the expert is at a Q-ROC, he becomes part of the cell, and the benefits of close communication and instant feedback are derived.

Cross-Functional Training

The case of the tooling expert leads to another common concern, which relates to cross-functional training. Employees worry about potential failure, fearing that they may be unable to absorb enough of another skill to make good decisions in that area. On the other hand, managers worry about the cost and extent of the training needed to get a Q-ROC up and running. In my experience, however, the tooling expert situation tends to be an exception. The key to success in cross-training for Q-ROCs lies in the following observation, which is similar to one in Chapter 5.

> QRM Nugget: In designing a cross-training program for a new Q-ROC, do not aim to train any worker in depth on a given skill that is needed in the Q-ROC. Rather, focus the training effort on the limited portions of that skill that are required by the worker to deal with orders for the specific FTMS served by this Q-ROC.

For instance, if a cell worker who is an engineer needs to be trained in cost estimating, the traditional approach would be to train this person on all the tasks a cost estimator might do in the functional organization. Such training would be quite extensive and would take a long time, and it is still unlikely that this person would be a good cost estimator without several years of on-the-job experience as well. However, if the cell only deals with aluminum products with limited options, the cross-training can be limited to teaching this person how to deal with cost estimating for these products and options. The Ingersoll case study that follows will demonstrate both the feasibility and the success of this focused approach.

Compensation and Professional Development

In the functional organization, senior experts in a special field can mentor the novices—who will be their mentor in the cross-functional cell? In the traditional organization, performance measures and compensation are well established and understood by all—what will be the new measures and rewards in the Q-ROC environment? For example, who will decide on pay raises for the cell workers? All these issues will be addressed in detail in Chapter 16.

Lack of Upper Management Support

A final concern relates to lack of upper management support for the Q-ROC.[5] Will it get the resources it needs? Will performance measures be changed to accommodate this new structure? Will management invest in the requisite amount of training to ensure success? These are important concerns, and they are tackled in depth in Part Five.

From Weeks to Hours at Ingersoll Cutting Tool Company

Based in Rockford, IL, Ingersoll Cutting Tool Company (ICTC) manufactures a wide range of indexable cutting tools and inserts (see Photo 12-1). It is a wholly owned subsidiary of Ingersoll International. The Ingersoll family of companies is well known for its leadership in the machine tool industry.

In early 1992, ICTC management was concerned about the increasing number of late shipments to customers, as well as the growing competitive pressures to produce nonstock specialized cutters and inserts with short delivery times. Around that time, some managers at ICTC heard about QRM; they asked to meet with me to evaluate whether this approach might help them improve delivery performance and reduce lead times. As a result of this meeting ICTC decided to engage in two QRM projects.

This case study focuses on one of those projects,[6] targeted at reducing the order processing lead time for nonstock (customized) items. Since these were nonstandard items, the order processing lead time wasn't comprised only of clerical data processing steps such as order entry, but required technical work too, such as design engineering and cost estimating. This project was conducted by University of Wisconsin graduate student David Johns, supervised jointly by Mike Wayman, team leader for steel products at ICTC, and myself.

Photo 12-1. Sample of Cutting Tools and Inserts Made by ICTC

The goal of David's project was to study the current order processing system and provide recommendations for reducing order processing lead time by at least 60 percent. (Although we now routinely set targets of 75 percent reduction and even more, at that time the target of 60 percent reduction appeared very aggressive to ICTC's management.)

Missing Information

As David and I launched into the project, we quickly realized that two key pieces of information were missing. The first was the sequence of steps in the current order processing system—specifically, ICTC did not have a clearly understood and accepted process for handling orders for nonstock items. The second was hard data on actual order processing lead times. (As discussed earlier, it is typical for a company to have poor knowledge of both these items.)

In order to get the first piece of information (sequence of steps), David applied the technique of process mapping. This included interviews with people in all departments related to the order processing activity, namely, accounting, design engineering, inside sales, inventory control, manufacturing, N/C and lathe programming, OEM, production control, proposal engineering, purchasing, quality, routing, and shipping. After a few weeks of work, he was able to produce a flowchart documenting the entire process, and we presented it to ICTC management.

(continued)

Managers at ICTC were amazed. The flowchart included more than 80 steps and, even with small print, extended to three pages. Figure 12-4 contains a summary schematic of the flowchart—details of processing steps are omitted, but icons are displayed to indicate the number of steps and decision points. Although most people at ICTC suspected that the process was complicated, each one knew only what was done in their own area and no one had seen it all compiled on one diagram before; thus the flowchart was far more involved than anyone expected. Presenting the flowchart to ICTC managers and staff served an important purpose: It laid the groundwork for subsequent improvements. Until this point in the project, there had been many skeptics about our ability to contribute anything to the processing at ICTC—how could an inexperienced graduate student and a theory-oriented professor know the practical complexities of tool design? But once we displayed the three-page flowchart, managers and workers alike agreed that there was plenty of opportunity for improving the process.

To get the second piece of information (time spent in various activities), we used the method of tagging. David designed a form to be attached to all order folders, to be filled out by each person who handled the order. A critical step in conducting the tagging was the groundwork by ICTC managers to assure employees that this was not a traditional "timesheet," and that ours was not an "efficiency" study or a search for people making mistakes. After two months a sufficient number of tagging sheets had been collected for analysis.

Tagging Sheets Reveal Multiple Opportunities

The first outcome of the tagging effort was a hard statistic showing ICTC typically spent 10 working days—two full weeks—handling an order before it was released to manufacturing. This represented a significant portion of the total lead time being quoted to customers; although staff in manufacturing had complained for years that orders were taking a long time to be processed in the office, in the absence of hard data no strong actions had been taken. As we eyed our newly derived statistic, it seemed like a wonderful opportunity for improvement, and we wondered how much of that time could be cut out.

A first-cut analysis of the tagging sheets showed that jobs spent about 70 percent of the 10 days being processed by some function at ICTC, and only 30 percent of their time waiting. This seemed too good to be true—usually one would expect the reverse—and disappointingly, it seemed to offer limited opportunity for improvement. Having worked on other QRM projects, however, I was skeptical of this naive analysis. I gave David a more thorough definition of value-added and instructed him to dig deeper into the tagging sheets with this definition in mind.

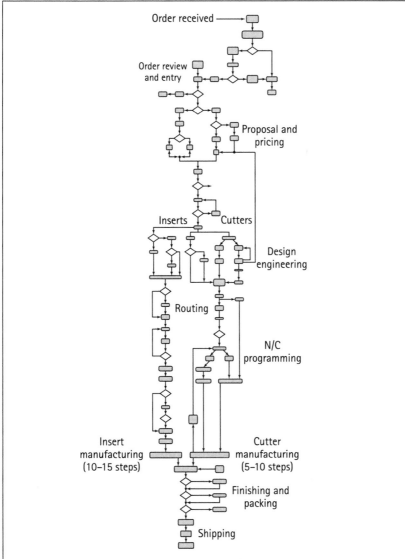

Order received

Order review and entry

Proposal and pricing

Inserts   Cutters

Design engineering

Routing

N/C programming

Insert manufacturing (10–15 steps)

Cutter manufacturing (5–10 steps)

Finishing and packing

Shipping

The original flow of orders consisted of over 80 steps and went through 12 departments. This is a schematic of the flow with details omitted. Also, since the purpose of this study was to focus on the nonmanufacturing portions of the process, the manufacturing steps are summarized in two boxes.

Figure 12–4. Complexity of Existing Order Processing System at ICTC

(continued)

This second analysis of the tagging sheets, supported by additional interviews, revealed several non-value-added activities buried in the data. Examples of these were: incorrect or mismatched information such as incorrect part numbers or missing dates; batched order folders; complex procedures to complete orders, requiring many departments to provide inputs; and orders returning to a department up to four times during the course of their handling. When this analysis was completed, it showed that 70 percent of the time could indeed be attributed to non-value-added activities. Publicizing this statistic at ICTC gave even more credence to our ability to find opportunities for improvement, and paved the way for our recommendations described next.

The Opportunity for Modified Standard Cutters

As I discussed in Chapter 11, the first step in redesigning office operations is to find an appropriate FTMS. ICTC already had such an FTMS in mind, which they termed "modified-standard cutters." These were requests from customers for cutters that resembled standard cutters, but with changes to one or two features such as diameter, length, or spindle mounting adaption.

At that time, these orders were handled in the same way as those for customized cutters with complex features, and so they had long lead times as well as a high cost. As a result, customers needing such a cutting tool often turned to an inferior alternative involving a cheaper and immediately available stock cutter supplied by a competitor. ICTC's sales staff felt that if both the lead time and cost could be reduced, they would pick up many more orders for such cutters.

In order to serve this FTMS, we recommended implementing a Q-ROC, with the QRM principles of multifunctional, cross-trained, collocated, closed-loop workers in the cell.

We also made specific suggestions to eliminate several of the non-value-added steps. One example was the use of cross-training to eliminate checking and return steps—for instance, the routing department would generate a routing that the quality department would check and then return to the routing department with notes and modifications. By cross-training a routing or quality person, the whole task could be done in one step.

Another example was creating mechanisms for the Q-ROC to access information across departmental lines—for instance, folders would circulate between departments just for simple data on prices and credit information to be inserted. Such data could be easily inserted by one trained person if it were available on a common database. Implementation of this idea, however, would require tearing down organizational walls, since traditionally information that belonged to a department could only be accessed by that department.

ICTC Takes the Recommendations Further

We presented our recommendations to ICTC management. Our initial estimate was that the Q-ROC would reduce order processing lead time for modified-standard cutters to about three days. Merle Clewett, president of ICTC, was enthusiastic about the recommendations, but felt that the opportunity for improvement was even greater. So in March of 1993 he formed a study team within the company to further refine and implement our suggestions.

The team came up with a number of additional ideas, of which three were key. The first was to replace the traditional detailed cost estimating process—which involved costing out each of the manufacturing steps and tooling needed to make each cutter—by a table-driven estimate based on key features of the cutters (as described in Chapter 11).

The second involved empowering the field sales force. To start with, they were given the pricing tables to use at the customer site. These tables included upper and lower limits for dimensions of a tool as well as any permissible modifications for the cutter to remain in the modified-standard category (see Figure 12-5 and Photo 12-2). But beyond that, since the sales personnel had a high level of technical knowledge, it was also decided to allow them to make the product application and manufacturing lead time decisions themselves. As a result, the previous one to two weeks required to respond to a customer inquiry was virtually eliminated, and the price and delivery time could be quoted to the customer on the spot.

The third key idea was to carry the cross-training even further than David and I had envisioned—the study team was able to recommend a Q-ROC with only two people in it. Imagine the shock when management saw an 80-step process involving 12 departments reduced to a process that *two* people could handle from start to finish. Figure 12-6 summarizes the final recommendation by this study team, a recommendation that the president implemented in the summer of 1993.

Result on Lead Time and Sales

Once the Q-ROC was implemented, it continued to hone its procedures. A year after implementation, it had reduced the processing time from more than ten days to less than 24 hours for most modified-standard orders, in many cases processing an order in as little as *four* hours.

Elimination of the quote generation time, along with the dramatic reduction in order processing time, brought ICTC's delivery times down from eight weeks to under four weeks (see the example in Photo 12-3). As we had hoped, the target market strategy also proved to be effective, and this lead time reduction had a significant impact on ICTC's business. According to Wayman, by 1994 orders for their modified-standard line of cutters had

(continued)

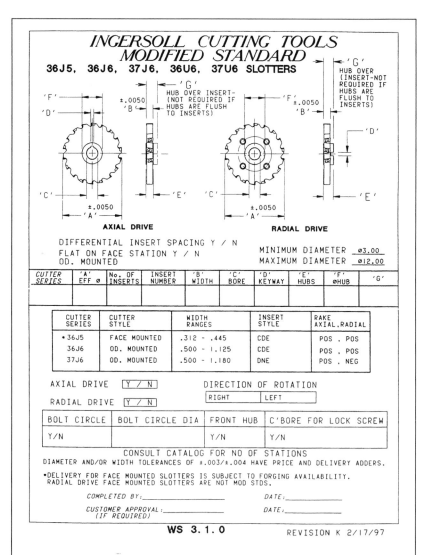

ICTC's field-based personnel were able to quote prices and delivery dates for specific modified-standard tools while filling out a simple worksheet right in front of the customer. This is an example of such a worksheet for modified-standard slotters.

Figure 12-5. Example of a Worksheet for Customized Slotters

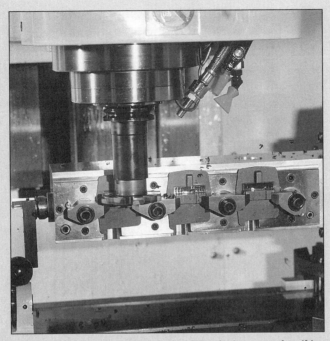

Ordering the customized slotting cutter for this job was simply a matter of specifying the size, the drive, and the insert radius.

Photo 12-2. Example of a Modified-Standard Cutter for Slotting

increased threefold, while by 1996, they had grown to fully 500 percent over their 1992 numbers. Also, many of the orders being obtained were with first-time customers, opening up the door to increased business in other products lines as well.[7]

Factors Contributing to Success

An important factor in the success of the project was early recognition that putting together a Q-ROC would radically change the organizational structure, which in itself might cause managers at ICTC to be skeptical of the proposal. So, to set the stage for our recommendations, we gave several presentations to ICTC managers, engineers, and office staff on the principles of QRM, as well as the rationale behind those principles. Such "mind-set seminars" have been key to the success of other lead time reduction efforts as well, such as the projects at Beloit Corporation. An additional technique we used in these presentations was to soften up the audience by showing them the complex flowchart for order processing, and

Photo 12-3. A Custom End Mill Delivered in Under Four Weeks

our analysis demonstrating the high percentage of non-value-added time. This left no room for people to claim that the current system was working well, and opened up the opportunity for us to propose major changes.

Another factor that made the two-person Q-ROC viable was the realization, early in the project implementation, that some modified-standard orders would not conveniently fit into the scheme being planned. To avoid stagnation of progress toward their goals, the study team agreed to focus on the 80 percent of orders that would fit, and to leave the difficult orders for a future implementation.

As the final, and perhaps most critical, contributor to success, Wayman stresses that changes at ICTC would not have been accomplished if the president had not been so supportive. "Merle Clewett championed the cause of implementing Quick Response Manufacturing," Wayman says. "Without his involvement, it would have stalled early on." The president's support was not just a matter of lip service. Key actions by Clewett included:

- Holding meetings and seminars on the subject, which he himself attended.

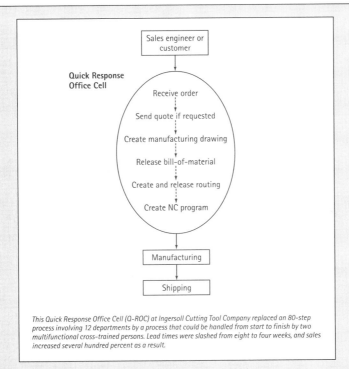

This Quick Response Office Cell (Q-ROC) at Ingersoll Cutting Tool Company replaced an 80-step process involving 12 departments by a process that could be handled from start to finish by two multifunctional cross-trained persons. Lead times were slashed from eight to four weeks, and sales increased several hundred percent as a result.

Figure 12-6. Modified-Standard Order Processing Flow

- Being present at the initial launch meetings of the study team that developed our recommendations further.
- Breaking down organizational barriers such as who could access which information.
- Giving up traditional detailed cost estimating in favor of an unproven and possibly risky table-driven pricing strategy.
- Supporting the creation of new job titles for the cross-trained personnel in the Q-ROC.

Long-Term Impact on ICTC

The huge success of the modified-standards project has had a lasting and deep impact on ICTC as a whole. This success spawned other projects to create additional Fast Track Teams (as ICTC now calls them). By 1997, more than 50 percent of ICTC's total nonstock cutter business was being handled through such Fast Track Teams, and plans were underway to expand to yet more segments of the business. There is one clear indicator of the impact of this whole new way at ICTC—in a market that has remained relatively flat over the past several years, ICTC's sales have continued to grow at a significant pace.

Main QRM Lessons

- Information handling principles include implementing the at most once rule for all information items, providing local access to information, and exploiting several electronic commerce technologies.

- Tools to assist in implementing Q-ROCs include process mapping, tagging, and value-added charting. Several fine points need to be observed for success of a tagging exercise.

- A detailed case study from Ingersoll Cutting Tool Company illustrates the principles in this chapter and Chapter 11, as well as the magnitude of success that can be achieved through a QRM strategy.

13

System Dynamics Principles for Quick Response

THE ORGANIZATIONAL PRINCIPLES described in the last two chapters, while critical for attacking lead time problems in the office, are by themselves not sufficient for achieving quick response. In fact, companies have implemented these principles, but lead time reduction has been minimal, or much less than expected. Similar problems have been experienced with reengineering efforts.[1] There are three reasons for these failures:

1. *Lack of use of system dynamics principles.*
2. *Management mind-set.* The survival of traditional management attitudes and values will negate most benefits of the reorganization.
3. *Obsolete performance measures.* If performance measures do not reflect the changes in office structure, once again benefits will be limited.

As you learned in Chapter 7, in the context of shopfloor operations, you cannot simply reorganize into cells; you need to understand and exploit the dynamics of complex systems to truly reduce your lead time. Similarly, you can best reorganize your office and achieve quick response by applying certain system dynamic principles, which are summarized in the "Main QRM Lessons" at the end of this chapter.

The very first principle in these lessons, *simplify sequence of tasks and reduce total number of tasks, using organizational principles and information handling principles*, is a reminder that you must begin by redesigning the system by applying the principles in the last two chapters. Although this may appear redundant, I have deliberately placed this in the main lessons because in my experience many "systems experts" will attack and change a system using all the other principles, but fail to begin by questioning the basic structure of the system. Redesigning the system must be an integral part of any system dynamics analysis. Also, the principles in these main lessons are worded such that you can apply them to achieve quick response on the shop floor as well. In fact you may wish to take a few moments and read each principle to see how it translates into the lessons of Chapters 4 through 7.

STRATEGICALLY PLAN FOR IDLE CAPACITY

To begin the review of the basis for idle capacity in the context of office operations we return to the second question in our QRM Quiz:

> 2. To get jobs out fast, we must keep our machines and people busy all the time.
>
> ❑ True   ❑ False

You know from the discussion in Chapter 7 that there is a penalty for high utilizations, namely long lead times. However, it's important to be reminded why the above belief pervades business operations. The root of the problem is the desire to minimize cost at each step of the operation. Chapter 3 gives two anecdotes that help you understand this extreme emphasis on keeping resources busy—one about a VP-manufacturing proudly displaying machine utilizations (Figure 3-2), and another regarding the manager of an order entry department. You also know from personal experience with supermarkets, banks, and highways, and from the analysis in Chapter 7, that as resources get busier, the waiting time for customers increases. The QRM principle replacing the above belief is:

> QRM Principle: Strategically plan to operate with sufficient idle capacity to meet your lead time targets. (Figure 13-1 contrasts the traditional and QRM approaches.)

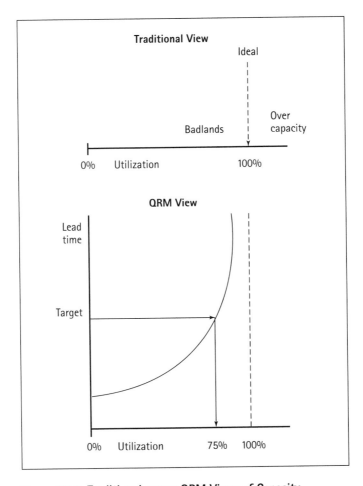

Figure 13-1. Traditional versus QRM Views of Capacity

Invariably, two questions arise when I present this QRM principle to managers:

1. *Won't this increase my cost of resources?* Let us understand this managerial concern further. If a department is running with five people working at 72 percent utilization, you could run the same department with four people working at 90 percent utilization instead (5 × 0.72 = 3.60 = 4 × 0.90). The manager could thus

reduce the department's headcount by operating at a higher utilization. (A similar argument can be made for machines on the shop floor.) The QRM counter arguments are many. First, through the product focus that is inherent in cells, the actual output per person typically exceeds the output per person in a functional organization. Second, even if initially it seems that more resources are needed, this will soon be countered through productivity increases (see all the benefits of cells, discussed in Chapters 4, 11, and 12). Putting these two in the context of the numerical example above, it may be that you can replace four people working at 90 percent utilization with four (not five) people working at 72 percent utilization and still get the same or higher output, yet with shorter lead time. Third, you reduce or eliminate all the ills of the Response Time Spiral and the cost-based organization—these include expediting, rework, and other forms of waste—with a resultant reduction in cost which will help to offset any increased cost of resources. Fourth, you are also angling for higher market share and growth through this quick response strategy. Thus you should see the increased cost of resources (if any) as an investment that will pay for itself many times over.

2. *How do I know the amount of idle capacity to plan for?* Once managers accept the concept of idle capacity, there is still the question of how much is needed. System modeling can help in answering this. The aim is to derive the trade-off curve in Figure 13-1, so you can relate a target lead time to a cap on resource utilization. For simple systems with a few processing steps, you can use the equations in Chapter 7 to derive this graph. For complex systems, you can apply modeling software that is based on more sophisticated implementations of the mathematical techniques in Chapter 7. Since these packages provide very fast answers relative to other capacity planning systems—typically they need just seconds on a PC—the method is also called rapid modeling technology or RMT.[2]

REPLACE TRADITIONAL EFFICIENCY MEASURES

Hand-in-hand with the traditional emphasis on high resource utilization is the focus on maximizing efficiency. Witness the third item in the QRM Quiz:

> 3. In order to reduce our lead times, we have to improve our efficiencies.
> ❑ True   ❑ False

As I discussed in Chapter 8, the problem lies not in the *concept* of efficiency, but in the fact that most *measures* of efficiency work counter to lead time reduction. For example, in a drafting department that creates detailed blueprints, a measure of efficiency might focus on the number of blueprints made by each employee. In this case, there are several incentives in place, such as:

- It is better to do one's own work than to help another person in order to keep a job moving.
- It pays to create a new job rather than reuse an old job. For example, if a blueprint already exists for a similar part, but it would take some time to find, it is better to create a new part and get credit for it. However, this hurts the rest of the organization because of the need for new routings and the resulting part proliferation.
- Doing work in batches helps to increase efficiency. As one instance, employees will wait until they have a stack of folders in their out tray before walking them to their destinations, or they save all computer entry tasks for one session on the computer.

All these incentives degrade the responsiveness of the organization. So what should you measure instead of efficiency? The answer is straightforward: *If you want to reduce lead time, then measure and reward lead time reduction.* Chapter 16 will show you how to do this effectively.[3]

ELIMINATE VARIABILITY

The formulas in Chapter 7 demonstrated the impact of variability on lead time. In our context, variability arises from two sources: (1) task arrivals, and (2) task times. Figure 13-2 repeats a graph used in Chapter 7 that illustrates the impact of variability, coupled with utilization, on the lead time of a job. This graph shows that for a resource working at 70 percent utilization and low variability, the lead time might be just 9 hours. However, with higher variability, the lead time could be 18 hours even though the utilization of the resource remains the same.

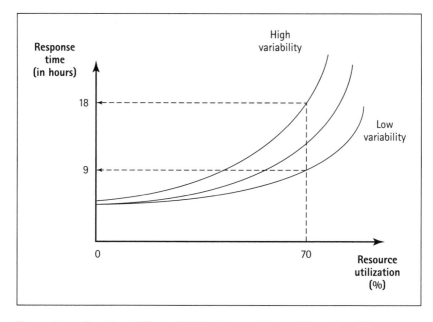

Figure 13-2. Combined Effect of Utilization and Variability on Lead Time

In terms of task design, the practical implications for reducing variability means you should attempt to make each task as predictable as possible, with minimum variation in task times. Techniques for doing this include:

- Standardizing procedures
- Standardizing forms and task routes
- Eliminating rework
- Separating complex from simple tasks

With regard to reducing variability in inputs, methods to do this include:

- Regularizing inputs through upstream controls.
- Using routine or periodic work as a cushion to even out uncontrolled workload.
- Reducing variability of upstream operations.

The last item relates to the ripple effects of variability. It uses the fact, demonstrated in Chapter 7, that variability in the task times of an operation shows up as variability in its outputs. These outputs, in turn, become inputs for downstream workstations. Hence variability can ripple through a system, resulting in large lead times. Conversely, attempts to reduce variability at each point in the system are doubly worthwhile because they also have systemwide benefits.

USE RESOURCE POOLING

Most of us have seen the principle of resource pooling used in banks and some fast-food chains. The idea is that instead of separate lines going to individual servers, you have a common line that is served by multiple servers (see Figure 13-3).

In the context of both office and shopfloor operations, however, the implications of this system redesign are more significant than might appear from the simple bank teller example. There are two prerequisites to achieving the pooled resources and common queue of Figure 13-3(b).

1. *Resources need to be flexible in order to serve the common queue.* Consider an order processing department where inquiries from customers need either a price check, a verification of purchased part costs, or a query about delivery date. The department could have three specialized staff, one who does cost estimating for quotes and subsequent price checks on incoming orders, one who deals with suppliers to get information on their latest prices, and one who does the job scheduling and commits to delivery dates. Then an incoming query would be routed to only one of these people, depending on its nature, as in Figure 13-3(a). If that person had a queue of jobs already waiting, the customer would have to wait for an answer. On the other hand, if the staff were cross-trained to handle any of three types of queries, as in Figure 13-3(b), then the response time for customers would be much shorter.

   You can have a similar discussion for machines in the factory. You can have three milling machines with different capabilities, or you can invest in three more flexible NC milling machines that can do all the operations needed. It may be that the investment is greater in the latter case, but the reduced response time and other QRM benefits may assist in offsetting this cost.

   One observation that managers have, as they look at Figure 13-3(b), is: "Aren't we going backward here, to the functional organization with pooled resources? I thought we were supposed to

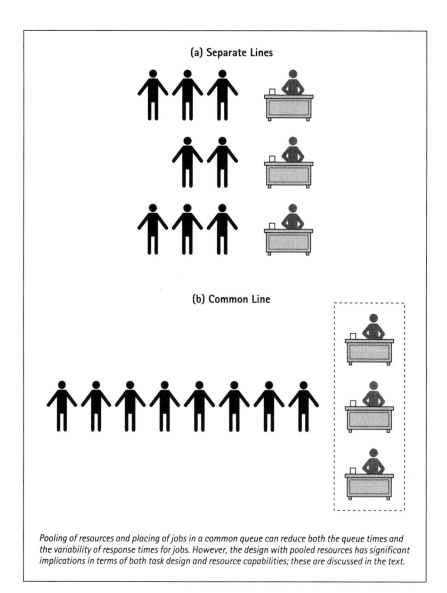

Pooling of resources and placing of jobs in a common queue can reduce both the queue times and the variability of response times for jobs. However, the design with pooled resources has significant implications in terms of both task design and resource capabilities; these are discussed in the text.

Figure 13-3. Alternative System Designs: Separate versus Pooled Resources

create cells with dedicated resources." This is indeed a good observation, and worthy of comment. To take the managers' observation further, one could argue that it would make sense to pull three grinders out of three cells and create a grinding department—a big step backward indeed. The counter argument involves all the reasons discussed so far on the disadvantages of the functional organization, such as poor quality, high WIP, long lead times, and all the other ills of the Response Time Spiral. If you wish to achieve quick response, then it is necessary to create the product-focused cellular organization, and the dedication of resources to cells is necessary. However, once you've established the general parameters of each cell, then the more flexible the resources you select for that cell, the better.

2. *Tasks should be redesigned to allow flexibility.* For the above example of customer inquiries, suppose the company's expert on scheduling publishes a daily table of which delivery dates can be promised for each type of incoming job. Then the other staff in the team can handle these types of queries. As a manufacturing example, suppose a cell makes two types of parts, some requiring regular NC milling and others needing a precision milling machine. The volume is enough to justify one of each machine. Applying this principle, you can question the design of the second set of parts to see if you can relax those close tolerances. If so, then the cell can operate with two regular NC mills, thus achieving not only reduced cost of resources but also the desired resource pooling.

The moral of these examples is that a cell is not a mini-replica of your current operations. In designing a cell, whether in the office or on the shop floor, rethink all your decisions; in this instance, the specific decisions have to do with task design and resource choice.

CONVERT TASKS FROM SEQUENTIAL TO PARALLEL

This principle is well known to those who work on setup reduction on shopfloor machines. A key point in setup reduction is to convert *internal tasks*, tasks that prevent the machine from running while they are being performed, into *external tasks*, tasks that can be done while the machine is running. In other words, these external tasks can be done in parallel with ongoing production. In the same way, you can apply this principle in many areas of office operations. Stalk and Hout give an example where a credit check would hold up the processing of an order. In only a very small percentage of cases were there problems with a customer's credit, yet this checking step would hold up all orders for several days. It therefore made more sense to continue processing every order and perform the credit check in parallel. Only in very few cases would it then be necessary to stop work that was already in process.[4] Converting sequentially ordered tasks to ones that can be done in parallel can have a large impact on response time. Thus you should search carefully for such opportunities, even going so far as to redesign tasks to enable them to be done in parallel.

REDUCE TASK SETUP TIMES AND MINIMIZE BATCHING

As demonstrated by the examples in the section on efficiency, batching occurs on the shop floor as well as in the office. In addition, some batching is caused by the structure of information systems that are still in place. Witness a personal experience I had with this issue and resultant delays.

> Recently I was amazed by my insurance agent's description of a lengthy transaction, and even more amazed that it was still happening in the mid-1990s. I had requested changes in both my home and automobile insurance, and the two policies were interlinked. My agent explained that even though the transactions could be done by computer, it would take six days to complete the change. The reason was that the insurance company involved (and it was a major U.S. company) used overnight batch processing to implement each change, and only one change could be processed at a time in order to keep the record consistent. The particular transaction I had requested needed six such changes, and therefore would take six working days to complete.

The solution lies in converting computer system applications to enable on-line processing instead of batch processing. In this age of CPU power and inexpensive memory chips, there is no excuse for batch processing in the case of most transactions.

Another way to minimize batching is to locate resources closer together. Indeed, this is one of the benefits of cells. However, one can look at this across cells as well. Again, this applies to both the office and the shop floor.

Batching is encouraged when task setup times are significant. If a materials analyst needs to review a price over the Internet, and it takes a while to start up the needed application and log onto the Internet, the analyst will wait until a stack of such jobs needs to be done. In all such cases, brainstorm to find ways to reduce setup times. Use available technology as well as task standardization, templates, forms, and other methods to make the setup as fast and painless as possible. The immediate aim should be to minimizing batching,[5] with the ultimate goal of being able to process jobs one at a time in the order that they arrive.

Finally, complement these efforts by working on the mind-set of everyone in the organization: It should be more important to keep jobs flowing than to focus on batching and other methods that improve traditional measures of efficiency.

USE CAPACITY MANAGEMENT AND INPUT CONTROL

The goal of using capacity management and input control is to keep lead time targets achievable. Without adequate capacity management, a Q-ROC can be faced with impossible demands, resulting in discouragement, lower productivity, and even failure of the whole Q-ROC concept.

It is important to continually perform rough capacity management calculations to keep targets feasible. There are standard objections to this which state that good data is not available for the time involved in many office tasks, or that tasks such as design are highly variable in their needs, and thus capacity planning is not feasible. For QRM, though, some capacity estimation, however rough, is better than no capacity planning at all.[6]

You can obtain one form of rough capacity management by strict input control to the Q-ROCs. Let us review Little's Law. This says that for a given cell:

Average number of jobs in cell = Output Rate × Response Time

Let us see how you could use this law for capacity management. Suppose you know from past performance that the Q-ROC can process 7 jobs a day, and suppose that our response time target is two days. Then Little's Law tells us that, ideally, you should have 14 jobs in process at any time. You could use this knowledge to raise a flag whenever the number of jobs rises significantly above 14. (Because of variability in the system, the number might normally fluctuate around 14, so you need not be alarmed if it goes a little over.) If the number rises to 19, say, then you can start corrective action. Such action can include temporarily adding capacity to the cell, diverting jobs to other cells, or even not accepting new jobs. The first two actions are made possible by adopting the principles of a flexible organization, discussed below. The last action is a difficult one for sales people. The lesson they must learn is that maintaining your reputation for quick response means occasionally having to turn down jobs.

Although the above control on the number of jobs may seem like the pull system of JIT, there are some differences. A pull system has a fixed number of kanbans, thus the number of jobs has an upper limit; in that respect the two control mechanisms are similar. A minor difference may be that the upper limit for a Q-ROC does not have to be a hard limit as it is with kanbans, but just a flag that initiates corrective actions—the number might be exceeded until the corrective actions start to take effect. On the other hand, a larger difference is that a Q-ROC dealing with external customers—such as those requiring a price quote—does not have the luxury of pulling an order from a customer each time one job leaves the Q-ROC; customers call when they have a need. In this case the soft limit allows the Q-ROC the flexibility to reorganize based on demand fluctuations, while at the same time giving it a way of checking capacity.

In more complex cases, where products need to flow through multiple cells in different combinations, whether in the office or the shop floor, you can employ the POLCA control mechanism.

CREATE A FLEXIBLE ORGANIZATION

The principle of creating a flexible organization serves as a way to counteract one of the risks of cells. We discussed above that in the functional organization, one achieves maximum pooling of resources, while in the cellular organization, one must dedicate resources to cells, thereby "unpooling" them. The risk then is that unexpected surges in demand (or other forms of variability) can overwhelm the limited resources in a cell. You can minimize such risks by creating flexibility throughout the organization by: creating flexibility across cells; using floaters; using vertical migration strategy for seasonal demand.

Creating Flexibility Across Cells

Early in their process of putting in cells, companies should include in their strategy the creation of flexibility *across* cells, for occasional diversion of jobs between them (see Figure 13-4). This has two implications, along the same lines as an earlier discussion on resource pooling. First, it means that the various cells should have some overlapping capabilities. In terms of human resources this implies that once workers have been cross-trained *within* their cell, companies should cross-train workers *across* cells. In terms of machines, this means that there should be some overlap between the capabilities of machines in various cells. (See the discussion in Chapter 5 for more on this.) Second, it implies that, as much as possible, jobs should be designed to enable use of more than one cell if needed.

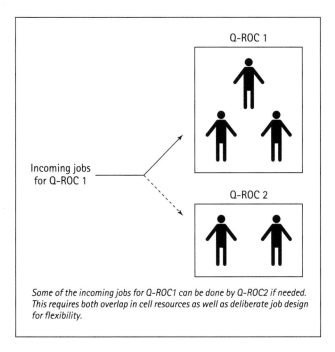

Some of the incoming jobs for Q-ROC1 can be done by Q-ROC2 if needed. This requires both overlap in cell resources as well as deliberate job design for flexibility.

Figure 13-4. Flexibility Across Cells

Using Floaters

As a second strategy, companies should train *floaters*–people who can work in several different areas of the business based on demand (see Figure 13-5). These are often the most experienced people in the company. One of their goals, in addition to just adding capacity in a cell by their presence, would be to train the cell workers on more advanced skills. While dealing with capacity surges, the company would also be spreading knowledge around its employee base.

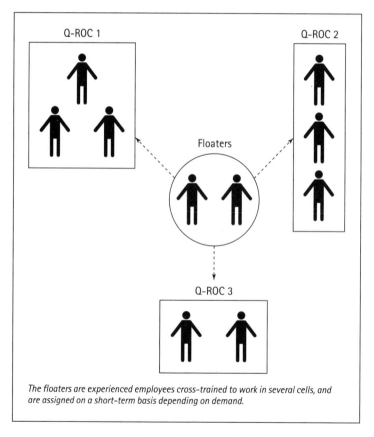

The floaters are experienced employees cross-trained to work in several cells, and are assigned on a short-term basis depending on demand.

Figure 13-5. Flexible Capacity Through "Floaters"

Using Vertical Migration Strategy for Seasonal Demand

For companies that face seasonal demand, where differences between peak and low demand can be more than 20 percent, the above strategies are not enough–those strategies work when demand shifts to different parts of the company, but overall demand stays about the same. In the case of seasonal demand, a company will face higher demand in practically all of its cells. Here a third strategy, which I call vertical migration, can help.

To motivate this strategy, let us consider one response to seasonal demand, typical of an inflexible organization, and that is to hire temporary workers for the various jobs that need to be done in the organization (see Figure 13-6). Some of those jobs are quite complex and take time to learn, so companies spend a lot of time in training temporary workers, or get poor quality output, or both.

An alternative is to bring on the temporary workers only for the low-skilled jobs in the company–jobs that they can learn quickly and perform reliably. But how do you accomplish the more skilled jobs, which are also suffering from lack of capacity? The trick is to have the permanent employees trained to perform jobs at multiple skill levels. During the off season for demand, these employees tend to do the jobs that demand lower skills. Then when the demand starts to surge, employees "migrate" to higher skill jobs. This happens at all levels of the organization, with some employees at each level migrating up and being replaced and supplemented by employees from a lower level, leaving the lowest skill jobs open to the temporary workers (see Figure 13-7).

A specific example of this is provided by a manufacturing cell, where during nine months of the year, two workers can meet the demand, as shown in Figure 13-8(a). However, during the three months of peak season, a third worker is added as shown in Figure 13-8(b). This worker's job involves bringing raw material into the cell; sawing it at the first machine; and unloading parts coming off the last machine in the cell (one of the skilled workers loads and runs this machine). Since the operations allocated to the third worker require low skill levels, this job can be learned quickly.

A seemingly obvious objection to the vertical migration strategy is that for a good part of the year the company, as a whole, will have people with higher skills doing jobs that could be performed by people with

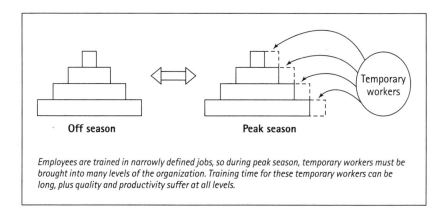

Off season   Peak season

Employees are trained in narrowly defined jobs, so during peak season, temporary workers must be brought into many levels of the organization. Training time for these temporary workers can be long, plus quality and productivity suffer at all levels.

Figure 13-6. How an Inflexible Organization Copes with Seasonal Demand

lower skills. This has implications for both payroll and job satisfaction. First let us discuss the payroll issue. It is reasonable to presume that during the whole year employees are paid at the scale for the highest skill they have, not for the job they are currently doing. This is consistent with current compensation practices, and besides, the company won't retain skilled employees very long if it doesn't abide by this rule. Note, however, that it is not necessarily more costly for the company to use the vertical migration strategy. True, during the off season the migration strategy costs more, but during the peak season the tradi-

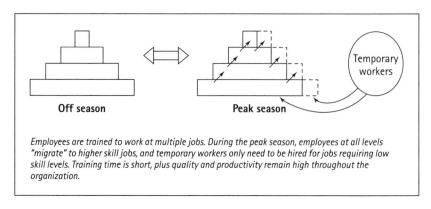

Off season   Peak season

Employees are trained to work at multiple jobs. During the peak season, employees at all levels "migrate" to higher skill jobs, and temporary workers only need to be hired for jobs requiring low skill levels. Training time is short, plus quality and productivity remain high throughout the organization.

Figure 13-7. The Vertical Migration Strategy for Coping with Seasonal Demand

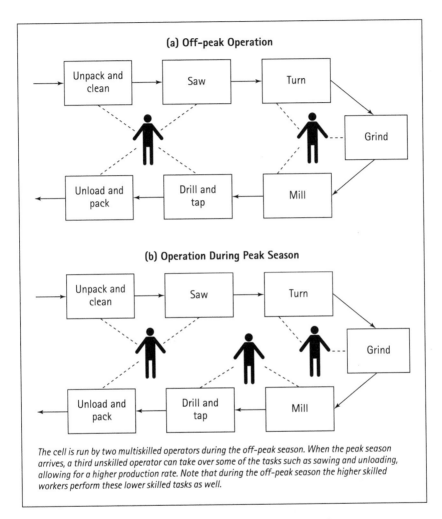

Figure 13-8. Operation of a Cell During Off-Peak and Peak Seasons

tional strategy costs more, because employees are hired at all skill levels. So there is a trade-off in payroll costs between the two strategies.

In the case where payroll cost for the migration strategy is indeed higher, the strategic issue to be resolved by management is the following: Is the added cost of this strategy outweighed by the benefits of improved performance, such as fewer rejects and less rework, during peak season?

In answering this, also consider the question: How much time do the permanent employees spend, in the current way of operation, trying to get the temporary hires trained and fixing their errors? Management should also keep in mind that present human resources strategies are encouraging increased skill levels throughout the organization; they may well have to pay for multiple skills for their employees, and given that, why not invest in a focused strategy for these skills?

The case for the migration strategy is further strengthened when you consider the cost of inventory. Many companies attempt to cope with seasonality by leveling production over the year—they build up inventory during the off-peak season, and consume it during the peak season. Companies do this in order to keep a stable work force occupied year round. If a company has a three-month peak season, this means that for nine months of the year it is building up inventory, and then working it off over the three months of peak demand. Thus the company must forecast as much as nine months into the future to decide what to build. Given the uncertainty in the actual demand, the company will be forced to build up safety stock for all its models. Inventory carrying costs for all these models, for up to nine months, can end up being substantial. A calculation for one company showed that the cost of such inventory, including interest, warehouse costs, handling costs, obsolescence, and other overhead, would have been the same as permanently hiring 50 percent more workers. This additional work force would have been sufficient to cope with the peak demand, with no need for inventory at all. In other words, the company could simply let these extra workers be idle for most of the year, use them during peak season, and still come out ahead of the level-loading strategy. Of course, we are not advocating this extreme approach; the point is that arguments based on inventory costs might help to justify the more reasonable migration strategy.

Turning next to the job satisfaction issue, several points can be made. During the off season, you should design jobs so that, as far as possible, each person does some of their higher skilled work along with their lower skilled work, such as the two employees in the cell in Figure 13-8(a). If some low-skilled jobs do not allow this flexibility because of physical or other constraints, there should be a timetable for people to rotate between the higher and lower skilled jobs that are in their range. With either of these solutions, you see that no one "gets stuck" doing the lower skilled job for the whole off season. As another point, many employees have shared with me their frustrations of having to deal with temporary workers during the peak season, taking weeks to get them to the level where they are performing, only to have them laid off just as they are becoming productive. Our vertical migration strategy will help to alleviate this frustration, thereby adding to job satisfaction. Finally, companies that make workers responsible for jobs at differing skill levels—for example machine shop workers who are also responsible for most of the janitorial and cleaning jobs within the area of their machines—find this creates a more holistic view of their job and the organization, and adds to their feelings of health. Thus, in more ways than one, the vertical migration strategy could add to, rather than subtract from, employee job satisfaction.

Main QRM Lessons

- Simplify sequence of tasks and reduce total number of tasks, using organizational principles and information handling principles.

- Strategically plan for idle capacity.

- Replace traditional efficiency measures with measures of lead time reduction.

- Eliminate or reduce all sources of variability in input load as well as in task execution times.

- Enable resource pooling by using flexible resources and redesigning tasks for flexible routing.

- Convert tasks from sequential to parallel.

- Minimize batching of work; find ways to reduce task setup times to enable this.

- Use capacity management and input control to keep lead time targets achievable.

- Create a flexible organization by designing flexibility across cells, using floaters, and using vertical migration to cope with seasonal demand.

Part Four

QRM for Rapid New Product Introduction

14

Extending Quick Response to New Product Introduction

DURING THE LAST TWO DECADES, a great deal of attention has been devoted to the product development process—and particularly so since the mid-1980s, as U.S. manufacturers have fought to regain market share in major industries such as electronics and automobiles. Researchers and practitioners alike have launched substantial efforts into rethinking traditional ways of developing and introducing new products. The results have been phenomenal: Entire new methodologies have been developed, taught, and practiced, and many companies have launched vigorous and successful attacks on the competition.

Given the vast amount of literature already available on the new product introduction (NPI) process, I will not attempt here to provide a detailed methodology for NPI. I will give a brief review of NPI techniques in the next section, but my primary aim in this chapter is to focus on the issue of implementing speed in the NPI process. You will find that many of the principles I laid down in preceding chapters can be translated to NPI. So I will emphasize those aspects of QRM principles that support an NPI methodology, and provide key insights related to speed, extending the QRM principles to enable rapid product development and introduction.

Since many companies make products that are routinely customized or even highly engineered, it is worth clarifying what I mean by *new product introduction*. If a product is based on customizing existing technologies and options for a given application, regardless of the amount of engineering involved, it's considered in the realm of *existing products*, a subject dealt with extensively in the preceding chapters. On the other hand, if a product is based on a new technology, or new design concepts, or substantially new options, it's a *new product*. Although this definition is somewhat subjective, two examples may help to drive home the difference:

- *Large capital equipment made by the Beloit Corporation.* Beloit Corporation manufactures paper-making machines. Every machine that Beloit sells is custom engineered to fit the space constraints and production needs of a particular customer. Despite all the engineering, though, normal orders would be considered "existing products" within our definition. On the other hand, if you look at the Concept IV-MH headbox introduced by Beloit in the early 1990s, you will see that this was based on a whole new technology in flow tube design, and you would classify it as an example of NPI.
- *Material-handling robots.* A company that supplies robots for material-handling applications will routinely customize their robots for each application. Although such customization requires some engineering, gripper design, software programming, and modification to existing controls, the robot is based mostly on existing technology. Implementing QRM for such products would be possible through the lessons of the preceding chapters. However, if this company decides to pursue a new market segment via a robot based on a new generation of software and controls, along with new concepts in grippers, I would consider this an NPI, and the lessons of this chapter apply.

These examples highlight the difference between NPI and previous chapters. Thus far we have considered work that was somewhat repetitive. Even if jobs were customized, most of the steps and processes were similar. I exploited these similarities in the QRM redesigns for both shopfloor and office operations. In contrast, NPI involves a one-time effort for each new product. Also, there is a much broader span of organization involved in NPI. Even though in previous chapters we created multifunctional teams and cut across departments, this was often confined to a few shopfloor departments or office functions. However, for NPI, most of the functions of a company are involved in the process.

BENEFITS OF RAPID NPI

Before describing the specific techniques for rapid product introduction, I'll review the benefits of speed in the NPI process. Chapter 2 described three of these benefits that will increase your sales and profits:

1. Beating the competition to market.
2. Always having newer technology in products that hit the market at the same time as the competition's.
3. Using fewer resources in the NPI process in addition to cutting NPI times.

One might be tempted to argue, in this age of JIT, TQM, and continuous improvement, that a firm can be very profitable by finding ways to manufacture products with increasingly higher quality and lower cost. Such an approach may be effective in the short term, but there is substantial evidence to show that frequent and rapid innovation is vital to long-term success of a firm. A study by a researcher at the University of Illinois found that for firms that were leaders in their industry, on average about 50 percent of their sales came from products introduced within the last five years. In contrast, for firms at the bottom third of their industry, this figure was around 10 percent.[1] Some cutting-edge firms have even more impressive numbers. In 1994, more than 50 percent of Hewlett-Packard's orders were for products introduced in the preceding two years.

Equally important as a motivator to rapidly innovate and introduce products, is understanding the penalty of being late in a product introduction. One supplier of parts to the computer industry stated, "Our customers make most of their profit in the first 90 days after product introduction. We have to support them in meeting their target date—or risk being the cause of substantial losses for them!"[2] Hewlett-Packard studied a high-growth market and found that if an NPI project was six months late, it could cause a 33 percent loss in profits, while if the project overran its cost budget by 50 percent but was completed on time, the loss was only 3.5 percent.[3] In other words, it is worth putting in a lot of effort to get an NPI project done on time. Fortunately, as you will see, it is not necessary to throw huge budgets at NPI projects. Rather, the key is in using proper methodologies along with modern tools, appropriate management methods, and QRM principles.

REVIEW OF NPI METHODOLOGIES

Although it is not my intention to discuss in detail the methodologies for NPI, it is useful to review the latest theories to set the stage for my subsequent discussion. Also, for readers who are not versed in the NPI literature, this section may serve as a guide to further reading. Note that, while I use the acronym NPI, I am including development in this process, i.e., in this chapter I refer to the entire process of new product development and introduction. Specifically, this means the process that begins with an idea for a product or market opportunity, and ends with the public launch of the product.

In a recent book, Ulrich and Eppinger provide a thorough analysis of the NPI process.[4] They first argue that, while the basis of invention is creativity, which may not appear to be controllable, in modern industrial corporations a well-defined NPI process is essential to the management and quality of NPI efforts. Then they suggest the following five-phase development process.

1. *Concept development.* In this phase, you identify the needs of the target market, generate and evaluate alternative product concepts, and select a single concept for further development.
2. *System-level design.* This involves definition of the product architecture, division of the product into subsystems and components, and the final assembly scheme for the production system.
3. *Detail design.* Here you develop a complete specification of all the parts, both fabricated and purchased. You develop process plans and design tooling for the fabricated parts, as well as for the assembly of the product.
4. *Testing and refinement.* Here you construct and evaluate multiple preproduction versions of the product to answer questions about basic functionality, aesthetics, performance, and reliability.
5. *Production ramp-up.* In this phase you make the product using the actual production system. Initially, the volume is low as you train the work force and work out any remaining problems in the product and process. Finally, the volume is high enough that you can make the product widely available, and you make a decision to formally "launch" the product.

Next, Ulrich and Eppinger provide a number of structured methodologies that can be used in the above five phases. The methodologies are not confined within distinct phases; in fact, most of them span several phases. These structured methodologies are:

- *Identifying customer needs.* The methodology ensures that these needs are expressed in terms of what the product has to do, not in terms of how you might implement the product.
- *Establishing product specifications.* The methodology here enables development teams to establish a set of specifications that spell out in precise, measurable detail what the product must do.
- *Concept generation.* This develops a number of alternative product concepts; each contains an approximate description of the technology to be used.
- *Concept selection.* This evaluates each of the candidate concepts with respect to customer needs and other criteria, and selects one or more concepts for further development.
- *Establishing product architecture.* This firms up the arrangement of the functional elements of the product into physical blocks, along with the interfaces to other blocks.
- *Industrial design.* This is the design of the aesthetics and ergonomics of the product.
- *Design for manufacturing and assembly (DFMA).* This reduces manufacturing and assembly costs, while improving product quality and reducing development time.
- *Effective prototyping.* A structured methodology supports the use of both analytical and physical prototypes during NPI.
- *Economic analysis of NPI projects.* This is an approach for both quantitative and qualitative analysis of a project, with emphasis on quick, approximate methods for supporting decision making.
- *Managing NPI projects.* This includes methodologies for both project planning and project execution.

Ulrich and Eppinger's framework and methodologies build on earlier work by several authors. For example, Wheelwright and Clark had earlier provided a methodology for NPI,[5] focusing on managerial issues such as teamwork, managing projects, learning from projects, and building organizational capabilities for effective NPI. Experience with implementing these ideas led the authors to the realization that senior managers needed to do more than simply launch these methodologies in their organizations; they needed to play a new role. NPI was not at the heart of managers' agenda and they did not lead this process. They only got involved when a project ran into problems. Wheelwright and Clark felt that senior managers needed to put product development at the heart of the business and shoulder the responsibility of leadership for this. Hence they wrote an accompanying book focusing on the role of senior management in leading the NPI process.[6] Key points that they mention include:

- Management must change its role from back-end problem resolution to leading front-end foundation activities, including such activities as decisions on product architecture—decisions that may have been left solely to the technical people in the past.[7]
- Senior managers must perform certain pivotal activities to ensure that teams do, in fact, work consistently and effectively. These include creating career paths that support teamwork; building the right capabilities in the organization; supporting teams in dealing with unforeseen contingencies; and creating an effective transition of responsibility for an NPI project from the team to the operating organization.
- Senior management needs to make changes in its own attitude and behavior. Changes include seeing NPI as a business process, focusing on the improvement journey and not just on the results, and exhibiting behavior that is consistent with the new values and principles that are being espoused. (I will have more to say on this topic in the next two chapters.)

In an approach he calls Lightning Strategies for Innovation, or LSFI, Zangwill also argues that the key to success in NPI is to think about product development strategically. He presents a seven-step process aimed at providing senior managers with the strategic thinking needed for effective and rapid NPI.[8] These steps are:

1. *Make innovation the strategy.* This means getting top management involved and committed to NPI as a core strategy.
2. *Establish foundation competencies.* Get risky research and invention done first, preventing delays during NPI.
3. *Eradicate fumbles.* Most NPI processes are riddled with rework, changes, delays and other "fumbles."
4. *Place customers first.* This is accomplished by using structured techniques to get more accurate information on what customers want.
5. *Develop a business strategy.* Ensure that the NPI makes sense from a total business standpoint.

6. *Design the product.* Zangwill identifies several tools, such as DFMA, to ensure cooperative development across functional lines.
7. *Improve continuously.* The key here is to create mechanisms that ensure knowledge gained from one NPI project is used to make future projects more effective.

Goldman, Nagel, and Preiss take a somewhat different tack in their description of the new form of commercial competition which they call *agile competition.* Their approach covers not just NPI, but also existing products, and customized products. Rather than providing a stepwise methodology, in their book they describe four dimensions of agile competition:[9]

1. *Enriching the customer.* The products of an agile company are seen by its customers as solutions to their individual problems. So the agile competitor can price the goods and services it provides a customer on the basis of the value of the solutions, rather than on some standard marketwide pricing.
2. *Cooperating to enhance competitiveness.* An agile competitor utilizes existing resources, regardless of where they are located or who owns them. So cooperating with other divisions, with suppliers, and even with competitors are all means employed to leverage the company's resources to achieve its goal.
3. *Organizing to master change and uncertainty.* Rapid reconfiguration of human and physical resources is needed to exploit changing market opportunities. Rapid decision making is enabled by distributing managerial authority.
4. *Leveraging the impact of people and information.* In an agile competitive environment, people and information are the differentiators between companies. The knowledge, motivation, and dedication of its work force are the key assets of an agile company. Such companies place great emphasis on work force education and training, as well as on motivation and rewards. Also, the collection, organization and distribution of information becomes a key infrastructure requirement for the company.

Sanderson and Uzumeri look at the rapid spread of product varieties in a different way.[10] They argue that there is a consistent pattern in the modern trend for companies to offer more product variety while subjecting individual designs to more frequent change. They call their analysis of this pattern the *variety-change framework.* By using these two variables, variety and rate of change, they are able to explain the interaction between product designs, manufacturing technologies, and market demands. Then they develop the NPI strategies that companies need to employ to respond to competition. These strategies include managing patterns of innovation to ensure design and model longevity, exploiting flexibility and agility in manufacturing to support the variety and change, and organizing the company's responses to dynamic competition.

The approaches in the above books are not distinct alternatives: Managers do not have to select one approach over the others. Rather, there is a strong overlap between the methodologies. For instance, the methodologies of Ulrich and Eppinger provide structured approaches for many of the steps in the other books. The variety-change framework of Sanderson and Uzumeri helps management understand the competitive environment within which they will execute their NPI project. In summary, if you are concerned with running an effective NPI process, you should be able to build on the lessons from all these books and integrate them in the context of your own operations, without being torn about making choices on which approach to use.

CONCURRENT ENGINEERING FOR NPI

A cornerstone of all the recent methodologies for NPI is the formation of a cross-functional project team. In manufacturing cells and office cells, the teams have usually spanned the functionalities of a few traditional departments. An NPI team, however, is the most aggressive implementation of cross-functional teams for QRM. Ideally, it should have representatives of all the functions involved in taking the product from the concept stage to market launch. This includes functions such as marketing, sales, research and development, product design, process engineering, tooling, manufacturing, purchasing, accounting and finance, and even suppliers and customers. This approach is also known as concurrent engineering (CE).[11]

The motivation for CE is that a large proportion of the cost of a product, over its life cycle, is determined by decisions made early in the product development. So by spending more effort in the development phase, with a team that can anticipate all the downstream effects, one can have a significant impact on downstream costs such as tooling, manufacturing, and repairs. Also, CE greatly reduces the number of engineering changes made during the development stages of a product. The closer a product is to final launch, the greater the cost of a change. The risk does not end at the launch stage either; after launch, a safety problem may involve recalls which are extremely costly. CE ensures more analysis is done up front to avoid such downstream problems and costs.[12]

Obstacles to Teamwork for NPI

In recent years, there have been a large number of successes reported on the use of CE in the NPI process. You can find many case studies in the books on NPI that I cited earlier. And yet, despite the publicity afforded these successes, I've noted in my study of engineering processes at firms throughout the United States that CE has not been widely adopted. Yes, many companies can cite individual product launches that used CE, yet the same companies have not adopted CE routinely for all their new product introductions. Or companies have created incomplete implementations of CE, where people have been assigned to teams but still had homes in their original departments, with commitments that prevented them from devoting enough time to the CE project. Other companies still have not even experimented with CE. Why this lack of adoption or incomplete adoption of CE? *Because prevailing cost-based strategies create incentives that drive the different functional departments in conflicting directions* (see Table 14-1).

As a motivating example to understand Table 14-1, consider a design department whose "efficiency" is measured by the number of new designs created each month. An engineer in this department is in the midst of a design that requires a structural fastener. The engineer has little incentive to look in the files for an existing fastener with the right shape and strength characteristics. That may take a half hour or more. With modern CAD systems, it is faster for the engineer to design a new fastener. The productivity of the design department looks better too, since they have designed an additional part. On the other hand, this generates more work for the process planning, tooling, and manufacturing arms of the firm, and leads to part proliferation, with more work for

Table 14-1. Departmental Incentives Resulting from Cost-Based Strategies and their Effects*

Functional Department	Incentive	Effects
Design engineering	Create new designs rather than working within imposed constraints.	"Not invented here" (NIH) syndrome. Deters use of DFMA.
Purchasing	Choose cheapest supplier.	Supplier may be distant, may require large orders, or have poor quality. Lengthens new product development time.
Manufacturing	Increase "efficiency" and utilization of equipment and labor. Minimize unit cost.	Long lead times. Low labor skills resulting in poor quality, less flexibility, and longer learning curve for new products.
Marketing	Create successful marketing campaigns by revising product specifications or introducing new models.	Specifications may not be easily achievable. Wreaks havoc on engineering and manufacturing.
Sales	Accept all orders.	Manufacturing may not have capacity or capability. Expediting, trouble-shooting, and other eleventh-hour solutions.

*This table was developed jointly with Dr. Suzanne de Treville of Helsinki University of Technology.

the materials department and more inventory for the company to carry.[13] The same incentives also result in lack of enthusiasm for the designers to use parts if they are not designed by their department, and reduce their desire to engage in DFMA exercises (because these take a lot of time yet result in fewer parts being designed).

Similar detrimental effects are caused by incentives in other departments, as explained in Table 14-1. All these result in a lack of support for methodologies that would reduce the lead time for NPI. Worse, many of them are the cause of longer development times. Once again, you see that local optimization for the sake of cost-based measures results in poor global results.

The solution to this problem lies in changing the mind-set, organization structure, and performance measures in the firm. The two issues of organization structure and performance measures extend beyond NPI— the same issues pervade all QRM efforts. I am therefore devoting Chapter 16 to studying them. With regard to the issue of mind-set, there are two prerequisites for rapid NPI. The first is for management to shift from a cost-based to a time-based view. This too is a point relevant to QRM as a whole, and will be discussed in depth in the next chapter. The second prerequisite is for everyone in the organization to be kaizen-oriented.[14]

Kaizen and NPI

What is the connection between QRM, NPI, and kaizen? At first glance, kaizen and NPI would seem like opposites: The former derives its success from small incremental changes; the latter succeeds through radical innovation. So how can kaizen be important for NPI? An initial answer might be that kaizen creates a mind-set of change: Nothing is cast in concrete, everything has room for improvement. However, this is only the tip of the iceberg. The deeper answer comes from understanding the basis of kaizen.[15] Although the goal of kaizen is continuous improvement, a kaizen effort begins not by changing current processes, but by thoroughly understanding them. Only if you precisely understand how you do a task today, say the kaizen teachings, can you be productive in your attempts to improve it. Consider the following situation. A friend telephones you. He has lost his way somewhere in your town, and wants directions to get to your house. The first question you ask is, "Where are you?" If the friend replies, "I have no idea," your retort is bound to be, "Then I can't tell you how to get here!" The moral of this story is simple. If you don't know where you are, you cannot reliably get to your destination. Let us drive this moral home with a manufacturing example.

The Case of the Coater

A company operates a continuous coating machine, called a coater, that lays down a thin coat of polymer on a continuously moving sheet of metal. The quality of the output is highly variable. On some days the coat appears to be just fine: It is even in thickness and there are no breaks in the coating. On other days the coating is uneven. Without investigating this further, the operator decides to change a setting on the coater one day. The next day, the coating seems to be excellent. What did the company learn? At this point, nothing, since the coating could have improved because of the setting change, or it could be fine just as part of the normal variability of the process as in the past. Even if the coating is fine for three days, this may not be evidence enough. Suppose the coating quality depends primarily on ambient humidity—a factor that (say) the company is unaware of; on humid days it takes longer to dry and this allows it to remain more even, while on nonhumid days it dries too fast, which creates breaks and uneven coatings. If the factory gets three humid days in a row, the operator might make the mistake of assuming the setting change was the factor in improvement.

Now consider a more complex situation with this coater, in the context of NPI. The company is in the midst of a project to introduce a new product that involves thinner coatings using a different polymer. They happen to test the first prototype by running it through the coater on a day that is very dry. The coating is a disaster on that day. This could seriously delay the NPI project, as engineers struggle to understand why the new polymer is creating problems. They may spend a lot of time barking up the wrong tree. The moral of this more complex story is that if you have a thorough understanding of the current process, it will also enable you to be more effective in trouble-shooting when you introduce new products and processes. The net result will be shorter lead times for NPI projects.[16]

The case of the coater also resolves an age-old dilemma between discipline and creativity, at least in the context of the modern corporation. As I mentioned earlier, there is a strong feeling among designers and innovators that their creativity will be stifled if you attempt to instill too much discipline among their ranks. While this may be true in small entrepreneurial workshops, in larger modern corporations discipline is an asset. As de Treville states, "Just as a systematic, disciplined approach to manufacturing has replaced much craftsmanship, a systematic, disciplined approach to innovation needs to replace the trial-and-error methods that prevail today."[17] For example, the discipline instilled by kaizen efforts on a coater will help to quickly launch new products on that coater. Similar arguments are made by many of the authors quoted earlier, who provide disciplined methodologies for success in NPI. Companies such as 3M have demonstrated, however, that individual creativity can still thrive within this disciplined development environment.[18]

MANAGEMENT PRINCIPLES FOR ACCELERATING NPI

In the remainder of this chapter, I'll discuss the QRM-related principles that apply to the NPI process. I'll assume that the first step in this process is for management to create a cross-functional team for the particular NPI project at hand. I'll also assume that this team follows a structured approach to NPI along the lines of Ulrich and Eppinger's methodology. In this way I'm aiming my comments at supporting principles that need to be in place for this team to achieve *speed* in the NPI process. The principles are laid out in three main categories.

1. Focusing on aspects related to management of the project.
2. Discussing principles related to the design and manufacturing procedures used in NPI.
3. Reviewing the principles from Chapters 11, 12, and 13, and showing how they can be applied to NPI.

Create a Sense of Urgency Early in the Project

Imagine a product launch that is anticipated for June, but is delayed three months. If the company was expecting substantial revenue from the product, this may have a big negative impact on performance for that year; perhaps heads will roll. On the other hand, often during the early stages of an NPI project it takes weeks just to get people to meet. In 1996, I was asked to advise a task force charged with lead time reduction in NPI. It took over two weeks for the coordinator to get team members to respond to his request for available dates, and the first date that everyone could (apparently) make was another month away. So six weeks were lost before the first meeting even took place. The same company has a seasonal product, and if it were to miss a product launch date by six weeks, it would probably lose half its sales for that model.

How does one create such a sense of urgency when the project is just in its infancy? The best way is to create a timeline back from the launch date, with intermediate deadlines, which reduces the "final urgency" to a set of "intermediate urgencies." Creating and using such a timeline is also assisted by project management techniques. These are described later. However, you must support such an approach by education—creating awareness throughout the company on the importance of generating speed in NPI—and the support of management leading the project through and emphasizing the criticality of deadlines.

Use Project Management and Critical Path Methods

Very early in the project, the team should conduct a critical path analysis (also known as PERT/CPM, for Program Evaluation and Review Technique and Critical Path Method).[19] This is a structured methodology for specifying task precedences and identifying which tasks may cause delays in a project. A tool you can use to generate the precedence relationships in the context of NPI projects is called the design structure matrix (DSM).[20] Once you have identified critical tasks, the team should focus efforts on ways to reduce time for those tasks. You can use many of the principles in preceding chapters to improve the performance of those tasks, such as increasing capacity, reducing variability, and other techniques of reducing delays. If you cannot easily eliminate capacity constraints, consider outsourcing some of the work or the prototype manufacturing. In this context, the strategy of committing to a certain amount of capacity at a supplier, in advance of knowing the specifics of the product, is also useful (see Chapter 10).

Also, teams should brainstorm to see if there is any way they can overlap tasks that are sequential in the critical path. In fact, teams should even see whether they can eliminate some critical tasks altogether. For example, in one NPI for turbine blades, the team realized that they could completely eliminate a test and refine cycle because they could obtain all its benefits in a later test and refine cycle. Finally, the NPI team should also use the strategy of releasing work in smaller batches, which I'll discuss in a later section.

Although the preceding methods focus on planning, an important aspect of project management relates to project review. There should be measurable ways of assessing project status along the way to ensure that the project is on target,[21] or to take corrective action if needed.

Control the Project Scope

Many companies use the term *specification creep* to describe the situation where new features and capabilities are added to the product as the project proceeds. NPI teams must have the discipline to freeze the design at some point and leave all subsequent enhancements to the next update of the product. Management must support this approach, and in particular, help NPI teams enforce the frozen design against pressures from other departments to incorporate "just one more change."

Encourage More Frequent, Shorter Iterations

The traditional, functional operation encourages people to complete a task in considerable detail before releasing it to the next step. For rapid NPI, management should encourage an atmosphere of multiple short iteration cycles. Not only does this improve communication, but it allows you to make improvements through cross-functional brainstorming early in the project. It also supports the small batch policy with consequent reductions in lead time for each step.

Create the Infrastructure to Support Information Exchange

The NPI process requires frequent and often voluminous exchange of information, such as large numbers of drawings or marketing databases. Management should anticipate this need and support it in several ways. First, management should consider collocation of team members—this is particularly effective with smaller teams. Second, for larger teams, blocks of time and space (such as meeting rooms or even seminar facilities) should be dedicated to the team. Third, management should invest in creating an infrastructure of electronic technologies to support teamwork. Examples of such an infrastructure are shared databases, computer networks, electronic mail, electronic data interchange (EDI) including computerized tools for communicating designs and design-related information, internet-based tools, and "groupware" such as Lotus Notes. This type of infrastructure has also been termed *collaborative technologies*.[22] Note that you should not confine such an infrastructure just to the company; managers should consider extending these electronic links to suppliers and customers as much as possible.

Partner with Suppliers and Customers

Beyond just communication links, the principles of supplier and customer relations in Chapter 10, should be carried over to the NPI process as well. The key difference here is, while Chapter 10 focused on *delivering* products, the emphasis during NPI is *involvement* of suppliers and customers during the design and development phase. Many companies include representatives of key suppliers in the NPI teams. Even customers can be an integral part of NPI, as Ingersoll Milling Machine Company discovered. In one development project where customers were closely involved, Ingersoll found only 7 formal engineering changes had to be made after the design was finalized, while a previous project of a similar nature had 62 such changes.[23]

As an extension of this concept, NPI teams should consider visiting suppliers to understand their capabilities and constraints, and visiting customers to understand their needs and see how the company's products will be used in practice.

There is an unexpected synergy between implementing QRM for existing products and the NPI process. One company switched from make-to-stock to make-to-order, in the process agreeing to minor customizations of some of its products. In doing so, the company found that interacting with customers revealed new cost-free ways of adding value for the customer. More interestingly, after a little time, the company found that its NPI time became shorter as a result of the active customer involvement that building to order encouraged.[24]

The NPI process has its own version of the bullwhip effect on the supply chain, described in Chapter 10. The designers create initial specifications for parts, then purchasing tightens them to make sure suppliers' parts conform to the specifications with a high probability. If a supplier subcontracts some of the parts, it further tightens the specifications to its subcontractor, to make sure the parts conform, and on and on down the supply chain. The further down in the supply chain, the tighter the specifications. Not only does this mean parts are overdesigned and the supply chain is incurring an unnecessary expense, but worse, if there are problems with the end product and the customer complains, the whip crack is felt all the way through the chain, with companies at bottom being hit the hardest.

This bullwhip effect is easy to fix. It simply involves cutting out the intermediate levels and having suppliers deal directly with the designer or even the end customer during team meetings. As an extension of this communication, companies are now experimenting with making product data such as process information and bills of materials available to suppliers via the Web. With such tools, suppliers could keep abreast of any design changes, which could help them shave time off their customer's design cycles.[25]

Even more innovative interactions with customers are evolving through the maturing of the Internet. Many companies are establishing World Wide Web home pages for marketing their products. This seems like a common idea today. But there is more to it than the mere implementation suggests: These home pages have a hidden characteristic, leading to a novel benefit for NPI. As customers perform searches, looking for certain products and specifications, they leave behind a trail of valuable information that market research departments can analyze to ascertain what customers are interested in, new trends, and also, when customers did not find what they were looking for. A few leading-edge companies are learning how to use this information in shortening the time for identifying new market niches and developing new products.

Use Virtual Organizations and Webs

One logical extreme of the above cooperation between suppliers and customers is the so-called virtual organization, in which many companies, often traditional competitors, link up to provide a product or service. More formally, a virtual organization is "an opportunistic alliance of core competencies distributed among a number of distinct operating entities."[26] The aim is to replace mutual competition with time-limited mutual cooperation, in order to serve a given market opportunity. The primary motivator is that cooperation can enhance competitiveness, but there are other reasons too. The strategic benefits of such an organization are:

- Sharing infrastructure.
- Sharing R&D, risk, and costs.
- Linking complementary core competencies.
- Reducing concept to cash time through sharing.
- Increasing facilities and apparent size.
- Gaining access to markets and sharing customer loyalty.
- Migrating from selling products to selling solutions.

A support structure that enables the formation of virtual organizations is the virtual web. In order for virtual organizations to form and operate on a regular basis, there needs to be a large set of qualified firms to draw upon. This is the intent of the web: It is a collection of prequalified partners that agree to form a pool of potential members of virtual organizations. Virtual organizations and webs are still new and implementations are underway that will provide lessons on how to make this approach successful. However, you should keep an eye on this structure as a means of rapid and successful NPI, especially as this approach becomes better understood.

Learn From Each NPI Project

For a company to become increasingly fast in product introduction, it should have mechanisms for organizational learning as a result of each completed NPI project. *Postmortem project evaluations* are one such mechanism. During such evaluations, team members and others in the company engage in discussion of the strengths and weaknesses of the entire project. In particular, the aim of this exercise should be to document lessons learned and recommendations for future projects. The resulting document should be widely distributed, and also be readily available for project teams in the future.

DESIGN AND MANUFACTURING PRINCIPLES

If the recent books on NPI could fill a pond, then the literature on principles of design in relation to manufacturing would fill a large lake! Rather than cover these principles, I'll focus on providing insights that will be useful to managers and members of cross-functional teams engaging in NPI. I will discuss those insights relating to speed in particular. The specialists on the teams can refer to the detailed texts on this subject.

Determining Architecture and Platforms

Product architecture is a key decision you make early in the NPI process. This firms up the arrangement of the functional elements of the product into physical blocks, along with their interfaces to the other blocks. A structured methodology for deciding on product architecture is provided by Ulrich and Eppinger.[27]

In this context, I'll mention one type of architecture that has become recently popular: the use of *platforms*. A platform is a core technological subsystem—such as the tape transport mechanism in the Sony Walkman—upon which is based a broad set of product offerings. In general, the chosen platform can be an existing technological subsystem, or it can be developed as part of the NPI strategy. Platform projects offer significant competitive advantage when done right, because they can be leveraged into a broad family of products that remain in the marketplace for a long time. Further, the reuse of the platform in all the family products enables frequent and rapid new product introductions into the market.

The platform approach also plays a role in the area where the QRM strategy comes into its own, that is, for one-of-a-kind custom products. Here, a well-thought platform can provide a solid base upon which products can be custom-designed for individual customer needs. In a later section I'll propose a whole new way of thinking about platforms by using lead time (not just design) as a guiding principle.

Quality Function Deployment

Quality Function Deployment (QFD) is a powerful tool for use at the early stages of an NPI effort. It is a structured methodology for accomplishing four goals: (1) it helps to identify product attributes that are critical from the customers' point of view; (2) it relates customer desires to quantitative engineering characteristics; (3) it helps rank the company's current products against the competitors' offerings in the context of these characteristics; and (4) it highlights the interactions between different engineering characteristics.[28] Since the charts used during QFD exercises have the shape of a house, this is also known as "The House of Quality" approach. Benefits of using QFD include, early in the project, getting the whole team to understand the NPI task and arrive at a consensus on the fundamentals, and later in the project, a reduction in the number of engineering changes as well as a reduction in customer complaints after product launch.

An interesting extension of QFD is described by Anderson in the context of customized products. The QFD approach is applied simultaneously to a range of definitions that represent various combinations of standard parts and modules, and customized parts and options.[29]

Standardization

The call for standardization in design has been so often repeated in the last two decades that it probably falls on deaf ears at this point. And yet, standardization continues to be a critical enabler of rapid NPI. But what exactly does standardization imply, especially as you think of custom-designed products in low volume, or even one-of quantities?

An extensive treatment of the use of standardization to support custom-designed products can be found in the book by Anderson.[30] He demonstrates that standardization is not limited to the idea of using the same part. There are many other areas where standardization can be effective, including components, subassemblies, fixturing geometry, design features, tooling, processes, and raw materials. Anderson also makes the point that the best way to encourage standardization is to educate people in the organization on costs incurred from proliferation of parts, tools, and raw materials.

Another way to support standardization is to change performance measures away from counts of new designs output by the design department. As mentioned earlier, the counting measures encourage designers to create new parts rather than spend time looking for existing parts that might do the job. Anderson cites an example of a revised measure: GE Lighting measures the percentage of existing parts that are used in a new design—the goal is to hit 90 percent or better.

Group technology (GT) is another technique that can assist in efforts for standardization. GT exploits similarities in the characteristics of parts for the purpose of performing various tasks. While GT can be used in many different areas of the enterprise,[31] I'll discuss here its application to NPI.

For GT to be effectively applied, a formal system must be in place for identifying part similarities. Such systems are commonly called classification and coding (CC) systems, where key part attributes are represented by alphanumeric codes. Although the purchase or in-house development of such systems may be costly, their merits are by now well accepted. For instance, an article written more than a decade ago documented several benefits of CC systems via research at a number of U.S. manufacturers.[32] The CC systems were used not only for fabricated parts, but also for purchased parts, subassemblies, end products, tooling, and raw materials. Benefits obtained from these systems included up to 75 percent reduction in the time to create a new design, up to 80 percent reduction in the number of unnecessary designs, and up to 50 percent reduction in the number of design errors.

In addition, most GT systems also include computer-assisted process planning (CAPP) systems, which enable you to produce process plans rapidly with just a little assistance from the designer. The benefits of the GT systems included up to 80 percent reduction in time to create process plans, up to 95 percent reduction in the number of process plans created from scratch, and up to 80 percent reduction in the total number of process plans (implying an increase in the use of standard parts and processes). Several of the companies in the sample also used GT to estimate new part costs by relying on the codes to find similar parts that had been made in the past. The result of this was fast and accurate cost estimates. In a similar fashion, GT was used for new equipment sizing. One company reported that the time for a typical analysis fell from three months to two hours. Another reported using GT to plan the entire range of machine tools and processes for a new plant. Note that such accelerated cost estimating and equipment planning can contribute to a substantial reduction in NPI time.

The GT concept can also leverage off cells that have been created for existing parts. If the GT system indicates a new part is similar to an existing part that is already being made in a cell, then the new part is also a candidate for that cell. The cell team could even be involved in the prototyping stage. By taking advantage of the knowledge embodied in the cell team, the NPI team can ensure that their new part will work correctly and also be easy to manufacture. The manufacturing ramp-up phase is likely to be shorter as well.

It would appear the goal of standardization goes against the grain of QRM, since we want small quantities of customized products. However, there are a number of creative ways to customize products without the drawback of part proliferation or extended design cycles. The strategies include modular, adjustable, and dimensional customization, and are described in detail by Anderson.[33]

Exploiting the Interplay between Product Design and Bill-of-Materials

There are several clever ways to exploit the way that the product architecture is implemented via the actual bill-of-materials (BOM). One strategy is called *delayed differentiation.* The idea here is that the production of multiple versions of a product—for different geographical markets or different user segments—begins with a single core assembly. Throughout the production process, different subassemblies are added, leading to multiple versions of the end product (see Figure 14-1). By delaying product differentiation, one postpones the time at which different product versions assume their unique identities, thereby gaining the greatest possible flexibility to meet changing customer demands. This also improves the effectiveness of the supply chain, since you can stock inventories in their predifferentiation forms, resulting in fewer unique items that need to be stocked.

An extreme example of such delayed differentiation is provided by the strategy used by Hewlett-Packard for one of its printers. By redesigning the printer, HP was able to have its distribution centers customize the printer for the local countries they served. This was no easy feat, since it required redesigning the product so that the power supply module could be plugged in externally instead of being internal to the product.[34] This example clearly illustrates the interplay between product design and materials management.

Another strategy under this topic is called *product enrichment.* Here, instead of providing optional add-ons to a product for a price, the

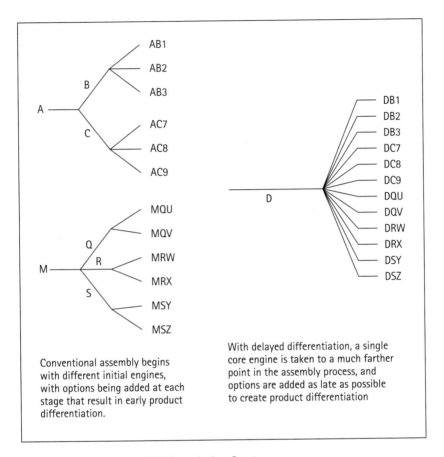

Conventional assembly begins with different initial engines, with options being added at each stage that result in early product differentiation.

With delayed differentiation, a single core engine is taken to a much farther point in the assembly process, and options are added as late as possible to create product differentiation

Figure 14-1. The Delayed Differentiation Strategy

manufacturer decides to include the most popular options as part of the basic offering for little or no extra cost. Although it may appear the company is losing money, it turns out in many cases that the savings—from not having to manage multiple inventories of options throughout the distribution channel—outweigh the cost of including the options in the product. This argument is further strengthened by the fact that designing the options into the basic product allows maximum economies of design and production at the factory—as compared to doing the work at the distribution point where adding the options may cost much more. A classic example of companies using product enrichment was the Japanese automobile makers who began including stereo equipment as standard on their cars, as compared with U.S. manufacturers who had such equipment installed at the dealership, with a heavy add-on price for the customer.

A simpler example of this approach, which also helps support the strategy of standardization, is to design components with multiple functionalities instead of different components for each job. For example, if a cover needs to have four holes drilled and tapped in it for one application, and three different threaded holes for another application you could design the cover with all seven holes in it (provided the extra holes were benign to the application where they were not needed). On the surface, the cost of making seven threaded holes is higher than the cost of making three or four, thus "pure" cost-based routing approaches will recommend making separate parts for the two applications. However, under the QRM approach of looking at total system costs, the advantages of using one part often outweigh the extra costs of fabrication.

This example doesn't show the deeper benefits of product enrichment, which—as the name implies—gives the product increased functionality. Carrying the previous example further, if you have a choice of fabricating several different options into a product, it might be better to fabricate all of them into every product. The total cost of carrying a large number of separate product types and managing the production, routings, and inventory for all of them could well exceed the cost of having one standard product with all the options built into it. This provides a powerful marketing strategy: You can provide the customer with all the functionalities for the same price at which the competition is offering only one functionality.

Rethinking the BOM is another way of exploiting the interplay discussed here. The aim is to flatten the BOM, and also to eliminate fabrication or assembly operations. Some of this can be accomplished through DFMA methods (see below). Chapter 8 discussed four specific strategies to support such rethinking. One extension of those strategies in the context of NPI is for teams to consider outsourcing an entire subsystem, rather than just individual parts.

Using Lead Time as a Driver for Designing Platform and Delayed Differentiation Strategies

Now you can combine several of the previously described strategies in a novel way with QRM as the focus, giving you a whole new way of thinking about platforms. Consider a toy manufacturer that is dependent upon a supplier of electric motors that provides a host of different motors for the various toys. The supplier's product has excellent quality, and no other supplier's product comes close in performance; the drawback is that this supplier has a lead time of several months. Despite its best efforts, the toy company has not been able to influence this supplier, who "owns" the market and has little incentive to change. So the toy manufacturer has to forecast many months into the future regarding its requirements for motors. Given the finicky nature of toy sales, the toy manufacturer often finds itself with excess inventory, or else with lost sales if it can't keep up with sales of a popular toy.

Traditionally, this company has designed its motorized toys and forecast its sales of them in terms of types of toys, e.g., airplanes, tanks, dolls, and the like, and each category uses many different types of motors depending on the size of toy and its functionality. Now I'll combine the three strategies of using a platform, delayed differentiation, and standardization to change the way this company thinks about its products. *The key is to realize that the long lead time item causing problems, namely the motors, should become the platform for design.* The end products that share the same motor can then become the delayed differentiation. Suppose, for example, that a particular tank, doll, and car use the same motor. Then these three products should be considered as "options," with the motor being the platform. After you have identified these new families of products, much effort should be put into standardizing the types of motors used by rationalizing the families further.

Next, by pooling the demand for each family and looking at the aggregate forecasts, you can convert the motors into a commodity item that is always in the pipeline for regular delivery each week. Why is this better than the original situation of forecasts? It is due to the cumulation of several effects. First, by pooling the demand for several items into one forecast, the variability of the forecast error goes down (this can be statistically proven). Second, a key idea here is to combine very different products into one platform. This way it is unlikely that all the end-items for a given motor will be popular at the same time. In fact, it is more likely that forecast errors for these very different items will cancel out. And third, by looking at these families and rationalizing them, you have reduced the number of different motors, which causes the demand for each one to be higher, increases the pooling effect, and allows a regular delivery schedule of the motors as a commodity item.

Using Prototyping

Many of the problems and delays in NPI arise from a difference in theoretical concepts (put forth in the design phase) and actual product performance (once it is built). The purpose of prototyping is to attempt to identify such problems by going through one or more design-build-test cycles. Although prototyping itself is an age-old concept—inventors often went directly to building prototypes—in the modern enterprise many methodologies and tools can help to speed up the design-build-test cycle and also make each cycle more productive.

First, there are a host of computer-aided tools that assist the designer long before any physical prototypes are constructed. These tools allow visualization, such as 3-D computer-aided design (CAD) systems and simulation tools, and analysis, such as computer-aided engineering (CAE) tools. Designs or models created and analyzed with these tools are often called *analytical prototypes*. I'll have more to say about the effective use of such tools under the subject of "Design for Analysis" later.

In building physical prototypes, the new rapid prototyping (RP) technologies can be extremely helpful. These are methods for creating complex parts (typically made from polymers) by downloading their CAD representations to an RP machine. This allows designers, manufacturing personnel, suppliers, and even customers to look at the parts and evaluate them. Even though the parts are typically not made from the final material, they can still be evaluated for a number of factors, such as assembly fit, aesthetics, basic correctness, and ease of manufacture. While in the past it might have taken days or even weeks to have a prototype made, now it is only a matter of hours. This can greatly speed up portions of the design-build-test cycle. Also, you can build prototypes with varying amount of functionality, and the type of prototype built should be consistent with the objectives of the current test phase.

The testing phase can be leveraged by putting considerable thought into it prior to the beginning of tests. Deciding the objectives of the test—exactly what key factors must be tested, and which decisions are thought to be influenced by a given set of tests—can ensure that the testing is productive. It can also prevent the need to go back and redo a test phase. Use of statistical methodologies such as design of experiments can help to increase the productivity of a battery of tests.[35] Quality function deployment (QFD) can also be used up front to help set the objectives of the prototyping and testing phases.

The new CAD, CAE, and RP technologies enable much faster and more cost-effective use of prototyping than was previously possible. Wheelwright and Clark recommend replacing the traditional prototyping cycles by a pattern of several shorter, more integrated prototyping cycles.[36] They also make a strong point that prototyping is not simply a technical tool; it is a key management tool that can be used to speed up NPI. So managers should be actively involved in each design-build-test cycle.

The availability of high-definition CAD systems is allowing some companies to do away with prototyping altogether. The apparel industry is using CAD systems to produce high-quality images of new designs in lieu of producing actual samples of fabric or garments for buyers' or mechandisers' approval. Since the sample production was a long lead time activity, eliminating it allows garment manufacturers to plan products much closer to the retail selling season, with fewer speculative forecasts.

In fact, you can extend the use of fast design-build-test cycles to the point where you can actually carry out the tests in the marketplace. The apparel industry has begun performing consumer preference tests and limited introductions to pre-test and fine-tune specific style, color, and size options prior to higher volume production.[37] For this approach to be effective, however, the manufacturer must be able to respond quickly to the particular options that are most popular. This implies instilling QRM abilities throughout the supply chain.

Design for Manufacturing and Assembly, and DFX

No discussion of modern NPI strategy would be complete without at least a mention of DFMA, an approach that has been tremendously successful in the last decade. I use the term DFMA to imply the integration of both design for manufacturing (DFM) and design for assembly (DFA) techniques. DFM focuses on the use of selected materials and manufacturing processes for the parts of an assembly, finding the most effective design by considering manufacturability as an integral design criterion. DFA then homes in on how to produce assemblies with fewer parts, as well as on how to make the parts easier to orient and assemble. (Traditional product design practices include deep-seated habits which led to creation of single-function parts with many fasteners.) DFMA results in dramatic reductions in costs and numbers of parts, and typically 50 percent reduction in NPI time.[38] Extending the concept of DFM and DFA to other criteria has resulted in a host of related approaches, commonly called design for X or DFX methods. Here X can be quality, testability, reliability, and other desired outcomes. These should all be explored as tools by the NPI team. One version of these, called design for analysis, has particular implications for NPI, and will be discussed in more detail in the next section.

Given that DFMA techniques are now well established, and entire handbooks are devoted to the subject,[39] I will not dwell on any specifics of the technique but instead make some observations that are relevant to speed in NPI.

NPI teams should consider using DFMA early in the project. One researcher notes that in the United States most companies use DFMA only after the details of the design are known, and sometimes only after prototypes of all parts are available. In contrast, in Japan he found that DFMA analysis may be conducted even at the concept design phase, and DFMA results are used in eliminating concepts or selecting them for further development.[40]

Note that the use of DFMA is not always beneficial to rapid NPI. Sometimes the application of DFMA guidelines may result in parts that are so complex that their design, or the design and procurement of their tooling, becomes detrimental to the speed of the development effort. So the apparent cost benefits of the DFMA redesign may not be worth the delay in project duration when you consider the overall project value.

Teams should therefore apply DFMA judiciously, by also considering the complexity of the resulting parts and tooling and engaging in trade-off discussions during the DFMA exercise, rather than accepting each DFMA result as a hard and fast recommendation.[41]

Design for Analysis

Computer-based analysis tools are becoming increasingly important for design. Commonly used CAD tools include 3-D solid modeling systems and parametric modeling. CAE tools include stress analysis, thermal analysis, fluid flow simulations, finite elements analysis, and analysis of how molds and castings fill and cool. Computer simulation tools include animations of robot and machine tool movements, product functions, worker movements, and recently, virtual reality systems that replicate an entire environment in 3-D. All these tools enable the use of *analytical prototypes* in place of physical prototypes. Usually you can quickly construct analytical prototypes with less cost and easily modify them for "what-ifs," thus tightening the feedback loop and speeding up the NPI process. So, they can serve the earlier stated goal of having many more, fast prototyping cycles.

In practice, however, it is common to encounter difficulties in going through this analytical prototyping process. These problems can arise from two sources. One is where the analysis procedure itself is so complex that it requires specialists in the organization. The design must then be sent to the specialists, who are typically located in a separate department. Not only does this consume time, but it also removes the designers from the analysis phase, which means you lose the benefits of immediate feedback and redesign. The second problem arises when the design is complicated enough that it is beyond the scope of existing analytical tools. In this case, the designers have no choice but to proceed with physical prototyping in order to evaluate the design. Again, this adds time and cost to the NPI project.

The effects of these drawbacks are more significant than one might think at first. At one U.S. car company, a designer complained to me that he would often wait for more than 10 weeks for a staff group to perform a finite element analysis for a simple component. This meant that he could only make one iteration of the design even with a

20-week lead time for his task. In cases where problems remained after the first iteration, he would be in the uncomfortable position of having to explain why it was taking him so long to design a component as simple as a window crank. With such difficulty for a simple component, what were the chances of the whole automobile being designed in less than four years?

Having observed such difficulties in many different companies, I felt that design practices needed to be changed to take full advantage of modern computer-based design tools. With this in mind, I proposed a design methodology for speeding up the NPI process. This methodology is called *Design for Analysis* (D/A). I use D/A since the acronym DFA is commonly used to denote design for assembly. D/A was developed further in collaboration with Masami Shimizu of Mitsubishi.[42] In order to describe D/A, the issue of good design tools must be clarified.

First, in this context, by *analysis* I mean any procedure that assists in determining whether a design will meet certain specified objectives. Next, I define an *effective design tool* (EDT) as any tool that assists designers in performing design and analysis activities and has three particular characteristics:

1. It should be easy to learn and use, requiring little training or specialized knowledge beyond that of a typical designer in the firm.
2. It should provide results of analysis quickly, preferably in a matter of seconds, or at most, within a couple of minutes.
3. It should be able to answer key design questions.

It is clear that an EDT enables designers to test their designs rapidly and also encourages them to explore many designs. In reality, as you saw from the examples above, in many cases either tools are too complex to qualify as effective, or else the designs are too complex to take advantage of EDTs. In looking at this situation, an idea emerged: Rather than passively accepting the need for a complex and time-consuming tool to evaluate the design—or worse still, giving up on analyzing the design entirely and waiting till the physical prototype is ready—what about actively influencing the design phase itself, modifying the designs so that the designer can use EDTs? This leads us to the formal statement of the D/A strategy.

....
QRM Nugget: The Design for Analysis Principle.
For rapid NPI, designers should be constrained to work with
only those designs that can be analyzed by EDTs.
....

This concept might appear backwards to many engineers who would view analysis as a tool that must serve the needs of designers. "Won't this stifle the creativity of good designers?" and "Won't the use of D/A eliminate many 'good' designs?"–these are concerns that I hear from many engineers when I first explain this strategy.

To address these concerns, consider the success of DFMA. This too is a strategy that constrains designers and eliminates several designs right off the bat. Yet from experience we know that many apparently good designs are not so good when it comes to looking at how the parts will be fabricated and assembled. In a similar way, one can see that if an (apparently) good design is hard to analyze, and this delays an NPI, then maybe it isn't really "good" after all. In fact, our research has shown that D/A can be a powerful strategy along many dimensions. D/A results in several benefits:[43]

- *Simplicity.* D/A leads to simpler products and systems; this is usually coupled with elegance of solution and concepts.
- *Robustness.* The strategy leads to designs that work well over a wide range of parameters. This is a result of the ability to explore many design alternatives and conduct wide-ranging sensitivity analyses.
- *Responsiveness.* D/A leads to reduction in NPI time. This benefit occurs via the rapid analysis of designs, early elimination of problems, and the stronger linkage between design and analysis through use of the EDTs by the designers themselves.
- *Ease of operation.* D/A creates designs of systems and products that are easy to manage, operate, and manufacture.
- *Strategic focus.* D/A forces designers to look at strategic issues (the forest, not the trees) and helps them achieve the strategic objectives of the design. This occurs as a result of working with tools that keep designers at a simpler level, leaving more time to think about the overall picture, instead of being encumbered by many complex details.

Several case studies of actual design processes help to demonstrate these properties.[44] For example, designers were asked to redesign parts for a Xerox copier using a very simple CAD system that did not allow the modeling of various complex features of the parts. The designers were able to accomplish the redesign for 30 percent of the parts and still maintain their functionality. The first observation, then, is that the complexity of those parts was not functionally necessary. More important is that the redesigned parts were cheaper to manufacture, with the savings flowing from lower costs for fixturing, inspection, handling, and assembly. So now you have a result that counters the quotes from engineers above: You didn't eliminate a good design; on the contrary, you constrained the designer and got a *better* design.

A more sophisticated example is provided by a CAD system developed by Toyota for its car body styling design process. Car bodies are composed of subtly expressed aesthetic surfaces. Traditionally, the design of such a body has depended on repeated trial and error, consuming many staff hours, trial clay models, and stamping dies, taking more than two years to complete the whole design. In order for its CAD system to be able to create highlight lines on a display of the car body, Toyota constrained the types of curves designers could employ. This enabled application of a modeling approach that provided quick response of the CAD system in generating the highlight lines. The results of this constrained design process were remarkable: Toyota's body styling designers could complete the entire design process, from initial input to generation of a shape, within a few days.

The D/A strategy is not restricted to product design. Designs of manufacturing systems benefit from the approach as well. Two examples from our research are an automated guided vehicle (AGV) system and a flexible plating line for multiple products. In fact, application of D/A to the plating line design not only enhanced the design but led to new systems concepts for plating lines.[45]

The results of these studies suggest that D/A can be more than just a design approach to speed up NPI; it can be a powerful strategic weapon that can provide an NPI team with substantial advantages for the entire product development and introduction process.

Investing in Manufacturing Process Development Early in the Project

Most executives have the impression that the benefit of focusing on process development is confined to lower unit costs in manufacturing.[46] In actual fact, an early investment in process development can result in accelerated time-to-market, enhanced product functionality, and rapid production ramp-up.[47] Because these factors affect market penetration and market share, they have much greater financial leverage than simply looking at unit manufacturing cost. Also, rapid ramp-up means you can free critical resources from troubleshooting and move them to support other NPI projects.

As part of this early analysis of manufacturing processes, initial modeling of manufacturing capacity and lead times can have substantial impact on product development, as illustrated in the following case study.

Early Analysis of Capacity and Lead Time
Assists New Product Development and Introduction

In most NPI projects, manufacturing capacity analysis is left to a later stage, after most of the product parameters have been decided. However, our experiences have shown that early "rough-cut" analysis of issues such as manufacturing capacity and lead times can make a substantial contribution to the success of NPI. You can easily and quickly accomplish such rough-cut analysis using the rapid modeling tools discussed in earlier chapters. In fact, the ease of use and speed of such tools enables their interactive use by NPI teams during team meetings.

In 1990, I was engaged in a project with Dr. Suzanne de Treville of the Helsinki University of Technology, assisting a Finnish company to introduce a new consumer electronics product.[48] A key strategy for success of the product was going to be a short lead time. The marketing director at the company explained to us that their main competitor had a four-week lead time. "Guaranteeing a delivery time of five days is key to our capturing orders from distributors," she told us. At the same time, however, she was concerned that not enough time was being spent up front to ensure that manufacturing actually would achieve the five-day lead time. Our role was to conduct a quick analysis of this issue.

Dr. de Treville and I arrived at the company in the morning, and using the input screens of a rapid modeling tool[49] as a communication medium,

---

(continued)

worked with managers of marketing, manufacturing, and technology to gather data on the planned manufacturing processes. By noon we had built a simple model that captured the essential features of the processes. Learning of this, the president of the company decided to drop by to "look at a couple of outputs." He had already decided to go from a one-shift operation to two shifts when the new product was launched. Our first run of the model showed that this would barely meet targets for the first month after launch, and that sales growth in subsequent months would render even the two-shift operation inadequate.

By now the president was deeply interested in the "what-ifs." Pushing us aside, he grabbed hold of the keyboard and asked his production manager what other options were open to them. This manager had already anticipated the capacity shortfall, and had developed a pet solution that used inexpensive automation in a novel way. The president himself entered the new parameters, and the model quickly showed that while this solution did indeed provide enough capacity, the lead times were still unacceptable. Once again, the marketing director emphasized the importance of the five-day lead time. At this point the technology manager jumped in with a suggestion: "Well, there is that Fuji machine which costs about three million Finn marks [about $800,000], but we will have to change some product designs if we want to use it..." Swallowing hard, the president entered data on the machine. In seconds, the model showed the five-day lead times could be nearly achieved. We worked for another hour, with the manufacturing and technology managers coming up with additional ideas, such as setup reductions and improvements in the burn-in ovens. Finally we arrived at a solution that, at the cost of about $1 million, would survive the sales growth for the next six months. Without loss of momentum, the president turned the meeting into a discussion of what design changes would be needed for the Fuji machine, and who would be responsible for implementing those and the various other improvements we had identified.

By analyzing the manufacturing processes early in the project, we were able to influence product design, enabling a manufacturing technology to be used that would ensure the short lead times. At the same time, the modeling exercise itself proved to be an important communication vehicle for the team members, and also served as justification to the president that the investment in the expensive machine was necessary. In the absence of our analysis, the company may well have found itself unable to meet the promised lead times, in the bargain losing the trust of dealers and then being unable to gain a foothold in the market. As it was, the company went to market with its product, gained a reputation for fast delivery, and captured a large portion of the market away from an industry giant.

Similar advantages of early process analysis have been reported by Alcoa, IBM, and other companies. In a published study, analysts at IBM reported that use of rapid modeling early in a project clued them in to certain process parameters and design features that needed improvement if they were to stay within budget for their factory.[50] An engineer at Alcoa similarly reported the influence of early modeling on process design issues such as expected yields and equipment reliabilities.[51] Our experience with these and other case studies has also shown that the transition to product launch and ramp-up is much smoother if you have conducted such analysis early in the NPI process.

ORGANIZATIONAL AND SYSTEM DYNAMICS PRINCIPLES APPLIED TO NPI

You can and should bring all the principles of Chapters 11, 12, and 13 to bear on the NPI process. I'll illustrate the application of a few of those principles with examples that are specific to NPI.

Applying Organizational Principles

First you should review the organizational principles in Chapter 11 for application to each NPI project that you're considering. A basic principle of QRM is to aim at a focused target market segment (FTMS) while keeping the product definition as simple as possible. This principle is equally applicable to NPI. Another core principle in Chapter 11 is the formation of a cross-functional team. NPI projects require the ultimate cross-functional team, in the sense that almost every function in the business needs to be represented. To the extent possible, collocation of the team members greatly enhances the speed of completion and chances of success. Also, many NPI projects are burdened with approval processes. Although it is important to have major approval stages to decide which concepts will survive, many of the minor approval steps should be eliminated altogether by empowerment of the NPI team. In the same way as I recommend rethinking of processing steps NPI teams should rethink the phases of the product development process, in particular to find ways to save time. Changing the order of steps, eliminating dependencies between steps, and even eliminating whole steps may be solutions that emerge from such brainstorming.

The concept of studying value-added work is applicable to NPI too. Zangwill[52] cites several instances of design teams where only 5–15 percent of each person's time was spent on value-added work. He makes the point that in a team where 15 percent of time is spent on value-added tasks, finding ways to add another 15 percent of value-added time has tremendous leverage: For just 15 percent of additional work you actually double the productivity of the team. Thus the techniques discussed in Chapter 12 for studying value-added time and improving it can be valuable for NPI teams.

Applying System Dynamics Principles

Next, you should review each of the system dynamics principles in Chapter 13 and design the NPI process accordingly. I'll provide a few concrete examples.

The need to plan for idle capacity is just as important with development resources as it is with the shop floor and the office. In fact, overcommitment of capacity is more often a problem in development activities than it is in manufacturing, because there are few standards on how much time an engineer needs to spend on a given task. In the absence of such standards, projects are initiated and heaped upon the development groups. One company was experiencing so many delays in NPI that its management decided to conduct a thorough study of all the active projects. They found that there were so many development projects in process that the deadlines for those projects would have required an average utilization of around 300 percent for each of the people involved.[53]

Another company had a similar problem and solved it using the principle of input control, described in Chapter 13. The company found that on many occasions it had more than 15 open NPI projects. Management imposed a limit of five NPI projects open at any time. The company immediately realized a dramatic reduction in lead time. More surprisingly, rather than reducing the company's development capacity, this input control strategy also *increased* the capacity of the development team. There were two reasons for this.

1. Efforts to reduce lead time result in numerous improvements and reductions in non-value-added work, all of which increase productivity. This is the fundamental reason laid down in this book.

2. By limiting the number of assignments, the amount of value-added work goes up significantly. This has its basis in the fact that when people are assigned to too many activities at once—this being management's attempt to increase their utilization and reduce their "dead time"—they end up spending a lot of time on tasks such as scheduling, coordinating, switching between tasks, traveling between assignments, and tracking down information for different and unrelated tasks.

The concept of resource pooling can help too. In one company, the development engineers were overcommitted on projects. Additional study showed that application engineers and technicians on the teams could take over some of the development engineers' more routine tasks at busy times. This would give partial pooling of several skill categories. Analysis of this pooling policy demonstrated that NPI lead time could be reduced by 50 percent.[54] Note that this is also an example of the vertical migration strategy of Chapter 13, since in slack times, the higher paid development engineers revert to doing the more routine tasks.

Surprising as it may seem, the ideas of transfer batching and reducing lot size are also applicable to NPI. The aim is to increase the overlap between sequential phases of the process. This is usually perceived as risky, because it requires release of preliminary information from one phase to the next. If this information changes substantially, it can create a great deal of rework for the downstream phase. Hence, in designing such overlap, the ideal characteristics to look for are:[55]

- Information that is unlikely to change.
- Information that will have minimal impact on downstream work if it does change.
- Information that can be used to launch a significant effort in downstream phases.

Companies should spend up-front time in the NPI process to identify such information. This involves breaking up typical design process groupings into smaller lots, determining key precedence relationships, and then rearranging decision sequences to allow release of qualifying information earlier.

As a concrete example, consider the design of a jet engine. In preparing to make the shafts, equipment designers need to know the length of the shaft and its maximum outside diameter. Then they can tell if existing machines can make it, or if they will need new ones. Only much later do they need to know more details of the shaft design such as the inside diameter and details of the outside profile. Similarly, if the shaft length and outside diameter are released, designers of blades that attach to the shaft can begin their design, since those dimensions do not depend on the inside diameter. On the other hand, if the shaft design were to be released only after the detailed design had been determined, since the details require lengthy engineering analyses to determine the best inside diameter and detailed outside profile, the equipment and blade designers would need to wait for a considerable amount of additional time.

RAPID NPI AND YOUR COMPANY'S FUTURE

Since many of the CAD/CAE/RP technologies are available to every firm, simply investing in these technologies does not guarantee superiority in NPI. Companies must develop internal advantages in their NPI process. To this end, companies must leverage as much as possible the lessons learned from each NPI project, using them to develop proprietary knowledge that is obtained from, and depends on, the technical characteristics of that company's products and its people. You can even use this to create proprietary software applications for use by designers and NPI teams. For example, such software could contain rules identifying those chains of design, analysis, and process creation steps that are essential to product performance.[56] Another possibility for such proprietary software would be in developing reusable modules across designs. The use of modular components and physical elements is already understood and used in design. A novel suggestion is to modularize functional elements and reuse them across new product offerings. This idea has been introduced under the term *virtual design* by Sanderson, who envisions these functional elements being used even across very different products, with tools in place to make their reuse quick and effective.[57]

A U.S. researcher who spent a year studying design approaches at Japanese companies noted that among the advantages of shortening NPI lead time are that designers learn to do design better and engineers get more practice at engineering. These improvements compound, adding up to a growing lead.[58] A practitioner in the United States sees this trend leading to a new dimension in business competition, marked by a much higher degree of inventive capability. He envisions the "rapid, constant, efficient creation of new innovative products—not just customized versions of existing products," and he labels this approach *continuous invention*.[59] So it would appear that companies interested in maintaining a lead over their competition will need to get better and better at rapid NPI. It is not too early to begin this journey. The time to investigate your NPI processes and speed them up using the principles in this chapter is now.

However, as I've mentioned before, changing the entrenched processes for product development and introduction is not trivial. Establishing a firm foundation for rapid NPI involves changing the management mind-set and creating a new organization structure along with revised performance metrics. These issues pervade all QRM efforts—on the shop floor, in the office, and in new product introduction—and they will be the subject of the last portion of this book.

Main QRM Lessons

- There is a high correlation between firms that lead their industry and those that are good at NPI.

- Being late in a new product introduction carries heavier penalties than being ontime and over budget.

- Review of the latest methodologies for NPI.

- Why cost-based strategies deter speed in the NPI process.

- The connection between kaizen and NPI.

- Management principles for accelerating NPI include adding speed to the beginning of the project, creating an infrastructure for exchange of information, partnering with suppliers and customers, using frequent short design iterations, and learning from each NPI project.

- Design and manufacturing principles for rapid NPI include standardization, use of rapid analytical and physical prototypes, exploiting the interplay between design and bill-of-materials, DFMA, design for analysis, and developing manufacturing processes early in the project.

- You can apply the organizational and system dynamics principles of Chapters 11, 12, and 13 to speed up the NPI process.

Part Five

Creating the QRM Enterprise

15

Management Mind–Set
to Support QRM

THROUGHOUT THIS BOOK you have seen that QRM requires a new way of thinking about fundamental issues in a manufacturing corporation. Traditional management approaches are rooted in economies of scale, leading to a cost-based mind-set. On the other hand, for QRM you need to be thinking in terms of minimizing time or maximizing speed. In a sense, then, to apply QRM you need to first identify the mind-set management must adopt to achieve *economies of speed.* Just as the desire for economies of scale led managers to measure and minimize cost, it is my view that the desire for economies of speed should lead them to the measurement and minimization of time. However, for companies to shift their emphasis from cost-based to time-based management there are several rules of thumb they must modify, rules of thumb that highlight to what extent they need to change their mind-set.[1]

The purpose of this chapter is to underscore the numerous changes that must occur in management thinking in order to implement QRM effectively. To show this I will consider typical management decisions that you must make in key strategic areas of the company. These include decisions about capacity, facilities, quality, personnel and organization structure, production planning and scheduling, supplier management, and customer relations. In each case, I will contrast the cost-based decisions with the ones required for quick response, illustrating the differences with specific examples. I'll begin by contrasting cost-based and time-based decisions in the area of capacity planning and facilities organization.

CAPACITY AND FACILITIES DECISIONS

Table 15-1 shows the difference between making decisions about capacity under cost-based and time-based approaches. Under the cost-based view, management wants to maximize asset utilization, meaning a well-run factory is one in which you utilize every machine or person 100 percent of the time. The time-based approach, on the other hand, recognizes that there is a trade-off between utilization and lead time; thus you determine the ideal utilization by the lead time target required for competitiveness. Instead of focusing on utilization as the driver and lead time as the result, we home in on lead time as the driver and let utilization be the result.

For similar reasons, in the cost-based view, management is upset whenever it sees idle machines or people. On the other hand, in the time-based approach, you recognize that idle capacity is strategically necessary. Therefore, there will be times when there is no work for a given machine or a given group of operators. However, the QRM approach, which depends on continuous improvement efforts, sees such "slack" periods as opportunities for employees to participate in such activities as setup reduction, quality improvement, preventive maintenance, and other kaizen activities. You can also use these slow periods for training and cross-training your employees, an essential aspect of QRM. You can educate them in basic skills such as statistics and data

Table 15-1. Comparison of Capacity Decisions

Cost-Based	Time-Based
Good factory is one in which every machine and person is 100% utilized.	Sufficient capacity is available to keep orders moving and to achieve lead time targets.
Management is nervous about idle machines or idle staff.	Idle time of machines and people is an opportunity—activities such as preventive maintenance, quality circles, and other kaizen meetings, and prototyping can occur.
Capacity investment lags demand growth; avoids risk of excess capacity, but long lead times result.	Plan ahead and install enough capacity so that lead times stay sufficiently short; take risk and capture market share.

analysis, which in turn support quality and kaizen activities. In addition, you can use these idle periods to support new product development activities such as prototyping, and for experiments with new settings on the machines.

The third row of the table states that in growing markets, the cost-based (and usually risk-averse) decisions result in a "lag strategy" for investment in additional capacity (see Figure 15-1). With this approach, since the enterprise always has less capacity than needed, lead times are very long. On the other hand, the QRM approach views short lead times as a way to capture additional market share. So capacity is purchased at a sufficient rate to ensure that lead times remain short enough to maintain competitive advantage. (See the set of steps labeled "Lead Strategy" in Figure 15-1—the terms *lag* and *lead* denote the fact that capacity investment either lags the market or leads the market.) At each stage in the lead strategy, the enterprise has some excess capacity. QRM views this as an investment that enables the company to get to the next level of growth and market share.

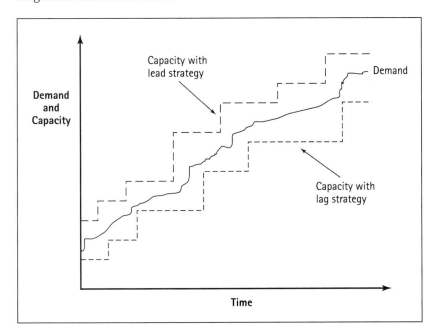

Figure 15-1. Comparison of Lag and Lead Strategies for Capacity Investment

With regard to organization of facilities the cost-based approach groups together all the resources of a similar type (see Table 15-2). The aim is to minimize the total number of resources, thereby (it is assumed) minimizing cost. However, this results in the Response Time Spiral with all its ill effects. The time-based approach, instead, organizes around product flow, grouping together all the resources needed to complete the orders for a given family, with less regard for utilization of individual machines.

The second row of Table 15-2 is less obvious. What could possibly be wrong with investing in high-speed machines that minimize the unit cost of production? An anecdote will help to drive home the point. In 1994, I was asked to review the operations of a company that made large electric motors. Electric motors contain stators and rotors, which are composed of many separate layers of thin sheet metal in special shapes. As a result, one important step of the manufacturing process at this company involved stamping out large numbers of sheet metal parts. To accomplish this in the most economic manner, the company had purchased a fast, heavy-duty press that could stamp parts out at the rate of more than 100 a minute. The only problem was, when it was necessary to switch from one part type to the next, changing the dies in the press took very long—often eight hours or more. This was because the power of the press required that the dies be very heavy and its high speed of operation meant there was little tolerance for error, so a lot of adjustments were required before the changeover was completed. As a result of the long changeover time, the materials department calculated that the stampings had to be made in batches of around 10,000, in order for the setup cost to be amortized over the parts.

Table 15-2. Comparison of Facilities Decisions

Cost–Based	Time–Based
Group similar processes together to minimize the total number of resources of each type.	Group together all the processes required for one product family to ensure orders move quickly through the system.
Use specialized, finely tuned, high-speed machines to minimize the unit cost of each operation; this forces the use of large lot sizes.	Use flexible equipment to maximize speed of changeover; speed of each operation is less of a concern.

Typical customer orders, on the other hand, were for batches of one to five motors. Each motor might use 50–100 stamped pieces in one stator. So a customer order translated into at most 500 stamped pieces for a given type of stator. With a batch size of 10,000, it was clear that the company ended up stocking large numbers of stampings. There was another problem too. The batch size of 10,000, along with the long setup time at the press, meant that lead times for stators and rotors were often a month or more. If the company received an order for a type of motor for which the stock of stampings had been exhausted, it could take two or three months to complete the motors. Since the quoted delivery time of the company (which assumed stampings were in stock) was four weeks, this type of order would initiate a cycle of expediting as they rushed stampings through the press. Of course, this meant that other parts got delayed, and the familiar Response Time Spiral hit the company in spades.

The attempt to economize on unit cost by investing in high-speed machines thus leads to yet another type of Response Time Spiral, shown in Figure 15-2. The only way out of this spiral is to use lead time targets as the primary driver, with unit cost calculations being a secondary measure at the time of equipment purchase. This tends to drive the purchase of flexible equipment with fast changeover times. Such machines are often slower, but they only have to make a few parts at a time between changeover. They also lead the company away from make-to-stock operations toward make-to-order operation. Finally, such flexible equipment also survives changes in product models, and even in entire product lines. As companies strive toward more platform-based products, such flexible lines will accommodate evolutions of the product line without requiring new investments in machines and tooling.

To help firm up these ideas, let us give some concrete examples of high-speed versus flexible equipment. In the case of the stamped parts above, an alternate technology could be a laser-cutting machine. This is programmable, doesn't require a die change, and the changeover time is limited to loading and fixturing the sheet metal and downloading the appropriate NC program. Admittedly, the time to cut 1 piece will now be on the order of several seconds, instead of multiple pieces per second. On the other hand, you could use such a machine

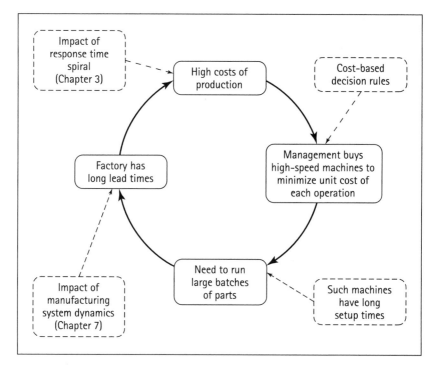

Figure 15-2. Response Time Spiral Resulting from High-Speed Specialized Machines

to cut only the number of pieces needed for an order: 50 or 10 or even only 1 piece.

A different example is provided by automated assembly systems. A high-speed assembly system, such as one used to assemble appliances, pens, or switches, contains many so-called "hard-automated" devices, which are particular to the type of part being assembled. A limited amount of flexibility is possible by loading different components into the part feeders or by having stations in the machine that are visited by only some products. Although such machines can operate extremely fast, with cycle times of seconds at each station, having to change the components in the part feeders can take from half an hour to several hours. If a model change requires changing the stations, this could be a major investment—easily $50,000 per station. An alternative approach is to use "soft-automated" assembly through the use of robots and other pick-and-place devices. You can program them to accommodate widely differing products. The cycle times are much longer and the unit assembly cost is apparently much higher, but when you look over the life cycle of a variety of product offerings, such systems may actually be more economical. Plus they allow production in small batches, whereas the changeover time in the hard-automated systems necessitates the running of large batches. When Sony introduced its legendary Walkman line of products, it made up-front investments in soft-automated systems with the anticipation of the need to ease model changeover and to accommodate modifications in products. As a result, it was able to use the line from one Walkman product offering to the next, even when design changes were significant. Indeed, its assembly system was so flexible that Sony even used it to assemble VCRs.[2] The versatility of its manufacturing system allowed Sony to keep introducing a wide range of Walkman products to the market, continuously staving off the competition and resulting in a highly profitable offering spanning over a decade.[3]

A story to the contrary also helps to instill an air of caution here: Simply investing in flexible automation does not automatically ensure success. In the early 1980s, General Motors spent more than $40 billion on modernizing its facilities, with much of this sum spent on flexible automation. Yet its factories did not experience the gains that would bring them to a par with world-class factories in their industry.[4] There are two points to be made here. One, as I've repeatedly stated, you must focus the QRM strategy on a target market, and the choice of technology must support serving this market. Two, as I'll discuss at the end of this chapter, technology is neither necessary nor sufficient for success in QRM. Without substantially changing management principles and the mind-set of the whole organization, no technology will bring a company into the QRM era.

QUALITY STRATEGIES

The traditional "inspection approach" to quality, as found in older texts on quality control, emphasized a trade-off between quality and cost of inspection. The aim was to increase the degree of inspection until the marginal cost of additional inspection exceeded the loss from bad product. This approach has been replaced by modern quality methods that show the cost of poor quality far exceeds those traditional calculations, and instead emphasize building quality into the operations by educating workers on basic techniques of quality control and quality improvement.

Indeed, these modern quality strategies take this approach much further. They emphasize relentless quality improvement: goals such as six sigma quality encourage workers to seek amazingly high quality levels. While laudable, such goals also lose credibility when workers fail to see the connection between such extremely high levels of quality and the functionality of the product.

Using the methods of previous chapters to find the root causes of long lead times, Table 15-3 shows the QRM approach as more pragmatic. Where you determine quality problems and defects to be the source of significant delays—due to rework or the need to make new parts—you tackle those processes via modern quality approaches such as SPC or TQM. This enables workers to see the need for specific quality targets at those operations. Another advantage of this approach is that you do not waste organizational resources on improving processes that are already optimal and are no longer in need of improvement. Not that achieving higher quality is bad; in fact, there are long-term benefits for an organization that pursues ever-higher quality targets. For example, the organization is ready to leap to new levels of technological sophistication, ahead of the competition. But you have to balance this against another pragmatic criterion. The total organizational energy available at any time is limited. If you focus this energy where it is not needed, you lose the opportunity to utilize it in more critical areas. Using the yardstick of time (as in Chapter 8), the QRM approach tells us where to focus

Table 15-3. Three Approaches to Quality Strategy

Cost-Based	Modern Quality Methods	Time-Based
Evaluate quality-cost trade-off (increase inspection to point of no marginal return).	Relentless quality improvement—six sigma—but risks loss of credibility with workers.	Use time as yardstick to determine where to focus quality improvement efforts and to set quality targets.

the quality improvement efforts for best results in terms of quick response to the customer.

ORGANIZATIONAL STRUCTURE AND PERSONNEL DECISIONS

The cost-based approach for hiring personnel is to use specialized staff for each job function. The aim is to keep wages low and minimize the total wage bill. The time-based view is to create a cadre of multiskilled people in the organization (see Table 15-4). Although you will pay such people more, the return on this investment is seen from three different angles.

1. You have the ability to move such multiskilled people, at short notice, to wherever a problem exists and jobs are being delayed. By adding capacity and reducing lead times in that area, you avoid all the costs of long lead times.
2. Using such cross-trained workers also reduces the amount of scrap, rework and non-value-added work, through their holistic understanding of the set of tasks to be completed.
3. Cross-trained workers contribute a continuous stream of ideas for improvement, again through their knowledge of multiple steps of the process.

Table 15-4. Comparison of Personnel and Organization Decisions

Cost-Based	Time-Based
Use specialized staff to minimize the wage bill.	Use multiskilled staff to add capacity and keep orders moving wherever needed; also higher productivity in long run.
Group similar functions into departments to minimize total number of resources and maximize efficiency of operation.	Group multiskilled staff capable of completing all tasks required to serve a given market segment.
Hierarchical structure.	Flat, team-based organization.
Supervision to enforce compliance.	Compliance through delegation of responsibility and empowerment of teams.
Communication is vertical through management structure.	Horizontal communication between task performers.

The second feature of the cost-based organization is the creation of departments containing staff that specialize in one function. The aim is to maximize the efficiency of the staff through specialized knowledge of that function and through supervision. At the same time, the goal is to minimize the number of staff. These strategies are part of an overall attempt to minimize the total wage bill. However, this organization actually results in long lead times, poor quality, and poor productivity, and it is not even clear that cost is being minimized any more. The QRM approach is to organize by market segment, putting together in one area the multiskilled staff needed to complete all the work to serve jobs for that segment.

The specialized department structure requires a hierarchical organization for two reasons. One, because goals are localized by department, there is a need for supervision in each department to maintain the efficiency of the work force. And two, since staff in each department do not have knowledge of other functions in the company, when problems arise the resolution needs to be accomplished through higher level managers. Thus communication goes vertically before it goes horizontally. The person detecting the problem and the person who caused the problem may never talk to each other directly. In the QRM organization, which is team-based with minimal levels of hierarchy, you do not need to enforce compliance. Instead, teams accept responsibility and accomplish their goals through empowerment. Communication is directly between the people actually doing the work. Problem solving is more productive and also results in continuous improvement.

PRODUCTION PLANNING AND SCHEDULING DECISIONS

As shown in the first two rows of Table 15-5, the traditional view considers a large order book (or backlog) healthy. This view is not confined to upper management—it pervades the present free market system. For example, a market analyst discussing a stock's value, often cites the company's order book as evidence that the stock is worth an investment. As a result of such thinking, there is every incentive for the company to accept all the orders it gets. Manufacturing then has tremendous pressure to get orders out, to satisfy customer due dates. However, even if manufacturing can't keep up, it is not entirely bad since the backlog makes the company look good. The QRM approach recognizes that in order to maintain short lead times one cannot have much of a backlog. Also, limits on capacity—imposed by analyses as in Chapter 7 that determine the maximum utilization that can be tolerated for the target lead times—are respected, and orders are not accepted if capacity analysis indicates that lead times are in jeopardy. Of course, this sounds like heresy to most executives—"What? Turn down an order?" they exclaim in disbelief. "What kind of business strategy is that? How will we gain market share?" And yet, accepting orders when you clearly do not have capacity only helps to prove the "Pay me now, or pay me later" dictum.

Let us understand this in more detail. If you accept an order with a due date for which you clearly don't have capacity, you set several unavoidable events in motion. Like a ship that has lost its rudder, stuck in a current driving it toward a reef, disaster is now inevitable. The pressure on manufacturing to meet the due date results in expediting, schedule changes, and bouts of overtime. Expediting creates non-value-added work for many people, schedule changes only mean other orders suffer instead, and overtime means you incur extra cost. You also know the longer term ill effects of such operation from Chapters 2 and 3. The orders for this customer or other customers—or even this customer *and* other customers—are shipped late, sometimes very late. In many cases

Table 15-5. Comparison of Production Planning and Scheduling Decisions

Cost-Based	Time-Based
Accept all orders—manufacturing will "somehow" manage to get them out.	Order acceptance matched with manufacturing capacity; don't accept if lead time jeopardized.
Large order backlog is sign of healthy company.	For short lead times, little or no backlog is neccessary.
Make economic batch size—even if no orders for most of the parts in batch—stock the rest of the parts.	Make to order—only the quantity needed for firm orders in hand.
Release jobs to keep machines and people busy—build ahead or make parts to stock.	Idle machines or people are not a schedule driver; release only what is needed for shipping; use idle time for learning and improvement.

these customers incur substantial consequences, even financial loss due to this late delivery. The customers are extremely unhappy and may never return anyway. In fact, this performance may remain a stigma that prevents your sales department from closing a sale with those customers for years to come. Instead, would it have been better to be up front with the customer, and hope that they would come back next month with another order? Would it have saved the enterprise a great deal of *angst* and unnecessary cost?

In contrast to the first two rows in Table 15-5, which deal with the situation where there are more orders than the company can fill on time, the last two rows concern traditional strategies that advocate making parts even when there are no firm orders for them. In fact, juxtaposing these opposite situations serves to bring out a critical point: Whenever you make parts that are not needed, you are taking capacity away from some firm order that may end up being delayed. As one customer service representative said to me, pointing to a large storage area while she showed me around her company's factory: "In this mountain of parts are buried most of the reasons why we shipped jobs late."

This is not to say you should ignore the impact of repeated setups on capacity. Clearly, too much setup also takes capacity away from production of parts for firm orders. However, the approach to lot sizing should be based on the methods in Chapter 7, focusing on lead time minimization, and complementing the lot size choice with appropriate setup reduction on key resources. This path will gradually improve the situation, eventually taking you to the point where you can make parts in lot sizes that correspond to firm orders.

In situations where workers or machines are idle, I have known supervisors to look ahead in the schedule—sometimes as far as six months ahead—and find work that needs to be done, just to keep the resources busy. In other cases, materials managers may look for a stock item whose inventory is somewhat low and build up its inventory—again, just to keep resources busy. "But this is reasonable!" you might exclaim. "We would have to do that work anyway. Why not now so that our resources are available to do other jobs in the future?" One reason for "why not now" is that if a short lead time job appears, but a machine has already been set up after a long and complex procedure and is now in the midst of a long run of a stock part, workers will be reluctant to break the setup. The stock part will run, the customer's job will wait. Another reason is all the costs of carrying inventory. For example, one company discovered after a year that it had a growing inventory of obsolete parts. Upon investigation it found that in one area whenever employees ran out of work, a supervisor had them make parts to stock from readily available bar stock and other materials. This, of course, meant that his area had the highest efficiency ratings in the company. Unfortunately, some of the parts he chose to make had, unbeknownst to him, become obsolete. The real tragedy of this story is not so much what the supervisor did, as the management behind him: Until that point management had considered him a very effective person who was helping the company make money. This tragedy only serves to underscore why you need to change the mind-set of management in preparing for QRM.

SUPPLIER AND CUSTOMER DECISIONS

Management's concern with unit cost has always been a primary driver of supplier relations. In recent years, the importance of quality and on-time delivery have been recognized, and these have been added to the criteria used to select suppliers (see Table 15-6). The QRM approach departs considerably from these criteria. The primary measure used under the QRM approach should be lead time reduction by the supplier (the next chapter will discuss how to measure this in an effective way). Cost, quality, and delivery should become secondary measures—still applied, but with a lower weighting.

Also, in the context of many suppliers bidding competitively for contracts, a piece of sage advice in purchasing departments has been, "Don't get too close to your supplier." The thought is that there may be a conflict of interest if you get too pally with a supplier. Similarly, if you share any corporate information with a supplier, it may give that company an unfair advantage over the other suppliers. On the other hand, the QRM approach is to develop partnerships with a few key suppliers. The idea of partnering has been put forth by JIT and quality movements as well. Some of the ideas are the same. The key difference for QRM is that partnering implies, first, helping the supplier implement QRM. This provides results of greater responsiveness, better quality, and

Table 15-6. Comparison of Supplier Management Approaches

Cost-Based	Time-Based
Unit cost is primary measure, also quality and delivery performance used to judge supplier.	Responsiveness is primary measure; lead time reduction used to judge supplier; quality, cost and delivery are secondary measures.
Use distant suppliers if price is lower.	Proximity is an important criterion.
Many suppliers compete for contracts.	Partner with a few key suppliers.
Maintain arm's-length relationship with suppliers.	As partnership, help suppliers implement QRM.
Order large quantities to get discounts.	Suppliers' QRM programs enable small quantity orders at low cost, with high quality and quick response.
Don't share internal planning or product information with suppliers.	Share forecast and schedule data; ask for suppliers' input on new products and processes.

lower costs. Second, it implies choosing suppliers that are close, not only because of faster delivery, but to help foster the relationship and communication between the organizations. Finally, partnering means sharing information about current schedules, forecasts for coming months, and even vital information about new products and processes. Companies have found that supplier feedback from such information can assist with development and prevent problems from occurring in the future.

In dealing with customers, management often sees QRM as a way to further the aim mentioned in one of the QRM quiz items:

> 9. The reason for implementing QRM is so that we can charge our customers more for rush jobs.
>
> ❑ True   ❑ False

This seems reasonable to many. After all, in certain situations customers do pay more for speed. Consider a company whose main production line is broken, and the company is losing thousands of dollars a day. Such a company would happily pay a premium for a spare part to be fabricated and delivered quickly. As another example that is closer to home, many of us willingly pay a courier $10 to deliver a letter that you could mail for a few cents, because the benefit of the speedy delivery is worth the cost to us. So why *not* charge customers more? The answer lies in a longer term view.

....
QRM Principle: The reason for embarking on the QRM journey is that it leads to a company with a secure future.
....

This security is founded upon three fundamentals.

1. *A QRM strategy must be based on creating value for the customer.* As I discussed in the chapters on cells, both shopfloor and office cells should focus on market segments where customers desire short lead times. Indeed, George Stalk, Jr.–the same author who coined the term *time-based competition*–warned, in a subsequent article with Weber, of the dangers of implementing a time-based strategy without considering whether it is creating value for the customer.[5] My own experience has been similar: Where companies undertake developing cells or other QRM initiatives without clear focus on specific customer needs, they often flounder and even abandon their efforts altogether–reverting to their previous method of operation.

2. *Implementing QRM leads to improved integration of the whole enterprise.* A QRM strategy results in less waste, higher quality, lower WIP, higher productivity, lower costs, and thus higher profitability. The faster delivery times coupled with higher quality and low costs result in increased sales. Hence companies achieve both growth and profitability with a redoubled effect on their shareholder value.

3. *The resulting QRM-based company is hard to beat.* With a large market share, fast response time, little waste in the system, and rapid introduction of new products, such a company gives its competition virtually no chance of getting a foothold in the marketplace.

There is another point to note regarding the warnings in the article by Stalk and Weber. They mention that in certain domestic Japanese markets there are so many time-based competitors that the rewards simply are not there for pushing this strategy any further. This is not a concern for most companies in the rest of the world. In my work in the United States, Europe, and many Pacific Rim countries, I have found that, as a rule, firms have not yet adopted QRM strategies. This observation is strengthened by the results of the QRM quiz, which showed that a majority of managers in the United States still do not subscribe to QRM strategies. The results are similar in Europe and Pacific Rim countries as well. If managers are not yet convinced about the basics, how can their companies have completed the QRM journey? So there is still considerable opportunity for those companies that are first off the starting block toward the QRM destination.

CREATING THE QRM MIND-SET

When they hear the words *quick response*, the first thought many managers have is that they will need to invest in state-of-the-art technologies such as group technology (GT), computer-aided process planning (CAPP), and rapid prototyping. Witness the last item in the QRM quiz:

> 10. Implementing QRM will require large investments in technology.
> ❑ True   ❑ False

Indeed, many technologies such as GT, CAPP, CAD/CAM, and rapid prototyping, offer great opportunities for lead time reduction.[6] Although these are important to consider, there are several steps that precede them. The first step lies in the following.

> QRM Principle: Recognize that the biggest obstacle to successful implementation of QRM is not technology, but mind-set.

My own experience with many firms, which is supported by research at the Center for Quick Response Manufacturing,[7] is that the most common reason why QRM efforts fail is not lack of investment in suitable technology, but lack of appropriate mind-set in the entire organization. Since QRM requires rethinking so many of the accepted rules about what is "right" in manufacturing and management, the mind-set of everyone in the organization—from the shop floor to the boardroom, from desk workers to senior management—must be realigned with QRM principles. Do not underestimate the magnitude of this task. The only way of combating the existing mind-set is through extensive education. *This must be the first step, or else other efforts will fail.* Successful QRM companies supplement their QRM projects with numerous training seminars for everyone in the firm, from shopfloor workers to top management. They also schedule visits by employees to other companies, to see first-hand the use of nontraditional strategies and work methods. They encourage their staff to go to national meetings and be inspired by the "whole new ways" that have succeeded in other firms.

The next step is to engage in lead time reductions that can be achieved through low-cost or no-cost solutions. Looking at the preceding chapters, much can be accomplished through rethinking current work methods and processes, without the need for large investments. Get a few of these successes under your belt and build on them, to further combat the mind-set issue.

Only when you have obtained substantial lead time reductions through such methods should the organization consider big-ticket technological solutions. These will eventually be necessary to stay ahead of the competition, but you bring them in at a later stage, after the initial successes have been achieved and you have explored the low-cost solutions thoroughly.

To bring about the mind-set change, however, while education must be the first step, it is not the last: You need to rethink performance measures. This is the subject of the next chapter.

Main QRM Lessons

- The biggest obstacle to implementation of QRM is not technology, but mind-set.

- The mind-set of everyone in the company—managers, office workers, and shopfloor workers—must be changed from thinking about economies of scale and cost reduction to focusing on economies of speed and time reduction.

• Such change affects decisions in all areas of the company; this chapter gives specific examples in the areas of capacity, facilities, quality, personnel, organization structure, production planning, scheduling, supplier management, and customer relations.

• The mind-set change is best accomplished through education, exposure to successes at other firms, and through low-cost QRM projects to demonstrate the validity of the concepts in your own organization.

16

Organizational Structure, Performance Measurement, and Cost Systems

I'VE DEVOTED CONSIDERABLE DISCUSSION to the need for a mind-set change to support the implementation of QRM. The preceding chapter identified the types of thinking that need to be changed and mechanisms for accomplishing the change, and the following chapter provides concrete steps to implementing QRM, including steps for creating the new mind-set in the company. In order to support such a mind-set change, however, you need to thoroughly rethink existing performance measures. Otherwise your QRM efforts are doomed to fail because your traditional measures will continue to reinforce the traditional mind-set. I've given examples on the pitfalls of traditional performance measures. This chapter provides more precise statements of those problems and presents new measures that are based squarely in the QRM approach.

Although one aim of performance measurement is to show people such as stockholders how a company is doing, another equally important aim is to motivate and reward the company's employees. Since QRM requires a new organizational structure, the incentive and reward schemes must also be consistent with this structure.

Performance measurement is also intimately tied to the cost accounting system. Again, this is for similar reasons: one, the company must have an accepted way of reporting its status to its stockholders; and two, managers need to have controls in place to ensure fiscal responsibility and against which their performance can be measured. So a third topic in this chapter concerns issues of accounting. However, this will not be another discussion on the need for whole new accounting systems such as activity-based costing (ABC). Instead, I will provide two simple remedies that you can use to supplement existing accounting systems to enable them to better support your QRM implementation.

From this discussion, you'll see that performance measurement, organization structure, and accounting systems are three facets of the same issue. This chapter looks at these facets and presents a unified approach for dealing with this trio—for you cannot tackle one facet without considering the others.

DRAWBACKS OF TRADITIONAL MEASURES

Traditional measures employed in manufacturing companies are either financial measures, or metrics derived from cost-based strategy. I've discussed at length the pitfalls, from a QRM point of view, of two such derivative measures—utilization and efficiency. Now let's discuss the drawbacks of traditional measures in general.

The purpose of financial accounting is to ensure that the basic fiduciary responsibility of the company is carried out for its stockholders, for government reporting, for potential investors, and for other people outside of the company. The aim of management accounting, on the other hand, is to provide information to people within the company to help them in running the company effectively. Both these aims seem reasonable; for example, if one is to invest in a company, then one wants reassurance that its financial health has been measured according to some generally accepted practices. Similarly, if one wants to hold management accountable for the health of the company, there should be measures by which you can judge if managers are doing their job. Ironically, our use of the metaphor "holding management accountable" serves to illustrate how deep in the marrow of our society—in the marrow of our language even—is the connection between accounting and measurement of managers! In principle, then, both these aims should serve companies and their stockholders well. In practice, there are problems with each.

The First Problem—Preoccupation with Financial Measures

The preoccupation with financial measures is exacerbated by the way the stock market works. In today's financial markets, institutional investors such as mutual funds own a large proportion of company stocks. The managers of these funds are driven (by *their* performance measures) to show short-term returns on their portfolio.[1] So they, in turn, exert pressure on the companies whose stock they own, for them to show returns. Since mutual funds have substantial shareholdings, their managers command a significant amount of attention at these companies. This means executives at the companies have an incentive to show short-term returns or else risk having their stock "dumped" with disastrous consequences for their stock price. This financial "bottom-line" view percolates down to each performance measure used at each level of the company. Every lower level measure is meant to tie in to some financial measure, and by motivating managers to improve these measures the company hopes to make its overall financial health look good.

The Second Problem—No Room for Manufacturing Strategy

The second problem is therefore a consequence of this tendency to be preoccupied with financial measures. Although management accounting is intended to help managers run the company effectively, traditional accounting practices require a strong link between the financial accounting reports and the management accounting numbers. As a result, the management accounting systems end up being driven by the bottom-line measures desired from the financial accounting systems.

The upshot of all this is that there is no room for long-term manufacturing strategy. Not only do employees have little incentive to work on efforts that have a long-term return, but even at the investor level, most decisions are made based on financial measures with little appreciation for manufacturing strategy. Two examples serve to drive each of these points home.

Example One: A President's Talk

In 1995 I was invited to speak to the managers of a company that made industrial equipment. I was instructed that the company was placing a great deal of emphasis on lead time reduction, and that my purpose was to help create the new mind-set that would enable them to apply QRM methods. Immediately prior to my talk, the president of the company spent half an hour showing the managers his vision for the company over the next decade. He displayed chart after chart with sales numbers on each product line, sales growth in different markets, market share, and profitability. Then he followed with charts about goals for individual department costs and profitability. Having been revved up about the importance of QRM for the company, I kept waiting for the charts with specific lead time goals to appear. They never did.

Example Two: An Annual Report

I gave this second example in Chapter 3 so I will repeat it briefly. In 1996 I was asked to deliver a training session for the management of a publicly owned company in the packaging industry. Again, I was told short delivery time was essential in their industry and all their divisions were working hard at reducing their lead times. Prior to the seminar I received the annual report for this company. Although it contained descriptions of the company strategy and many performance statistics, nothing in the report mentioned or measured responsiveness. Since annual reports are a key instrument used by potential investors, it is clear that companies, their managers, and their stockholders only want to see the bottom line in financial terms. On the other hand, as I've shown in many places in this book, many financial measures will only show short-term health if the company takes decisions that run counter to QRM.

Consequences of Traditional Measurement Systems

The preceding stories make the case that both management and financial accounting methods are not sufficiently synchronized with manufacturing strategy. It is not surprising, then, that statements such as the following abound in today's literature:

> The fundamental flaw in the use of management accounting reports for operational performance measurement is the assumption that financial reports are valid and relevant to the control of daily business operations. This assumption is wrong. Not only are financial reports irrelevant to daily operation, they are generally confusing, misleading, and in some cases, positively harmful to the business.[2]

The following points make concrete the drawbacks of traditional accounting methods and related performance measures:

- *Traditional cost-based performance measures motivate employees and managers to do the wrong things.* In the context of QRM, I've provided many examples of measures such as "efficiency," "unit cost," and "utilization," which promote behavior resulting in increasing lead times. Worse, even in an apparently enlightened company that implements a time-based strategy, the weight of the traditional entrenched measures can cause it to abandon that strategy and revert to cost-based manufacturing methods: "A stubborn refusal to abandon traditional performance measures put the brakes on an automotive supplier's efforts to compete."[3]
- *Traditional measures mislead company managers and result in poor strategic decisions.* Incorrect understanding of overhead costs, and thus invalid allocation of these costs, can result in cost distortion to the point where companies pursue product lines that actually cause them to lose money, while giving up their profitable product lines. In a similar way, a purely financial approach to measurement, combined with a superficial understanding of overhead costs, has led to companies closing down portions of their operation and outsourcing the production of those parts. This, in turn, led to more overhead being allocated to the production of fewer parts, which were then deemed to be too costly and, as a result, were outsourced. Eventually such a spiral led to the closure of whole divisions of companies.[4]
- *Traditional measures were designed for traditional management methods.* The traditional organization has a hierarchical structure with functional departments. Each department has cost goals and other goals based on measures like utilization and efficiency. Top managers in divisions have aggregate goals, which they translate into targets for their subordinates. These managers, in turn, translate the targets into goals for the people working for them, and so on. Traditional reporting systems are therefore designed to produce reports for each layer of this hierarchy to ensure that the goals at its level are being met.
- *Cost-based measures are not cell-oriented.* QRM's approach to the role of people in the organization is fundamentally different. The traditional approach views each person as a specialized resource that has to perform at a certain rate in order to justify its cost. However, QRM views the organization as composed of cellular

teams that own processes that deliver products to their customers. Existing management systems typically have little in terms of reports to measure the effectiveness of this type of organization in achieving these particular goals.

- *Cost-based measures are backward looking.* Effective manufacturing strategy today requires a clear vision of where a company should be headed. Yet most reporting systems, and financial systems in particular, only tell management where the company has been, not where it is going. To drive this point home, ask yourself, how effectively could you drive a car if you were only allowed to look in the rear-view mirror?

It is not my purpose in this book to reinvent accounting systems; considerable effort has been spent over the last decade in designing new systems that do a better job of allocating overhead. Also, much work has gone into rethinking financial justification methods. My goal is to make the reader aware of the drawbacks of traditional measures so that the company is willing to let go of its attachment to them, and second, to focus on new measurement systems that are relevant to the QRM effort.

NEW MEASURE OF PERFORMANCE—INTRODUCING THE QRM NUMBER

I've emphasized that if you are concerned about lead time you should measure lead time. This much is not new; others have made the observation that despite making noises about long lead times and late deliveries, management does not put sufficient emphasis on measurement of lead time.[5] However, I have an additional observation. QRM is not just about lead time and meeting certain lead time targets. QRM is about lead time *reduction*. So if management desires the benefits of QRM, then it should reward *lead time reduction*. To do this management needs a fair and equitable measure of lead time reduction. However, in the last decade I have not seen any measures that truly fit the bill. As Tom Malone, president of Miliken, said, "If you are not measuring it, then you can't be serious about it."[6] Therefore, in order to help companies be serious about lead time reduction, I have developed a new measure suited to this task. It is called the *QRM Number*. This number can be used to measure an individual cell, a set of cells, or a whole factory.

In order to establish the QRM Number for any entity, say a given cell, you first measure the lead time of that cell for jobs over an initial period, called the base period. (Even the task of just measuring lead time can be tricky; in the next section I will discuss more precisely how to measure lead time, along with potential pitfalls in this process.) The average lead time over the base period will be called the *base lead time*. Then, for any subsequent measurement period (e.g., each quarter), you also measure the average lead time during that period, which is called the *current lead time*. The QRM Number of the cell, for the current measurement period, is then defined as:

$$\text{Current QRM No.} = \frac{\text{Base Lead Time}}{\text{Current Lead Time}} \times 100$$

Note, in particular, that the current period's lead time is in the *denominator*. The reason for this choice will become clear.

To make this approach concrete, let us consider a newly established manufacturing cell. After the cell is in place and functioning, you measure the average lead time for the first quarter. Suppose this is 12 days. Thus the base lead time is established at 12 days. In the second quarter, through several obvious improvements put in place by the cell team, the average lead time drops to 10 days. By the third quarter, the team is working well together, many ideas for continuous improvement emerge and are implemented, and the team manages to bring the average lead time down another 2 days, to 8 days. In the fourth quarter, the pace of improvement seems to slow down. The team has detected most of the obvious problems and put in place the easy solutions, so in this quarter they only reduce the lead time by 1 more day, to 7 days, on average. How does this performance stack up using the QRM Number?

Table 16-1 summarizes the calculations for this cell. By putting the lead time numbers in the formula above you derive the QRM Number for each quarter. (The QRM Number for the first quarter, which is also the base, is 100. The base period will always have the value of 100, as is obvious from the formula.)

Desirable Properties of the QRM Number

At this point you might ask, why such a complicated procedure? Why not just record the average lead time, which is in the second column,

Table 16-1. QRM Number Calculations for the New Manufacturing Cell

Calendar Quarter	Average Lead Time	Details of Calculation	QRM Number
1. Jan.–Mar.	12 days	$\frac{12}{12} \times 100$	**100**
2. Apr.–Jun.	10 days	$\frac{12}{10} \times 100$	**120**
3. Jul.–Sep.	8 days	$\frac{12}{8} \times 100$	**150**
4. Oct.–Dec.	7 days	$\frac{12}{7} \times 100$	**171**

and make it the team's goal to bring this number down? The QRM Number has the following attractive properties:

- *The desired performance results in an increasing number.* A fundamental rule of performance measurement is that humans react better to graphs and numbers going up, as opposed to going down. As stated by one expert, "Psychologically, people are motivated more by seeing the results of their improvement as an increase in the measures, a graph going upward, rather than by seeing a reduction."[7] Clearly, if the desired performance is reduction in lead time, then graphing raw lead time (Figure 16-1) results in a graph going down when the team is doing well and going up when the team is doing poorly—this is the opposite of the expert rule. On the other hand, graphing the QRM Number (Figure 16-2) gives the desired properties.

- *An equal reduction of lead time in the future produces a larger increase in the measure.* To illustrate this property, consider the cell's improvement from the first to the second quarter. The team cut its lead time by 2 days, and the QRM Number went up by 20 (from 100 to 120). Now consider the change from the second to the third quarter. This was also a reduction of 2 days—the exact same amount—but the QRM Number went up by 30 (from 120 to 150). Why should this be, and is it reasonable?

  Intuitively, it is indeed reasonable to show a higher QRM Number increase for the second improvement. You would expect that, compared with the reduction from 12 to 10 days, it would be harder to

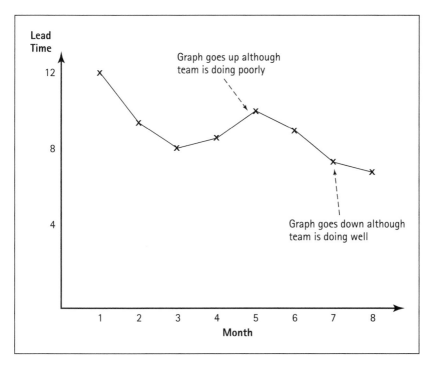

Figure 16-1. Average Lead Time for a Manufacturing Cell

reduce lead time from 10 to 8 days, for two reasons: one, the percentage reduction is greater for the second improvement, and two, there are usually fewer opportunities remaining the second time around—the easiest and most obvious steps are implemented right away, leaving the harder problems and more difficult solutions for later. So it seems fair to give the team more recognition for cutting the lead time by 2 days the second time around. The QRM Number does exactly that.

- *The QRM Number continues to motivate people, even when additional lead time reduction becomes difficult.* This is an extension of the previous observation. Consider a team that has reduced lead time from 12 to 6 days and appears to be approaching the limits of its ability to improve. A conventional graph, just plotting the raw lead time, would make it appear that improvements have almost stalled (Figure 16-3). Yet the QRM Number continues to motivate people, because even small improvements will create significant increases in the QRM Number graph (Figure 16-4).

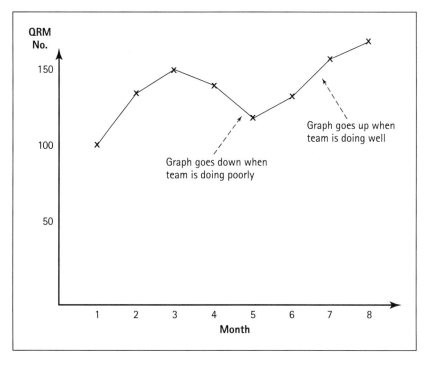

Figure 16-2. The QRM Number for the Same Manufacturing Cell

- *It provides a single unified measure that can be used throughout the organization, regardless of team size or type of work.* You can use the QRM Number to compare different teams across the organization: manufacturing cells, office cells, entire factories, and even whole organizations with each other. For example, consider a company that implements three cells: a manufacturing cell that has a base lead time of 14 days, an assembly cell with a base lead time of 6 days, and a Q-ROC with a base lead time of 3 days. While the manufacturing cell may cut its lead time by 1 day quite easily, the Q-ROC may struggle to take even half a day out of its lead time. In this environment, the QRM Number is the great equalizer. You can judge all three cells on a relatively equal footing by looking at their QRM numbers. In fact, management can use this approach to foster a spirit of friendly competition between teams across the company.

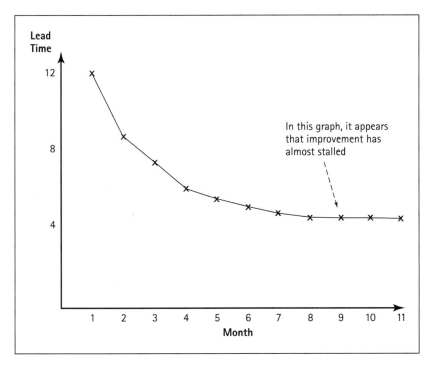

In this graph, it appears that improvement has almost stalled

Figure 16-3. The Average Lead Time Graph Plateaus

In addition to measuring individual teams, you can use the QRM Number to measure groups of teams to compare whole market segments or entire factories. Suppose, in the previous example, that the Q-ROC, the manufacturing cell and the assembly cell together serve a focused target market subsegment (FTMS), which was defined in Chapter 11. Customer orders arrive at the Q-ROC which does the order entry, along with some design work and preparation of manufacturing routings. Then the orders are released to the manufacturing cell for parts fabrication, and the final product is put together in the assembly cell. In this case, you could measure the three cells together by another QRM Number, which would help management track the improvement seen by the end customer. The overall team for this FTMS would start its QRM Number with a base lead time of 23 days (3 + 14 + 6). Management would not need to choose between this measure and the individual team measures.

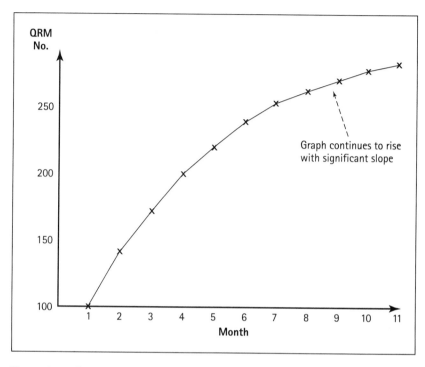

Figure 16-4. Increase in the QRM Number for the Same Time Period

They should publish both, because they serve different and complementary purposes. The individual team measure motivates improvement within each cell; the overall measure encourages improvement across cells. You could also use the overall measure to foster competition between this FTMS and groups of teams serving other FTMSs. In the same way, entire factories or organizations could be measured as well. Using this approach, we could compare a factory in Wisconsin making electric motors with one in North Carolina making complex parts for aircraft, and still feel that it was fair to compare these very different organizations.

Problems with Alternative Lead Time Measures

As the alert reader already will have deduced, the QRM Number is none other than a measure of speed, normalized to allow measurement across different teams. Other similar measures could be proposed; one such measure might be the percentage of lead time reduction, which would also be normalized across teams since it could be a percentage of the base lead time. It also has the property of being an increasing number: The more lead time is reduced, the higher is the percentage reduction. However, the problem with this measure is that after initial improvements, as teams face harder and harder targets, using this measure their progress will appear to have stalled.

Another measure used by some companies, including several leading electronics firms, is a cycle time ratio. Two alternatives have been proposed for this; one is simply the reciprocal of the other. The first measure, which we call Ratio A, is defined as the ratio of value-added time to total lead time (also called cycle time in these companies):

$$\text{Ratio A} = \frac{\text{Value-Added Time}}{\text{Lead Time}}$$

For example, if the value-added time in a manufacturing cell (e.g., the time a part is actually being machined) is 3 hours, while the lead time is 25 hours, then Ratio A has the value 0.12 for this cell. The aim in these companies is to increase this ratio, because they want to maximize the proportion of time that value is being added. However, there are two immediate drawbacks to using this ratio.

1. If a team discovers how to reduce value-added time, for example, by eliminating an operation or speeding up an operation, the team might end up being penalized for this improvement, because the ratio will go down unless the proportional reduction in lead time is at least as great. For instance, if an operator in the above cell eliminates a 15-minute deburring operation, bringing the value-added time down to 2.75 hours, but the lead time only goes down by 45 minutes to 24.25 hours (this will be the case if the deburring operation doesn't have much queue time associated with it), then the new value of Ratio A is 0.11. The operator has made an improvement and yet the measure has gotten worse!

2. This disadvantage is more subtle and is caused by the behavior just noted above. Teams soon find out that the ratio depends heavily on the value-added time, and they make every effort to increase the number that measures value-added time. Similar to traditional labor union approaches that attempted to maximize the standard

times associated with each operation, incentives have been created for operators to make each value-added operation appear as long as possible, at least for the sake of measurement.

What the two behaviors above show is the fundamental principle of performance measurement: By measuring something that is not clearly related to our goal, we've lost sight of the goal altogether. The goal is lead time reduction. It is not clear what the connection is between Ratio A and this goal.

The other cycle time ratio measure, which we call Ratio B, is simply the inverse of the previous ratio:

$$\text{Ratio B} = \frac{\text{Total Lead Time}}{\text{Value-Added Time}}$$

For the same example as above, the value of Ratio B is 25/3, which equals 8.3. Now the goal is to make this number smaller. You can use the exact same arguments as above to show the disadvantages of this ratio as well. The QRM Number, on the other hand, is directly related to our goal of lead time reduction and does not suffer from these unexpected consequences.

KEY ISSUES IN MEASURING LEAD TIME

Although the QRM Number seems relatively straightforward and simple, there are subtle points to consider when measuring lead time, as well as pitfalls to be avoided in implementing this measure.

Some Problems in Defining Lead Time

Since the QRM Number involves the ratio of a base lead time to a current lead time, the first issue is exactly what is meant by lead time. For instance, if a cell makes the same part frequently, it can stock extra quantities of that part and then supply them to the next operation on demand, with a lead time—the time from the next operation requesting it to the time it can ship the parts—that is virtually zero. (In fact, this is what happens with a pull system, as we pointed out in Chapter 9.) However, in measuring lead time for the QRM Number you should really measure the lead time for material to flow through the whole system being measured. The manifold benefits of QRM discussed in this book result from focusing on reduction of this total time. Thus in the preceding example, the lead time should not be zero, but the time from when the raw material arrived at the cell to the time when the finished part was eventually shipped to the next operation. This is also called *flow time* or *cycle time* by some companies.

However, even with this clarification, there are some complications in ascertaining lead time. Specifically, for each job, when does the clock start ticking and when does it stop? It is important to identify these checkpoints early in the project to avoid later pitfalls. Using two examples I'll illustrate these issues.

Example One: Office Lead Time

One company attempted to measure its lead time in order processing for a line of products, for which it had established a new Q-ROC. Data was recorded via computerized transactions at each step of the process. After they had collected several weeks of data I was asked to analyze the data to help establish a base lead time. Upon examining the data I became a little suspicious and spent some time talking to people at the front end of the process. It turned out that the first computer transaction only occurred when someone in the Q-ROC logged the order onto the computer system. However, a purchase order might arrive by mail or fax and sit in the company's mail room, the fax area, or the in tray of the Q-ROC for up to three days. This literally doubled the premanufacturing lead time for some orders. After discussions with the Q-ROC team, we agreed that it would be best to measure the lead time from the time the order first arrived at the mail room or at the fax. Even though the team knew that these processes were not under its control, it realized that from the customer's point of view, time was passing regardless of where the order was sitting. The team decided that in order to best serve the customer, it wanted to be responsible for the lead time as soon as the company received the written order by mail or fax.

As a result of this decision, several changes occurred in the company and in the behavior of the team. First, the team convinced management that mail room personnel should time-stamp each envelope as soon as it arrived. Second, instead of waiting for the mail room to deliver the mail, the delivery process was modified. The mail room sorted the mail into boxes for each major department. This had been done before, but now these boxes were made into pigeon holes accessible from both inside the mail room and outside. This enabled team members to visit the mail room several times a day and pick up their mail, while other departments still had the option of waiting for their mail to be delivered. Interestingly, addition of the time-stamping process, plus watching the team members pick up their mail regularly, also motivated the mail room workers to minimize delays in the process.

As regards the faxed orders, again two changes occurred. One, the team convinced management to give it a dedicated phone line and fax machine, and it publicized this phone number to its customers and related sales staff. Two, for customers that still used the company's main fax number, the Q-ROC ensured that several times a day someone from the team went to the main fax area. As a result of all these changes they reduced lead time by one to three days for most orders.

Example Two: Manufacturing Lead Time

I was asked to review computer records of lead time for some manufacturing cells at a consumer products company. Curiously, the cell lead time seemed short enough, and yet customer service people were complaining that lead time was actually much longer and resulting in unhappy customers. As I investigated the issue, analogous to the problem with the Q-ROC's lead time measure, I found that the measurement started when the first manufacturing operation was logged onto the computer. This was because the company was using a shopfloor control system that only "knew" a job had arrived when it was logged at the first operation. After that the system accurately tracked the job's status through the other shop operations. However, a job might be released from the office to the shop floor and sit for days—more than two weeks in some cases—before beginning its first operation. We only found this out by going back to the paper records, which had handwritten dates from the office, and comparing those with the first-logged computer dates. The discrepancy was quite large in some cases. Worse, in looking at the difference in dates between the last office operation and the first manufacturing operation, we could not identify why there was a long delay for some jobs. Was material not available? Was the first machine backed up with too much work? Was a machine broken? Or did the job just sit because the cell did not know it was there? Whatever the reason, the data gave us no insight.

Guidelines for Lead Time Measurement

The preceding examples crystallize some general rules that I have developed for measuring lead time.

- *Identify unambiguous start and end points for the lead time.* In particular, there should be no gaps in the measurement, particularly at points where there is a hand-off from one part of the company to another, or between the company and the outside. Also, be sure that the start and end points do a good job of measuring the total time for material to flow through the system, as discussed earlier in this section.
- *Ensure that responsibility for each segment of lead time is clearly identified.* As far as possible, give the responsibility to the Q-ROC or manufacturing cell that is handling those jobs. Even if it appears, at first glance, that this lead time is outside their control, gauge whether giving them responsibility for that segment might result in some improvements—you saw two examples of such improvements with the Q-ROC above. On the other hand, if the team is made responsible for a lead time segment over which it has little hope of influence—such as parts from a supplier whose delivery is managed by the materials department—then it could lead to frustration and even apathy regarding the lead time measure.
- *Use a "status indicator" in the recording template.* This is a data field that indicates what is happening with a job at any time. The purpose of this data item is to clarify the ownership of lead time for every moment of a job's progress through the company. You can record this manually on a sheet that accompanies each job, or better still, make it part of a computer transaction. I'll illustrate this with a manufacturing job involving housings, starting with the lead time measurement from the moment of releasing the job from the office to manufacturing. This occurs via electronic notification to the cell that a job is ready. I'll assume that all job data, such as blueprints and routing, are available on-line and no folder is needed. Table 16-2 shows the progress of the job and the way it is logged into the system, along with status indicators at each stage. The explanation in the table provides additional insight into use of this method.

Two important issues are brought out by the table. The first is that ownership of the lead time must be clear at each moment. The advantage of this is that as lead time statistics are analyzed in the future, the company can ascertain the contribution from each team or department to overall lead time. Knowing this, each of the support functions, such as materials, will also be diligent about serving the customer (in this case the cell). Also, since times such as when the team is waiting for a casting or input from engineering are not made part of the team's lead time, team members will not be frustrated that they are being measured on actions out of their control. The second is that even within the cell, it is useful to have some breakdown of lead time into components such as queue and operating. Additional status indicators such as "machine down" should also be added. These details will help the team's QRM detectives in their quest for continuous improvement. The issue of how much detail to include is to be discussed by the team. On the one hand, operators might want to measure details such as setup time separately; on the other, they might prefer some autonomy on how they do a job, and only want to be responsible for total time. Such an issue should be resolved by the team.

- *Think through the units and weighting to be used.* Consider a manufacturing cell that deals with orders ranging in size from one piece to 50 pieces. It is to be expected that the 50-piece order takes longer to complete. But should you compute the average lead time by weighting each order equally, or should it be weighted by the number of pieces? One could say that each order represents a customer, and each customer is important. Or one could say that large orders represent more business, or will be sent to more customers, and should receive greater weight. It could even be argued that you should weight orders by their cost or profitability—although this would lead right back to the debate on whether these numbers were fairly computed in the first place! Nevertheless, the question still remains, should the lead time for each order be counted equally, or should it be weighted in some way to account for the size or complexity of the job?

  The first point to note about this question is that simply raising the issue ahead of time is important, because it makes everyone think about the goals of the team. Second, the way to answer it is addressed by the next item in this list. Third, don't be shy to reexamine this issue after the cell has been in operation for a

Table 16-2. Example of the Status Indicator Use

Time	Status Indicator	Employee Coments	Explanation
Oct. 4 9:30 a.m.	Released	Notified housings cell via electronic message.	The last step in the office is to notify the appropriate cell, and change the status indicator to "Released" when this step is completed. As of 9:30 a.m., the cell now "owns" this job.
Oct. 4 11:14 a.m.	Review		The first chance the team leader gets to review the newly released job is at 11:14 a.m. As soon as she opens the record to look at the job, the status changes to "Review."
Oct. 4 11:27 a.m.	Materials	Castings not here. Left voicemail and electronic message.	Now the materials dept. "owns" the lead time until the castings are delivered.
Oct. 5 1:42 p.m.	Ready	Castings delivered to cell. Team leader notified verbally.	A material-handling operator delivers the castings and changes status when delivery is complete.
Oct. 5 2:28 p.m.	Review		The team leader again reviews the job, and assigns it to lathe #1066.
Oct. 5 2:33 p.m.	Queue 1066		However, since 1066 is busy, the leader puts the job in its queue.
Oct. 6 8:01 a.m.	Operating 1066		An operator starts the job on 1066. (The team has decided not to measure details such as setup and run, just the start and end of jobs.)
Oct. 6 11:52 a.m.	Queue 2001	Moved job to 2001, m/c is busy.	2001 is a CNC mill. As soon as 1066 is complete, the job becomes part of the mill's queue.
Oct. 7 7:02 a.m.	Operating 2001		An operator begins the job on the mill at 7:02 a.m.
Oct. 7 7:21 a.m.	Engineering	Question about tolerance. Left voicemail.	The operator has a question about the engineering drawing. Now engineering "owns" the lead time.

(continued)

Table 16-2. Example of the Status Indicator Use *(continued)*

Time	Status Indicator	Employee Coments	Explanation
Oct. 7 7:44 a.m.	Operating 2001		The question is answered, and the job continues.
Oct. 7 10:17 a.m.	Assembly	Delivered to assy.	A cell operator delivers the shafts to the assembly dept.

Notes: The "Date and Time" entry is logged automatically by the computer system when each entry is made. The "Status Indicator" value is set by the person entering data, by choosing from a list of acceptable values. The "Comments" field is for this person to add some information that may be useful when reviewing performance, or for process improvements. (In some cases this entry is simply left blank if there is nothing to be added.) The "Explanation" column is not part of the computer record. It has been added here to explain to the reader what is happening at each stage.

while. The team and management should meet and see whether the measure of lead time being used is helping the company head in the right direction in terms of QRM.

- *Use organizational goals to resolve issues of how and what to measure.* One company had an ongoing debate about partially completed orders. If a customer ordered 100 parts, and 95 were completed but 5 needed to be reworked, the company would ship the 95, and then ship the next 5 later, or with another order going to the customer. The rework often involved several days of time, so the company did not want to delay the rest of the batch. Let's say the 95 parts took 4 days and the rework took another 3 days, so that the remaining 5 parts had a 7-day lead time. The cell team wanted the lead time for the order to be measured as follows:

  Lead Time = $4 \times 0.95 + 7 \times 0.05 = 4.15$ days

  Customer service, on the other hand, wanted to measure the lead time based on when the order was finally completed. In this case, that would mean a lead time of 7 days. The difference between 4.15 and 7 was quite substantial and was giving rise to considerable tension between the two groups.

  To resolve this issue, management went back to the original goals of their QRM program. One goal was to remain a preferred supplier of their customer. Upon checking with their customer's materials department, they found that the customer did not record a shipment as received until it was complete. So this company's delivery performance, as recorded by the customer, was based on receipt of the final 5 parts. This left no ambiguity as to which measure the company should use.

Interestingly enough, once it was clear to the cell that their lead time was dependent on getting the full batch of parts completed, two things happened. One, the cell worked hard on improving quality and reducing the occurrence of rework, and two, the team put procedures in place to enable them to perform the rework very quickly, if it was needed. Hence our approach to this issue assisted both the company in pursuing its goals and the cell team in obtaining relevant performance improvements. You should use a similar approach of examining the organizational goals to resolve the issue of weighting of lead times, which was brought up in the preceding point.

Leveling the Playing Field

An issue with the QRM Number that you need to tackle periodically is called *renormalization*. Some teams that have been in place for a while and have been successful may have QRM numbers that are very high, easily 300 or more. Other teams that are relatively new or have not had as much success could have much lower numbers, around 150 or so. This might be discouraging to the point where these other teams could give up on improvements. This problem can be resolved in two ways.

First, you should set up reward systems so that the primary motivator for a team is improving *its own* QRM Number relative to *its* previous values. So the actual value of the QRM Number is less of a concern, as long as the team is working to make that value go up. Second, management should periodically renormalize all the QRM numbers. This means that at preset intervals, such as once a year, you should reset all QRM numbers to 100. You can do this by making the period with the best past performance the new base period, and the lead time during that period becomes the new base lead time. This will help to periodically level the playing field and give all teams a new and equal chance of competing with others.

Additional Attributes of Performance Measures for QRM

Although lead time reduction measured via the QRM Number is the primary measure used under QRM, secondary measures should also be used to help teams and management understand the progress of the enterprise. There are some general guidelines to be followed in the use of such measures:[8]

- *Supplement the primary measure by a small number of secondary measures.* The key word here is "small." Too many measures create confusion and deliver a mixed message. Many leading companies have a corporate policy of focusing on no more than five measures.
- *Ensure that the measures are directly relevant to your manufacturing strategy.* How does one choose these secondary measures? Let the major elements in your manufacturing strategy drive the choice.[9] For example, if your strategy involves achieving certain quality standards in order to capture a given market segment, then make one of the measures relevant to quality improvement. If cross-training is critical to your strategy, then you could measure the number of skills each person has. Since this measure is aimed at the process needed, not the specific results, such a measure is called a *process-oriented measure*, as opposed to traditional measures that tend to be "results-oriented." This is where one sees the difference between cost-based measures and the ones advocated here. At the minimum, cost-based measures have little connection to manufacturing strategy, and at worst, they work against the tactics needed to implement the strategy.

  In a similar way, one also has to be alert when implementing a lead time measure, because it doesn't necessarily result in the best manufacturing strategy, unless the measure is carefully thought through. At one factory that made cooking equipment for the fast-food industry, management was under pressure from its owners to reduce lead times to under eight days in order to retain market share. The factory had a fabrication area that made components and an assembly area to put together the final product. The total lead time, including fabrication of all components and assembly, was six weeks for a typical product. However, the assembly lead time was under five days. While there were too many end items to stock in finished goods, the number of unique fabricated components was smaller. Hence the factory manager decided to build up a large inventory of fabricated components. In this way, he could assemble to order and also achieve the desired lead time. However, the factory ended up with over a million dollars of component inventory, whose carrying cost ate into most of the profits of the business. In addition, there was always the problem that one component was not in stock when needed, which led to a cycle of expediting in the fabrication area. And finally, the long lead times in fabrication meant that the Response Time Spiral was impacting the operations in a big way, with its loss of quality and its other effects leading to high overhead costs.

What this anecdote illustrates is that a naive policy that measures lead time is just as perilous as a traditional cost-based measure. The solution in the above case would have been to institute three measures: one, lead time to respond to a customer order; two, lead time of material through fabrication and assembly; and three, the value of component inventory. Indeed, knowing that the best manufacturing strategy would be to have the ability to fabricate and assemble only to order, and that the major obstacle to this is fabrication lead time, reducing fabrication lead time should be made the primary measure (specifically, the QRM Number would work well for this), and the other two ought to be secondary measures.

• *Keep the measures simple.* Don't go for complicated measures involving a composite of many different individual measurements. These have two drawbacks. First, they obscure an individual person's contribution, thereby removing the incentive for that person to try harder. Second, they are subject to manipulation. Employees may attempt to maximize the index by focusing on portions of it, to the detriment of the rest of the operation. I witnessed an example of this a few years ago when I studied the operations of a capital equipment manufacturer. One area of the company had a very long lead time, and on closer examination I found that it had an unusual amount of machine down time. The workers in that area were measured on efficiency, and on a hunch, I asked what happened to the measure when machines were down. It turned out that many years before, the union had negotiated a clause stating that workers were credited a certain proportion of productive work hours whenever their machine was broken. The aim was not to penalize operators who couldn't make parts through no fault of their own. However, over the years as batch sizes had become smaller, workers in this area found that the amount of credit they received when a machine was broken was more than the average they could achieve with the setup and run time of a typical batch. It was no surprise, then, that for one reason or another, machines were down in this area! This example is a classic instance of the oft-cited dictum: "You get what you measure."

- *Display the measures clearly.* Some companies are reluctant to share performance details with workers, for fear that workers may want a greater piece of the pie if the company does well. The QRM approach involves complete openness with employees about the company's strategy and direction—after all, you are going to rely on these same employees and tap into their skills and suggestions to help you get there. As part of this openness, all areas of the company should be willing to share their key performance measures with others. This is true even when an area is performing poorly: Such an occasion should be an opportunity for others to rally around this area and assist them in rooting out problems to help the company improve as a whole. "If you want a goal to be met, share the progress with those that can make it happen," says Jim Riihl, manager of production planning at Beloit Corporation, who worked together with me for several years to train Beloit's employees on implementing QRM. In order to effectively display measures, company management should use charts, graphs, and other visual aids to communicate the measures in pictorial form, not just in reports with numbers. Also, you should prominently display these charts and graphs for all to see, in a central location, in each area, or both.
- *Supplement the QRM Number and other summary numbers with detailed and direct measures.* The QRM Number is a summary number, involving an average of a number of observations. Such a summary statistic can hide potential problems, such as occasional very long lead times, or when lead times fall into two main groups, or whether there is a trend during the observation period. So it also pays to plot the raw lead times on a chart (one point for each order, say). This helps the team focus on issues such as, why do we occasionally have a very long lead time? Also, if we consider that in the QRM Number formula, the base lead time and the "100" are essentially constants, then the QRM Number involves the reciprocal of the average lead time. Even though this is a simple calculation, there are benefits to showing some direct observations so people can see the actual lead times before they are aggregated and turned into a reciprocal. The same is true with other performance measures. In addition to the summary statistics for each period, show the individual detailed observations so that employees and managers can derive any insights possible and take corrective action where needed.

THE TEAM-BASED ORGANIZATION

You've seen that the core building block of the organization structure for QRM and the backbone of all QRM efforts involves cells, which in turn derive their benefits from teamwork. So QRM performance measures must be compatible with and supportive of this cellular team structure.

The first implication is that measures must be much more people-oriented than they have traditionally been. What is interesting is that even as far back as 1956, this was well recognized. More surprising (given all that I've said about accounting systems), the following statement came from a text on accounting: "People, not figures, get things done." The author goes on to say:

> An obvious fact about business organizations is that they consist of human beings. Anything that the business accomplishes is the result of the action of these people. Figures can assist the people in the organization in various ways, but the figures themselves are literally nothing but marks on pieces of paper; by themselves they accomplish nothing. It is surprising how often this point is overlooked. An accounting system may be beautifully designed and carefully operated, but this system is of no use to management unless it results in action by human beings.[10]

Or "unless it results in action *relevant* to the current manufacturing strategy," we might add as a friendly amendment to the last sentence.

QRM strategy refines this view of organizations versus people further—the organization is constructed of cells, and people come together in cell-based teams. However, transitioning entire companies to a team-based cellular structure is not going to be easy for U.S. companies, given several hundred years of American history, which values the independent spirit:

> American culture has long admired individual achievement. Despite paying lip service to teamwork and team players, our heroes are the lone cowboy, the hardy pioneer, the rugged individualist, and the self-made man. Surely football is a team sport; yet the highest honors (the Heisman Trophy, the Butkis Award, the Most Valuable Player Award, etc.) go to individual players rather than teams. Similarly,

reward systems in organizations are based on individual performance, not on contributions to a team effort.[11]

And:

...U.S. workers, despite their own self-image as team players, have little cultural experience to fall back on. Many have grown up in a society in which the Lone Ranger is an icon of popular culture. Individual competition has been encouraged for everything from school grades to job promotions.[12]

To make matters more difficult, even if workers were fully convinced, many managers are still not on board, despite all the publicized results on successes achieved via teams. As one instance of this, in the spring of 1996, a few weeks after we concluded a study for a Wisconsin division of an international company, a key recommendation being that they adopt a Q-ROC for a given line of products, the vice president wrote to us saying, "Many of your recommendations were going to be implemented; however, due to a change in upper management, all of this work was abandoned. Our new manager doesn't believe in a permanent team-oriented work environment."

Since teams, through their implementation in cells, are a vital element of a successful QRM effort, and employees and managers continue to have difficulty in accepting teams, it is worth devoting time to understanding more about what it takes to successfully implement teams.

Why Teams Fail

One way to enhance the chance of success in building a team-based organization is to have a good knowledge of why teams fail. Since teamwork is all the rage today, there has been an explosion in the literature on teamwork. In fact, there are several publications—from short articles to entire books—on the very topic of why teams fail.[13] This section summarizes the insights from such publications, as well as my own experiences in the context of QRM.

Among two primary reasons cited for the failure of teams are unclear goals and changing objectives. Fortunately, for QRM these two reasons are easy to combat. The clarity of purpose of QRM, in terms of its focus on lead time reduction—indeed, the relentless emphasis on lead time reduction that QRM calls for—can eliminate these two problems. Related to these are two other major reasons: lack of accountability and lack of role clarity. Again, by making lead time reduction the focus of the team and through measures such as the QRM Number, team accountability and the roles of team members both crystallize.

Other reasons for failure of teams include lack of management support and low priority of the team in the organization. These issues go deeper, because to tackle the first, one needs to change the mind-set of management. The issue of priority goes to the heart of the mind-set issue, because it really asks, is management truly committed to the lead time reduction it has stated? If it is, the team will not have a problem with being a priority. If, on the other hand, management is simply paying lip service to lead time reduction and is really focused on more traditional measures, then indeed the team will find itself a low priority item when requesting support or resources. This can be demoralizing, leaving employees disillusioned and eventually apathetic, thus completely destroying the reason for creating cells in the first place.

Another potential problem with teams occurs when companies institute teams but stick with existing individual compensation systems. This lack of team-based pay works counter to teamwork. Without team-based pay, the team members get a confusing message. While desiring the members to work together, the organization still rewards them based on individual accomplishment.

Teams also fail when their members don't incorporate a sufficient skill set or if lines of authority are fuzzy. These issues are addressed in the next section.

A major cause of failure in team implementation is fear. This emotion pervades the whole organization when the term *team* is used. Managers and supervisors fear that they will lose their jobs. In fact, you must not use team implementation as an excuse to fire supervisors and middle managers. People will begin equating team formation with downsizing and job elimination, and this will create resistance at all levels. In particular, experts note that teams fail 75 percent of the time if managers and supervisors are dead set against the idea.[14] Remember, as I've said and conveyed throughout this book, QRM is about growth, not downsizing.

Front-line employees fear the move to teams because of their ingrained mistrust of management's motives—maybe this is a way to get more work for the same wages. Likewise, unions fear the implementation of teams, seeing it as management's way to undermine their influence. Support departments and specialists fear that their entire function will disappear as the company disperses knowledge into teams via cross-training. In addition, these people worry that if they are assigned to a product-oriented team they will no longer have peers who understand their profession and can help them stay current. More important, who will be able to appreciate their technical progress and decide their pay raises? Who will be their technical leader, providing career guidance and mentoring in the field of their specialization? And what hopes do they have of a career path if their specialization has disappeared in the organization?

These fears and the other issues in this section will now be addressed with specific recommendations for QRM teams.

GUIDELINES FOR SUCCESSFUL QRM TEAMS

Let us emphasize that I am not advocating the type of team that is simply a group of people getting together periodically. That would be more of a committee than a QRM team. To review an issue raised in Chapter 11, many companies make the mistake of creating teams whose members remain located in their functional departments and report to their traditional functional managers. Such "teams" have to schedule their meetings in company conference rooms, and they do not approach the full potential of a cellular team that delivers products to a focused target market segment (FTMS). Indeed, let us revisit the misconception about teams:

> **Misconception: We can implement cross-functional teams while keeping our organizational structure intact.**

It is worth repeating—*a QRM team must be collocated, and it will cut across functional boundaries and change reporting structures.* Before proceeding with the issues related to implementing teams, readers may wish to reinforce their understanding of what constitutes a QRM team by revisiting the discussion on shopfloor cells in Chapters 4 and 5 for a definition of QRM teams, and looking at Chapters 11 and 12 for the issues involved in implementing Q-ROCs. This section discusses 10 guidelines for implementing effective QRM teams.

Creating Clear Focus and Goals

Studies on teams have repeatedly found that to ensure success, it is imperative that teams begin with clearly understood goals. Even better is when the focus is not very broad at all: "Many companies are finding that teams work best when their focus is narrow and goals clear."[15] Fortunately, QRM provides teams with a clear focus—investigate the causes of long lead times—and an unambiguous goal—reduce the lead time. Further, use of the QRM Number as the primary performance measure reinforces the clarity of the team's focus toward this goal.

Another rule of thumb that contributes to the success of teams is to give them challenging goals, often called "stretch goals." The aim is to make the goals such that you cannot achieve them via tweaking current methods, forcing teams to "think outside the box." At the same time, the goals should not be so far out that the team perceives them as completely impossible to attain. In fact, you should perform goal-setting in joint sessions between management and the team, to ensure buy-in from team members. Again, in the context of QRM, there is substantial experience that enables us to do this goal-setting effectively: goals of 50-75 percent lead time reduction are normally in the domain of stretch-yet-achievable.

Educating All Employees and Managers on the Need

Everyone in the organization has to begin the process of understanding why the organization must change, why the traditional methods are no longer good enough, and why there needs to be radical change. The only way to do this is through repeated training on the latest developments and exposure to what is happening at other companies. Although, despite all this education, you can expect resistance to such changes, sometimes there are refreshing surprises. In 1994, I was in the midst of a discussion with union members at an Illinois factory of a metalworking company, explaining to them the need for cross-training to improve the quality of a particular set of operations. How would their union leaders react to this suggestion, I asked them. I was not fully prepared for the depth of the answer a worker gave: "We have to change the way we work. We've been doing things the same way for over 50 years. We can't go on like this! How can we expect to compete with companies all over the world if we aren't willing to try new ideas? We will take this suggestion to our leaders."

It is critical to get not just the employees, but also managers and supervisors lined up behind the changes. This includes not only training to change their mind-set, but also acknowledging and addressing their fears. They should understand that a successful QRM program means there will be enough work for everyone. At the same time they need to understand that most everyone's roles and responsibilities will change. For example, many supervisors will turn into coaches with plenty to keep them busy. Managers may also end up with more rewarding jobs, being team trainers, quality trainers, facilitators, or champions, instead of being under pressure to always solve problems and make decisions (and to be right all the time). If managers can see such successes at other firms, it will help to bring them on board early in the program. Invest in visits by managers and supervisors to companies that have accomplished these changes. Also, be sure to involve them in planning the changes. Finally, make sure the people in the human resources department fully understand what is happening, and then get them to hold meetings and answer questions to allay fears at all levels.[16]

It is helpful to review the strategy used by U.S. Gauge, a division of AMETEK Inc., in implementing the concepts of work teams in its Barlow, FL plant. This division has been highly successful in implementing teams, although the success did not come easily at first:

> Our front-line supervisors had a great deal of difficulty relinquishing control, and this was compounded by management's insistence on holding the supervisors responsible for the actions of the teams ... This resulted in a power struggle between the supervisors and the teams ... We finally came to the conclusion that, in order to make this work, we had to re-focus our attention on the front-line supervisors

who were extremely frustrated in their new roles as well as confused as to what the future held for them. Management spent a great deal of time working with the supervisors assuring them they would still have a vital role to play in the success of the operations after Self-Directed Teams were implemented. Upper management and supervisors worked together to develop plans which would transition each supervisor into a new supporting role of his/her choice, e.g., training, process improvement, Total Quality Management, management associate, etc. It was only after this conflict was resolved that we noticed a willingness on the supervisors' part to play an active role in the implementation.[17]

Selecting a Team—Use Volunteers

Chapter 5 gives an example of the failure of a cell because the company decided who the cell members would be. In our experience, the best teams are formed when you use a selection process that involves people applying or volunteering for membership in the team. Companies should post job descriptions for the proposed members of the team and then engage in a selection process. However, there is a key step that precedes even the job postings, as demonstrated from an anecdote. When Ingersoll Cutting Tools Company was forming its Q-ROCs, the initial job postings got a very low response rate, because the company did not do a good job of communicating its "Fast Track Team" program to the rest of the organization. Hence management must start with a foundation of education for the whole company, explaining the need for the new structure, as well as the career opportunity for people moving into these teams. The idea is to get employees to buy in to the QRM vision, and then have them implement that vision. This education for the company is only the beginning of a great deal of training required along the QRM journey.

Training, Training, and More Training

In addition to providing general training for the whole organization regarding the journey to QRM, companies must invest in specific training for each team. Do not underestimate the amount of time your organization will have to invest in training. Successful companies note that employees spend up to 25 percent of their time in activities such as cross-training and learning new skills, short-term problem-solving meetings, and other team meetings for longer term improvement. As a QRM team is being put in place, team members will need training in a number of areas:

- *Team dynamics.* How to work together as a team, and how each person can be an effective team member. Often this involves outdoor physical activities where team members have to learn to trust each other in slightly dangerous or challenging situations. Many managers pooh-pooh this "touchy-feely" topic; yet every company I know that has invested in team dynamics training has stated that it was critical in getting the team off to a good start. It is not necessary, though, that the team be in a constant state of chumminess; some element of conflict is healthy for a team. There should be degrees of both friendship and combat along the way to achieving the best teamwork.[18]

- *Interpersonal and communication skills.* How to be a good listener, how to make an effective presentation, and how to interact with others in constructive ways. In the traditional organization, where many employees could do their part for the company without much interaction with others, these skills were not part of the job description. "In 1980 an immigrant without much English could just throw the right switch on the machine. Now he has to join 'team' meetings and read memos."[19] Employees will not magically change overnight once the team is formed; the company has to invest in training them in these skills.

- *Meeting management.* How to run meetings, and how to participate effectively. Again, employees may never have had to do this. Without specific training, a lot of time could be wasted in meetings where the team was just spinning its wheels. Basic skills like using an agenda, keeping a meeting focused, and ending meetings with action items need to be taught.[20]

- *Team leader/facilitator skills.* The role of the leader or facilitator is discussed below. However, everyone in the team should have some basic training in this area, under the assumption that this role will rotate around the team.

- *Skills enabling team improvement.* These involve problem-solving skills, kaizen methods, project management, and other tools to help the team find root causes of problems, brainstorm about solutions, and then manage their implementation and follow-up.

- *Multifunctional training (cross-training).* This involves both technical training to gain the skills needed for another job, as well as empathic or experiential training, which means actually doing that job for a while and experiencing it. Here are two specific examples.

  A U.S. design expert noted from his observations in Japan that deliberate efforts are made to broaden rather than narrow engineers: "At a camera company, mechanical design typically falls into four areas: film transport, mirror box, autofocus lens, and shutter. Even though shutters, for example, are very difficult to design, there are no shutter gurus. Instead, each designer rotates from area to area until he has worked on all of the elements."[21]

  I described at length the success that Ingersoll Cutting Tool Company had with implementation of a Q-ROC. In recruiting people for this Q-ROC, Ingersoll's management made cross-functional training part of the job description for potential Q-ROC members in two ways: Employees had to be willing to learn new skills, and they had to be willing to engage in training coworkers and sharing knowledge with others. Further, when they posted the jobs, looking for volunteers, they also posted a formal training matrix for each job description showing what skills would be needed. Finally, each person who joined the team had to fill out on this original matrix the skills that they already had, and which ones they needed to learn via training. This training was then done on the job, primarily by others in the team, and tended to be informal.

- *Miscellaneous skills.* This catch-all category is meant to indicate that there will always be additional skills that each particular team will need. For instance, a team may take on administrative jobs involving record-keeping that was typically handled by managers. Or it may need customer-service training; a team always has customers, whether internal or external, but in some cases the interaction may require special training. Some team members may need to be trained in computer skills to use the tools that the team relies on.

Determining Team Size and Structure

Effective teams seldom exceed 15 members. In my own experience, it often pays to set the limit even smaller, around 10 or even 7 people. If serving an FTMS requires more than 15 people, consider splitting the subsegment into two finer subsegments, or breaking the delivery process into a "front half" and "back half," with a team for each.

Regarding the leadership role in a team, there are several models, including those that suggest not having a leader, only a facilitator. My own preference, to be most consistent with QRM principles, is to create a position with the title of Team Leader. This person has three key roles.

1. *Acts as an interface with the rest of the organization.* Other teams and departments find it useful to have one person to call with questions, problems, complaints, etc. In this context, it is best for the person to have the title of leader, so that the other person feels they are talking to someone who can make things happen.

2. *Coordinates and facilitates team meetings and interactions.* It is important that the team leader not give directions at such events, but have the same decision-making authority as any other member. However, their role is to set meeting times, create agendas, move the meeting along, and compile action items at the end. It pays to train the team leader in both how to facilitate and how not to grab control.

3. *Advocates for the team, to management or to other parts of the organization.* This includes lobbying for resources that the team needs, or defending it against efforts to undermine it.

Another important part of this model is how the leader is chosen. Some experts suggest a voting process. My preference is to go with the rule that the position will rotate among team members, with a time period ranging from three months to a year, depending on the experience of the team with the startup time it takes for a leader to gel. The reason for my choice on this issue is that it creates a truly equal team; the leader is not a superior being, just a regular team member who is filling that role for this period. Also, this forces everyone in the team to learn the skills needed for this position and to appreciate what it takes to be in that position. In exceptional cases and under extenuating circumstances a team member can "pass" on their turn to be the leader, but this situation should be discussed in a team meeting and accepted by the team ahead of time. In the long term this means the team supports the leader better and the leader serves the team better.

Supporting Empowerment

Everyone talks about empowered teams these days, but what does empowerment really mean? I can best answer this in the context of cellular teams for QRM.

You must provide the team with some resources and authority to utilize those resources as it sees fit, e.g., a budget for training or tooling. One important resource to have available for the team is an experienced facilitator. This person is not part of the team but is available as required, to participate in occasional team meetings, to provide guidance, and to be available for consultation when team progress is stalled or a barrier arises.

In addition to providing the team with authority, management must clarify the boundaries of the team's authority; within these limits, the team should have full authority to make decisions. Examples of boundaries to be set are: To what extent can the team sign off on quotes or proposals to customers? Can it accept orders from customers? Can it decide vacation time? Can it quote delivery dates? And so on.

Empowerment also means encouraging risk-taking. If the team takes an initiative that fails, and management comes down hard on the team, this will be a sure-fire way to kill future improvements. Instead, management should convey the attitude: "We're happy that you tried something new. Too bad it didn't work. Why don't you try to understand why it didn't work and learn from this exercise, so that you can decide what your next initiative will be."

A good way to support the team's empowerment is to make sure it has a champion outside the team. This is someone with significant authority who can make things happen. For instance, this person can be an advocate for the team when organizational policy changes are required or barriers need to be dismantled.

When you create a team it is a good practice to state right off the bat that there will be periodic reviews of how the team is working together and how it is progressing toward its goals. By institutionalizing such reviews, they are not seen as signals that management feels something is wrong; rather, they become part of the regular team process. The expert facilitator can be helpful in conducting these reviews. Initially, when the team is starting up, these reviews could be as frequent as a one-hour meeting every week. Later, they could become much less frequent, say, a two-hour meeting every three months.

Providing the team with adequate support also means the occasional support it will request from existing specialized departments, such as engineering, materials, or purchasing, to get expert help on an issue. This is not trivial; these departments may be resentful of the team effort in general because they are losing personnel to teams, or because they feel the teams are taking over some of their jobs. Management must tackle this issue ahead of time in two ways, similar to the issue with supervisors. First, the support department personnel must understand why it is imperative for the company to change its mode of operation. Second, they should understand their long-term role and career paths—this issue is discussed separately below. In addition, during the transition period to the new organization structure, management should give support departments a clear signal that it considers the success of the newly formed teams to be critical, and that it expects the support departments to contribute in positive ways to this new direction.

Changing the Traditional Reporting Structure

Since it is the goal of QRM to eventually dismantle the functional structure of the organization, the reporting systems for the teams must be outside the traditional departments. Having a manufacturing cell under the milling department manager or the grinding department manager would be counterproductive. If you examine the QRM organization that I'm creating, you'll see that it is oriented toward market segments. This means cellular teams should ideally report up through an organization structure that reflects these segments. In the absence of such a market-segmented organization—during the transition period when the organization is still being restructured—cells should report directly to "neutral" (i.e., nondepartmental) higher-level managers such as the director or VP of operations.

Involving the Union

In cases where companies have labor unions, management can make several efforts to involve them in the progress toward the team-based organization. But first, it is worth making two observations. One, the presence of unions is not automatically an obstacle to success. Recall the observation that manufacturing cells have typically been more successful in union shops rather than nonunion shops. Two, in many cases unions have taken the initiative in forming teams, and approached management with the idea. I'll give more examples of this at the close of this chapter. To get a union lined up behind the team effort, Hitchcock and Willard offer four specific suggestions:[22]

1. Make the union a partner in the change process by involving the union early in the thinking. Include union representatives in visits to other companies and in the mind-set training.
2. Show unions that the team organization helps to further, not undermine, the very values for which unions were originally created.
3. Allow the union to share responsibility for implementing the changes, such as being involved in detailed decisions about the implementation and conducting some of the worker training sessions.
4. Put effort into helping the union redefine its own role. This isn't only something a company must do with managers and employees, it is just as crucial to ensure the union evolves in its role, or there may be backtracking later in the process.

Providing a Vision for the New Organizational Structure

As all these pieces come together, what does the new organization structure for QRM look like? It has three main characteristics. First, in place of functional departments there are FTMS-oriented teams.

Second, the large support department functions become dispersed within the teams; however, core groups of experts still exist in support functions. These experts are the repository of highly specialized knowledge in their area. Their key role is to support teams that occasionally need skills beyond what their members have; secondary roles include mentoring the specialists who are now actively on teams and providing training to team members who wish to learn some of their skills.

To elaborate on this, what you are creating is a company where your teams become experts on the whole set of processes required to serve the customer in their FTMS—as opposed to departments that are experts on one specialized task. However, the specialized expertise is

not completely eliminated; it is retained in the small support departments. The flow of orders, however, is not zig-zag through the functional departments, it is direct through one or two teams, and the
support departments are true to their name—they support, without
impeding the flow (see Figures 16-5 and 16-6).

Third, the organization realizes it will be in a constant state of flux
as markets change, customer demands shift, and you need to create or
disband cells to best serve the new segments. Unlike the apprentice who
joins the engineering department as a drafting person, works his way
up to being an expert designer, then a chief engineer, and maybe even
head of the engineering department—a career path of more than 20
years in one department—unlike this path, people in the QRM organization can expect to be in many different teams over their career span. As
an engineer (and team member) at Hewlett-Packard said to me when I
asked him about his future: "I don't know what I'll be doing two years
from now, but I do know that there will be a job for me, and I do know
that it will be challenging and rewarding."

What is the career path of people such as the apprentice above, in
the new organization? It involves quickly learning the key ways that
their specialized knowledge (e.g., design) can assist in serving the particular FTMS of their team. It involves demonstrating that the person
can move from one team to the next, and quickly adapt to the new circumstances. Instead of pay raises being decided by the head of the specialist department, team results and team members now decide them
(more on this below). And so it should be, if this individual is to contribute effectively to the organization's goals. The career path may
evolve in other ways too: It may involve becoming a trainer or facilitator. This designer may end up being part of the small central functional department that supports designers in many teams. The
challenges and rewards of this work environment are just as many as
in the previous one. The key difference is that career growth is not so
much upward as forward.

And what of the employees who do not want to cross-train or take
on responsibility? A manager quoted some of his employees during
their transition to cellular teams: "I don't want to be multiskilled—don't
ask me to learn new skills"; "Don't ask me to make decisions; tell me
what to do." It's not that these are poor workers, he explained to me. In

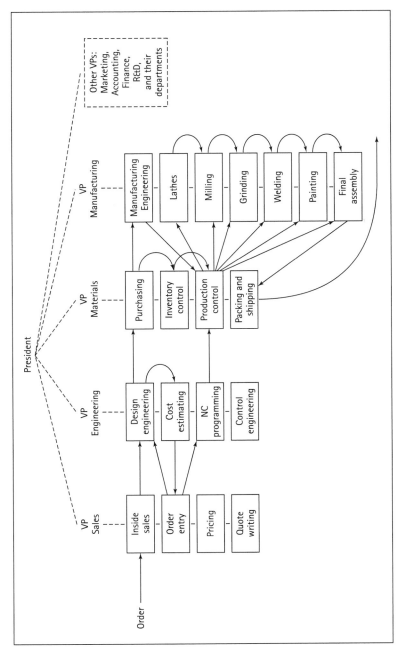

Figure 16–5. The Traditional Functional Organization Structure

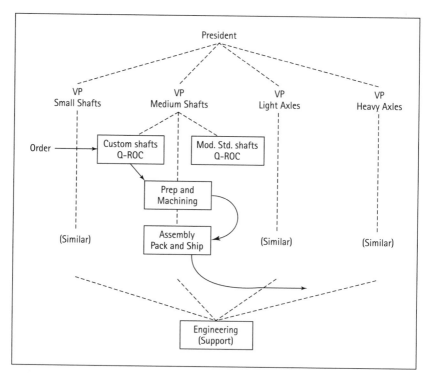

Figure 16-6. The Focused Market-Oriented QRM Organization

many cases they are excellent employees, but they just don't want to take on any more. The bittersweet realization is that in the QRM organization there will be limited scope for such dedicated workers who do not wish to expand their repertoire. Over time, such people will be replaced by those who do. Herein lies an important lesson for human resource departments. As they look to the future in terms of people to hire, they must bring in those who will be open to multiskilling, accepting responsibility, continuous improvement, and constant change.

One company adopted an even more stern approach. Rather than relying on attrition or voluntary departures of employees who couldn't stand the pressure, the company and its employees agreed to a "three strikes and you're out" policy. In this policy, team members had the opportunity to vote out a team participant after a few months, if that person had not gelled with the team and was perceived as not contributing sufficiently. Such a person would have three "lives." If they were voted off three teams, their employment would be terminated.

The changes I envision may seem too difficult to pursue. And yet, as company after company has shown, the rewards from these changes are clearly worth it. For example, as General Electric moved parts of its organization to team-based structures, the company noticed that "the more responsibility it gave the workers, the faster problems got solved and decisions were made."[23] Also, despite all the obstacles, fears, and transition problems, the grass-roots enthusiasm one eventually finds can be a refreshing change. As a shopfloor worker at an 80-year-old company that makes heat-transfer equipment said to me a few months after we implemented the first QRM team in the company's main plant, "This is the best thing that has happened at our company in a long time." Management should also realize that investing in the people and team-based organization isn't just a matter of making people happy; it is a core element of competitive strategy: "Very quickly, the raw materials, the production equipment … and the marketing techniques that are employed by a company … are available to its competitors. What cannot become quickly available are the knowledge, skills, initiative, motivation and dedication of its work force…"[24]

Making Incentive and Reward Systems Compatible with QRM

An important observation to make about incentive systems is that study after study shows that rewards are not synonymous with monetary compensation. In fact, nonmonetary rewards are perceived to have more lasting value. Examples of such nonmonetary rewards are public recognition via plaques, award ceremonies, and the like; picnics, field trips and other outings that include family members; and special parking privileges or other organizational privileges that are given to individuals for limited periods. The plant manager of one General Electric facility echoed this sentiment: "A monetary reward goes in their pocket and is forgotten in a week. We go for something that they can see and display for a longer time, such as hats, T-shirts, badges, and plaques."[25] Harley-Davidson has had a similar experience in their team-based Sportster assembly operation: "…but [their pay rate] is not substantially higher. Apparently, working on the Sportster line involves something intangible—like pride and identity."[26] Another expert on teams says: "... it is well known that many of the most successful reward programs are not monetary. A \$12 plaque awarded in front of the entire organization is often more effective than many times that amount given in the employee's check as bonus."[27]

Nevertheless, team members have to pay their mortgage, send their children to college, save for retirement, and have money with which to enjoy life, and so you must also carefully choose the mechanism for monetary rewards. For QRM, these rewards should consist of the following components:

- *Team-based rewards.* If you subscribe to all that has been said so far, then these should be the primary rewards. The QRM Number can form the backbone of measures used to set these rewards, with the secondary measures complementing them to a limited extent. To ensure that team goals indeed come first, the team-based measures should carry significant weight.
- *Team redistribution of rewards.* A majority of the team-earned reward should go to all team members. However, a small amount (e.g., 10–20 percent) should be kept in a pool and redistributed by team members after evaluating everyone in the team. This allows a recognition of those who worked particularly hard, and also provides an incentive for others not to slack off and shove their tasks onto their teammates.
- *Individual growth rewards.* The purpose of this portion is to encourage people to learn new skills. Skill-based pay schemes (also called "pay for knowledge") are appropriate for this, as are other schemes for people to set personal objectives and be compensated as they achieve them.[28]
- *Multiteam rewards.* Where a group of teams must work together to serve an end customer—such as an office team, multiple component fabrication teams, and an assembly team—a portion of the reward should be based on the performance of the group as a whole. Once again, you can use the QRM Number to measure the whole group.
- *Divisional rewards.* At the broadest level, where many teams serving different markets comprise a factory or division, it may be appropriate to make a small portion of the reward dependent on the performance of the broad division. This helps to keep some spirit of camaraderie among the various groups under the same roof. In this

case, it may be acceptable to base the reward on a financial measure. Even though I have concerns about the short-term or invalid perspective that results from financial measures, this might suffice for three reasons: (1) the teams and markets could be so disparate that something like the QRM Number may not be easily accepted across the board; (2) the systems are already in place and it will be simple to implement; and (3) this should only be a small part of the total compensation so that the financial systems will not significantly distort people's behavior.

Also, you should set up the compensation system so that team members do not have to take a pay cut in order to join the team. One misconception about teams is that you have to pay everyone the same amount. Not at all. It is entirely reasonable to have experienced members join the team at a higher pay scale than novices. However, what is important is that all team members see the distribution of rewards and pay raises as equitable, using the components listed above.

ADJUSTING THE ACCOUNTING SYSTEM FOR QRM

In several places in this book I've discussed many of the drawbacks of traditional accounting systems when it comes to implementing QRM. In the last decade, whole new accounting systems have been proposed to make them more compatible with modern manufacturing strategy. Prominent among these new approaches is activity-based costing (ABC). Traditional accounting systems allocate indirect costs based solely on direct labor hours or machine hours, and they use an allocation ratio that is fixed for all products. Instead, the ABC system estimates the demand for support resources based on numerous characteristics of each product and the processes it uses. This requires a substantial amount of data to be gathered on the "drivers" of cost for each activity performed by the company, as well as data on the demand for each activity generated by each product or process, again based on various characteristics of that product or process.[29]

However, it is not my intention to propose a whole new accounting system here. Instead, I'll present two quick fixes to existing accounting systems. While relatively simple, these two mechanisms could go a long way toward enabling QRM efforts.

Strategic Overhead Allocation Approaches

The ills of the traditional accounting system, and in particular its aspects that deter QRM implementation, stem mainly from overhead allocation schemes.[30] As I showed in Chapter 2, such allocation schemes can make cells look as if they are causing the company to lose money.

The problem arises because while cells actually reduce the overhead expenses for a company, they continue to be "burdened" with existing overhead rates. Therefore it is my proposal that management create a strategic overhead rate, lower than the regular one, for each new cell put in place. They should calculate this rate based on rough values of expected reductions in overhead activities required by the cell. (While this may sound like ABC, it is not, because QRM does not require detailed calculations to be carried out, just rough "ballpark" estimates, for reasons that become clear in the following discussion.) With this revised overhead in place, the cell's performance will appear more reasonable.

However, because this overhead rate is lower than the regular one, it is to be expected that when the cell is first put in place, there may well be some underabsorption of overhead. What this means is that the company may incur more overhead expenses than will be allocated to all its regular operations (which include the new cell plus all the remaining operations). The company should collect this underabsorbed amount in a pool that management should view as an investment—much as one would view the purchase of a new machine or a building. As time goes on, more cells are implemented, and the QRM efforts start to yield significant results, the actual overhead costs and material costs will decrease significantly. Since the costing system will continue to use the standard material costs for products, this plus the existing overhead rates will mean that costs will start being overabsorbed. The company can use this overabsorption amount to recover the costs that were put in the earlier pool, much as depreciation is used to recover the cost of a building.

In conducting this exercise, it is key that management not be under pressure to predict these changes precisely and exactly cancel out all the pools. This is not supposed to be an exact science. Rather, it is a way to retain some sanity while implementing QRM, without needing to revamp the entire accounting system. As you bring many programs on line and you realize many improvements, as long as the pools remain manageable (not too excessive) and roughly cancel out, you have achieved the goal of this exercise.

Knee-jerk reactions to the above description might include: "How can we guess the reductions in overhead that might take place?" Or: "This is entirely too risky a process!" Put these comments in perspective by examining the criticisms I've leveled against accounting systems: They are backward looking, and they don't encourage any risk taking. By creating the strategic overhead rates and pools, haven't we accomplished what we wanted, that is, forcing managers to look ahead and encouraging them to take risks?

Remember, too, that we are only concerned with the management accounting portion and are not proposing that you change the entire financial accounting system. The financial accounts will not be able to use this approach because government regulations require full absorption costing for financial reporting. However, my intention is that you can use this alternative approach for management accounting purposes, to support managers in driving the company in the right direction. It is not unreasonable to expect a divergence in these two systems. As Robert Kaplan, an expert on accounting and one of the inventors of ABC, once said:

> I have no argument with the use of financial accounting for external reporting to shareholders, tax authorities, creditors and the like. My complaint is that managers are using the same figures to make important decisions for which the data is unsuitable. Entirely different methods are required to give managers the information they need for operational control of the business.[31]

Why ABC Is Not the Answer

Kaplan, Cooper, and others have proposed ABC as a solution to this problem of misleading accounting reports. However, I now explain why ABC is not the answer to the issues that I have raised with respect to QRM implementation. There are four reasons.

1. As I explained earlier, implementing ABC requires a substantial data gathering and analysis effort, and this alone could impede the progress of a QRM project for months or even years. Your company

would be embroiled in the details of figuring out its cost alloca-
tions, while the competition continued to make advances in reducing its lead times. However, the reason why ABC is not the answer
goes much deeper, as we see from the next three more fundamental
points.

2. ABC attempts to provide an accurate analysis of indirect costs for
the current operation. However, when you implement QRM, the
structure and dynamics of your manufacturing enterprise change
substantially. Also, as we have seen from the numerous examples
in this book, the behavior of individuals and other organizations
such as suppliers also changes in significant ways. Entire feedback
loops are modified or eliminated. But the drivers in the ABC model
are based on the current system structure and dynamics, and thus
predictions of future costs are based on this structure. At first blush
ABC advocates would claim that their approach would enable us to
estimate the new demands for activities, and then the ABC system
would quickly calculate the new costs. But on deeper inspection,
we find that not only would demands for activities change, but the
whole structure of cost drivers would change as well. The only
proper way to apply ABC would be to recalculate all the major cost
drivers under the new situation, a task as daunting as the estab-
lishment of the original cost drivers. Plus, the values of the current
drivers are based on a lot of existing operating data such as, how
do the features of a product affect the number of engineering
change orders generated for it? Since the future scenario is being
hypothesized, it would take a great deal of discussion and difficult
judgment calls to guess the new values for the drivers in such
complex situations.

3. Even if one could quickly "guesstimate" the values for these drivers, there is another, fundamental problem to be overcome. We
saw in Chapter 7 that changes in setup time and lot size can have a
dramatic impact on lead time and WIP; further, this impact is nonlinear (recall the U-shaped graphs there). Similarly, a 10 percent
change in capacity utilization can double or triple lead time and
WIP. Hence WIP carrying costs and other related costs bear a nonlinear relationship to drivers such as lot size, setup time, and
capacity. But the ABC model is a linear model. What this means is
that all costs go up in a fixed proportion to the value of some driver. No linear model could ever predict the organizational impact
of changes that result from QRM decisions. In more rigorous terms,

we can say that a linear model such as ABC can only be meaningfully applied to "what-ifs" that are in the neighborhood of the current operating methods. QRM makes drastic changes in the operating methods, and thus goes beyond the applicability of the ABC model.[32]

4. The final reason is also fundamental. In a nation where a high proportion of top managers come from the accounting and finance side of the business, we have long assumed that fixing the accounting system will put a failing company back on the path to manufacturing competitiveness. Even well-known accountants have by now realized that nothing could be farther from the truth.[33] Instead, the prerequisite for manufacturing competitiveness is for top managers to understand the fundamentals of manufacturing enterprise behavior, and then to combine this understanding with a vision of how to exploit the behavior to create competitive advantage.

In the context of QRM, once managers have developed such an understanding and corresponding vision, as well as the conviction to accompany the vision, then the simple accounting fixes I propose here will be sufficient for them to carry out their decisions. There will be no need for implementation of complex accounting systems to support this effort.

QRM Projects Justification via Strategic Funding

The previous section looked at how to keep a QRM initiative alive after you've put it in place, by not having it burdened with unfair costs. Now we look at solving the up-front issue, namely, the pervasive problem of justifying QRM projects, given the obstacles of cost-based measures. (As examples of this, review the discussion of the $200,000 CNC machine required by MPC in Chapter 5, or the $15,000 printer needed by the breakdown department in Chapter 11.)

The QRM solution is to have management create strategic funds, which they will hand out based on *lead time reduction proposals*—not traditional cost justification calculations. Management itself should create the strategic fund pool through some high-level cost and return calculations; however, giving out the money is based on very different criteria. I'll illustrate this approach through the following example.

The QRM Criteria for Giving Money to a Project

A company has a complex product line that involves many different fabrications, assemblies, and purchased parts. Management wishes to increase market share in a particular segment of the market by dramatically reducing lead time. However, this isn't an effort that will take the mere implementation of one or two cells; to accomplish the lead time reduction will require many initiatives to pull together all the components from all the sources. Management estimates that if this strategy is successful, sales will increase substantially, as will profitability. Based on three years of projections, management decides that it can afford to invest $750,000 on this lead time reduction effort. I call this pot of money a strategic fund for QRM.

After management creates the strategic fund they also put in place a QRM strategy committee to manage this fund. As a first step, the committee should dip into this fund to educate its members in depth about QRM and also provide some basic education to all employees about the overall QRM strategy. During this process, the committee publicizes the fund and the company goals for the target market segment and gets employees to come up with proposals for lead time reduction. (The next chapter discusses in more detail how to accomplish this.) The fabrication area may come up with proposals for several cells. The office area might suggest one or two Q-ROCs. Engineering and manufacturing may suggest redesign of some parts or changes in material. Purchasing may suggest changing vendors to focus on responsiveness. Materials may suggest bringing some subcontracted operations in-house. All these proposals are presented to management with two key parameters identified: how much money it will take to implement the suggestion, and the anticipated reduction in lead time.

Then the QRM strategy committee evaluates proposals and allocates the funding based on return (in terms of lead time reduction!) versus the amount of funds needed. This does not have to be a one-time process. The committee could give feedback on proposals and have them be refined, altered, or combined with other proposals, to better fit the overall strategy. As the committee accepts the proposals, it can hand out the funds allocated (against certain timelines and milestones if needed) and enable projects to commence.

From an accounting point of view, the strategic fund for QRM approach is no more complex than, say, investing in a building for a new product line or in an R&D project for a potential new product. I do not envision any significant accounting system hurdles for companies that wish to adopt this approach. What is required is the management will to push for it.

The key point to note about this procedure is the criterion you use to select projects. Conventional justification methods use financial criteria such as payback period, return on investment (ROI), internal rate of return (IRR), or net present value (NPV). The QRM approach uses lead time reduction as the primary criterion for evaluating projects.

Not that the QRM approach is ignoring financial returns altogether. What it does accomplish is to have senior management engage in macrolevel planning to gauge the financial returns and justify an investment. But unlike cost-based management methods, QRM methods do not reduce these macrolevel financial goals to microlevel financial goals for individual departments. *Instead, the QRM approach reduces the macrolevel financial goals to microlevel lead time reduction targets.* Herein lies the key difference, and in my view, the strength of QRM—and note that this is key, not just in this one instance, but in all my examples in this book.

EPILOGUE: DEMOCRACY IN AMERICA (REVISITED)

Our society is coming full circle from the 150-year-old observations by the French historian Alexis de Tocqueville quoted in Chapter 3. To understand what this full circle is, first note that during the twentieth century, the United States and most of the free world has lived a contradiction: While fighting for the cause of democracy within our countries and around the world, we have accepted dictatorship in our workplace. Witness this observation:

> We vote for president of the United States, but accept that the board of directors will appoint one of their cronies as president of our organization. We disallow taxation without representation, but permit management to determine what is fair compensation. We make major financial decisions in our households, but then must get three managers' signatures to buy a Bic pen ... In a nation based on equality, we accept that ... inequality is embedded in our language. Managers are still called *bosses* and *superiors*.[34]

Amazingly, this divergence between our political and work systems was observed 150 years ago by de Tocqueville:

While the workman concentrates his faculties more and more upon the study of a single detail, the master surveys an extensive whole, and the mind of the latter is enlarged in proportion as that of the former is narrowed. In a short time, the one will require nothing but physical strength without intelligence; the other stands in need of science, and almost of genius, to ensure success. This man resembles more and more the administrator of a vast empire; that man, a brute ...The master and the workman have then here no similarity, and their differences increase every day ... the one is continually, closely, and necessarily dependent upon the other and seems as much born to obey as that other is to command. What is this but aristocracy?[35]

To put this all in perspective, and see the connecting thread, observe that most organizations have a four-way separation of fundamental roles.

1. Ownership is represented by the stockholders.
2. Decision-making is performed by management and supervision.
3. Productive work is done by the employees.
4. Protection of employee rights is represented by unions.

Much of the organizational dynamics and resulting nonproductive time spent by everyone comes from this separation, with each camp lobbying for its own goals. The key to understanding the team-based organization is that if you fuse these four roles together, the interests of all parties coincide. Everyone can focus their energy on making the company successful.

United Airlines demonstrated the huge impact of combining worker, manager, union, and stockholder roles when employees bought 55 percent of the company and acquired 3 of the 12 seats on the board of directors.[36] "Join the side you're on," stressed Chairman Gerald Greenwald to the employees, as he encouraged them to buy in to the company when it was in severe trouble. A union leader at United echoed the sentiment, "No one washes a rental car." In other words, you take care of what you own. In fact, while traditionally unions have wanted to increase the number of employees, United's pilots' and mechanics' unions have now had an incentive to minimize new hirings, and—even more radical—have encouraged the hiring of nonunion temporary workers. The results of the employee ownership were already phenomenal by the end of 1996. United's stock had more than doubled. Sick time and workers' compensation claims declined 17 percent. Operating revenue per worker rose faster than at American and Delta, even though United ended up hiring an additional 7,000 people. Merging the interests of capital, labor, and management is breaking new ground in the evolution of capitalistic society.

Distributing some ownership to employees does not necessarily mean that employees must own stock. In the QRM context, ownership is an attitude as much as it is a physical or legal term. While you can spread around some ownership through employee stock options, nonstock ways of distributing ownership include team-based rewards, profit sharing, allowing participation in decision-making, and giving teams "cradle-to-grave" responsibility for certain activities.

It isn't just management that is reaching out to workers. Unions have also taken some radical steps. In 1994, the International Association of Machinists (IAM) offered to help David Groetsch, president of an Alcoa division, create team systems and joint decision-making councils involving workers and managers at a Denver Alcoa plant. Said Groetsch in an interview: "If someone from [a consulting company] had said: 'We can improve product delivery, customer satisfaction, and profitability—hire me,' I would have yawned. But when the union walked in and said all that, it got my attention."[37] This approach by the IAM was not an anomaly at one plant, but part of a recent trend: "After years of skepticism, some unions are now advocating partnerships with management. A few examples: AFL-CIO, Laborers, Machinists, Needle Trades, Steelworkers...."

Lest my aim be misinterpreted, let me also clarify that the involvement of workers, employees, and unions in the management and ownership process is not a step toward some form of socialism or communism. This is not a trite observation, as a manager at Motorola pointed out to me in a written note while he was in a class at the University of Wisconsin-Madison:

> At the risk of stereotyping, manufacturing managers and executives tend to be a patriotic lot; their values include patriotism, democracy and love of freedom. At the same time, in your experience, have you ever made recommendations regarding employee involvement to a manager and gotten the impression that he considered you some kind of left-winger? Labor unions have often been called un-American. Without a doubt, communism and labor-activism shared some of the same leaders in the early part of this century. But I wish we could get beyond that and realize that employee involvement involves democracy, a traditional American value."[38]

These observations help to crystallize the point that the team-based organization is a natural step in our quest for complete democracy in our society. "Organizational democracy is this self-governance within an organization: of the employees, by the employees, for the employees."[39] Even an organization with strong authoritarian management styles, such as a military support organization, can change sufficiently to implement such democracy and benefit from work teams.[40]

However, democracy in the workplace alone will not guarantee a successful company. The democratic workplace must be designed with a viable business goal in mind. QRM offers just such a focused and highly viable goal, one that is ideally suited to the team-based democratic organization, and one that can launch us into the twenty-first century where our workplace and societal goals can be more equal than they ever were. The final chapter of this book shows you how to ensure the success of this groundbreaking venture.

Main QRM Lessons

- Performance measurement, organization structure, and accounting systems are three facets of the same issue and need to be tackled with a unified approach.

- Review of the drawbacks of traditional performance measures in the context of QRM.

- Definition of the QRM Number, a new measure for lead time reduction.

- Lead time measurement is not trivial; several tricky issues need to be resolved. Use of a status indicator for orders can go a long way to resolving lead time ownership issues.

- How to create successful cell teams for QRM.

- How to supplement cellular team creation with compatible incentive and reward systems.

- Two simple fixes to traditional accounting systems can go a long way toward supporting QRM: strategic overhead allocation and strategic funding.

- The new organization structure for QRM is less radical than you may think at first; it is simply an extension of our established democratic principles to the workplace.

17

Steps to Successful Implementation of a QRM Program

So you're convinced QRM will work for you. You've weighed the arguments for QRM, absorbed the case studies and success stories, and pondered the opportunities in your own marketplace. Having evaluated all this data you feel there is excellent potential for implementing QRM at your company. But how and where do you start? And given the difficulties and potential obstacles that you have read about in this book, how do you maximize the chances of success?

This chapter provides fifteen concrete steps for you to follow to ensure success of implementing QRM at your company. I will attempt to build on lessons I have learned over many years of implementing QRM at dozens of companies. Before I proceed, however, it should be clear in your mind that you cannot reduce lead time as a *tactic*. It has to be part of a *QRM strategy* led by top management. Let us understand this further.

Throughout this book you have seen that in order to significantly affect lead times, firms need to change many of their traditional ways of operating. This will necessarily have a major impact on the organization's structure. Such change has companywide implications and cannot be accomplished without total commitment from top management. It would be naive for senior managers to think that they can order their staff to cut lead times in half, delegate the responsibility, and expect that it will eventually happen.

This is why upper management must lead the company on the QRM journey. However, despite all the knowledge from experiences with prior implementations, there is no one sure-fire way for success—each

497

organization must learn its own best way of achieving QRM. This is because each company and its management have a unique dynamic. So, even while upper management realizes they must lead the company down the QRM path, it will not be obvious exactly how to do so. True, a number of principles can be stated (and indeed, these form the basis for this book), and these principles can help greatly in promoting success, but in the end each company must endure a certain amount of learning based on its own unique nature. In order to support this learning process and yet leverage off all the knowledge we already have about QRM, I have designed a generic implementation process for QRM efforts.

Other novel approaches to manufacturing strategy also advocate some similar implementation steps, but my approach is unique in a few ways. One, it doesn't involve just putting together a team for a project, but rather, it has stages that progress through several different teams, from a steering committee, to a planning team, to an implementation team. Two, although each implementation step is summarized below, it is based on a large number of details given in previous chapters, which the teams should use for their initial training and as a reference for their detailed work. Thus, the whole QRM approach to implementation is based on a set of documented and argued principles, as opposed to a number of loosely stated philosophies.

FIFTEEN SPECIFIC STEPS TO IMPLEMENTING QRM

These specific steps will not only assist you in planning and implementing a QRM program at your organization, but will also provide a roadmap to successful implementation, while at the same time allowing for the learning that must occur in each individual organization.

Step One: Get Top Management Commitment

To be successful in implementing QRM you will have to change many operating norms of the company, including the way you use financial measures and other metrics such as efficiency, existing reward systems, traditional reporting structures, and functional boundaries. Such changes will not occur without the backing of top management.

Failing to get this commitment up front will only mean that the project will stall later and may even be perceived as a failure, thereby undermining the chances of ever getting another shot at implementing QRM.

If getting total commitment from top management seems too daunting—because management is too deeply rooted in cost-based methods, for example—then let me share a little secret with you. This secret has helped me to achieve successes even in very traditional organizations. It is this: *Getting upper management's commitment is not to be seen as scaling a high wall in one attempt; rather, it should be seen as climbing a series of small steps over time.*

The reason for this is that in most cases, you don't know up front exactly what norms need to be changed first for a QRM project to succeed. Each company is a little different from the others, and some are very different. For similar reasons, top management often doesn't truly understand the extent and implications of what it is committing to when it gives the green light to a QRM effort. So you must begin your effort by getting an initial commitment from upper management, and then, as the effort progresses, return to it time and time again. It is best that this not be seen as coming back with a change in direction or a modification to the rules of engagement. Instead, each visit back with upper management should be seen as seeking additional definition and refinement of the original commitment.

It is not unusual for a typical session with a senior manager to go as follows.

Scene: *The executive vice president of a company has received a request from a QRM implementation team to change a long-standing company policy. He is currently in a meeting with one of the representatives of this team.*

Executive VP: "Did I really commit to that when I gave you the go ahead?"

QRM Implementer: "Not specifically, but you did give us an overall direction and goals when we began this project, and committed to supporting that. And in order to proceed in that direction, we will have to change this existing rule at our company."

Executive VP: "Are you sure we can't make some simple modifications or exceptions to this policy?"

QRM Implementer: "Unfortunately not. We have spent a good deal of time debating this in our team. Here are the reasons why the policy as a whole must be scrapped..." [she continues with detailed reasons, and then concludes] "...thus, if you still believe in pursuing the goals you laid out for us, the policy will have to go."

Note that there will typically be many such meetings between the QRM implementers and representatives of upper management.

There are only two sure-fire ways to get the steadfast commitment of senior managers. The first involves education: Have them attend seminars, workshops, or conferences, and have them read about QRM. The second involves their peers at other companies. Have your top managers call and speak with executives at other companies that have been down this path, or better still, have them visit those companies and talk to managers and employees there.[1]

Step Two: Steering Committee and Champion

After getting the initial and general commitment from top management, put together a steering committee for the QRM effort. This should include at least one senior manager who is willing to be a champion of this effort.

This committee will not run the project or even manage it. The main function of the committee and the champion is to identify an initial area for a potential QRM project and then to be available to clear obstacles in the path of studying this area further. Also, the committee or champion can be useful in strengthening top management's commitment to the project by helping out in situations like the scenario I described with the vice president and the need to eliminate the company policy.

Step Three: Pick a Potential Product and Set Rough Goals

The steering committee should home in on a product whose lead time is too long. (Throughout the following discussion, I will use the term *product* in the broad sense of Chapter 11. Remember that this usually will be not one actual product, but a family of related deliverables serving a given market segment.) Committee members should present several candidate products to the whole committee for consideration.

At this point it is worth reviewing the principles in Chapter 5 (if the product involves the shop floor) or Chapter 11 (if the product involves the office). In particular, some of the key points to be observed in picking a product are: Aim for a lead time reduction large enough so that it can't be obtained by tweaking the current system. If the desired lead time reduction is achieved, it should "make a splash" in terms of its effect on company performance or sales. The resulting project should have a reasonable chance of success (in other words, don't start your QRM implementation by choosing a project with an almost impossible goal). The project scope should not be too broad, so that the implementation team isn't spread too thin. And the scope should be such that there is a good chance that ownership can be had of all the processes required for the particular product chosen.

The above principles, as well as the supporting details in Chapters 5 and 11, have evolved from successes in getting many companies to adopt QRM. The basis of this success has been to start small and then build on the initial success with additional projects. In this respect, the QRM approach differs in a fundamental way from approaches such as reengineering (or BPR), which have often attempted to reorganize whole divisions or companies in one fell swoop, in many cases with disastrous results.

Step Four: Put Together the Planning Team

Based on its initial impression of the processes covered by the chosen product, the steering committee should put together a cross-functional team of people called the planning team. This team should have a representative from each functional area that has a substantial impact on the delivery of the given product. The role of this team is to study the QRM opportunity in detail and come up with specific implementation recommendations to management. The people on this team are not necessarily going to be the ones that end up in the cell (if one is recommended); however, there may be some that do. I'll discuss this issue later.

In creating the planning team, the steering committee should attempt to use, as much as possible, the QRM principle of seeking volunteers. If not enough people volunteer, both employees and managers should use the approaches outlined in the previous chapter to drum up more interest in participation.

Ideally, the planning team should be collocated and devoted to the study full time for a short period such as one or two months. However, many organizations may not be able to afford giving up an individual from each of several departments for this length of time. The alternative is to have regular times for the team to get together—e.g, every Tuesday and Thursday, from 8 A.M. to noon—and a space that is permanently allotted to the team, where they can meet, keep team records, have computer resources, and the like. It is extremely important to set regular and frequent meeting times, rather than having to put people's calendars together to schedule each meeting. Instituting regular meeting times will clarify the time commitment needed from all team members to their managers and also maintain a sense of urgency for the project. Note that, in order to put this team together and get people's time and the space, the steering committee needs to have sufficient authority over the rest of the company.

Step Five: Invest in Team Building

Even though the planning team does not have a long life, it is useful to engage in some team-building exercises along the lines detailed in the previous chapter. The only point to keep in mind is that these exercises can be less extensive than they would be for a team that may end up working together for several years. Still, the planning team needs to gel, in order to be able to work together and brainstorm effectively, so some investment in this step is important.

Step Six: Get Rough Measures of Current System Performance

The first significant task for the planning team, after it has completed the team-building exercises, is to get ballpark measures of current lead time performance for the given product. I've noted in several places that companies do not have good data on their actual lead times. In order for the team to set its goals and have an idea of what it takes to accomplish these goals, they should conduct some preliminary data gathering on current lead times. However, this step should not occupy too much time; more accurate and detailed data gathering comes at a later step.

Step Seven: Refine Scope and Set More Precise Goals

After the team has a better idea of the current lead time performance, it can establish more precisely the goals for the project and what processes or activities should be included (the scope). In setting the goals, note the points made above and in Chapter 16. At this point the team should also run the refined scope and goals past the steering committee and get its blessing on the project, or further refine the scope and goals based on feedback from the committee.

Step Eight: Conduct Detailed Data Gathering and Analysis

Now that the team has agreed upon the scope and goals, they are truly ready to sink their teeth into the problem of reducing lead time. At this stage, the team should use the methodologies detailed in this book. For example, if the project involves office lead time, then the team may wish to gather data using tagging and process mapping exercises. If it concerns a fabrication operation, then data on product volumes, routings, setup, and process times might be called for. Chapters 5 and 12 discuss in depth the types of data needed and the analysis methods that serve the team in this effort.

Step Nine: Brainstorm Solutions

In this step you apply the full power of QRM methodologies. The team begins to rethink the way jobs are done with the aim of minimizing lead time. In particular, a critical part of the brainstorming is for the team to home in on a focused target market subsegment (FTMS). Choosing the FTMS is important because all the details of the subsequent solution, as well as its impact, are dependent on this choice. After picking the FTMS, you should apply all the remaining principles discussed in preceding chapters. Also, team members should review

Chapter 13 to help them come up with additional ideas to reduce lead time.

In addition to generating potential solutions, the team should attempt to quantify their impact. Various analysis tools such as rapid modeling can help with both quantifying the lead time impact of decisions, as well as eliminating inferior solutions.[2] It is advisable for the team to generate not just one, but multiple solutions, with the quantitative and qualitative impact determined for each.

Step Ten: Present Recommendations

The planning team should next hold a formal presentation for the steering committee and upper management, to share its analysis and insights and to present its recommendations for implementation along with anticipated results. Data gathered during the team's work can help in this stage by showing management the extent of non-value-added time in the system and other forms of waste (see the Ingersoll case study in Chapter 12 for detailed examples of such analyses and how they helped). Also, the radical solutions proposed by the team, along with estimates of the dramatic lead time reduction potential, can help to open management's eyes to the need for change.

Step Eleven: Create the Implementation Team

Assuming management approves one of the solutions recommended by the planning team, the next step is to create the team that implements the solution. Since the QRM solution involves some type of cellular team, the implementation team will consist mainly of the people who will be in the cell. However, it may have some additional people allocated to it for an initial period, such as support people or specialists on a given topic, to help the team get off the ground. Ideally, there should be some overlap between the planning team and the implementation team. For example, if you make a few of the planning team members part of the implementation team, either as cell members or as part of the transitory support staff, it will help to maintain continuity of thinking from the planning stage to the implementation stage. You should keep this point in mind when choosing your planning team so you can choose some members who have a good chance of ending up on the implementation team.

At this stage you may ask, if we wanted overlap between these teams, why did we create two of them in the first place? Why not create just one team from the start? In fact, this is one of the key differences between my approach and those in other texts on related manufacturing strategies. The reason for two teams is that the implementation team will almost certainly differ significantly in size and skill base relative to the planning team. The planning team needs to have a sufficient knowledge base to understand all the processes involved, study QRM principles in detail, conduct appropriate analysis, brainstorm solutions, and make recommendations to management. The implementation team needs to have the skills to do the tasks needed in the actual cell. Consider a lead time reduction project for order processing, which results in the recommendation for a Q-ROC. The planning team could well involve 10 people from a host of departments. The implementation team might consist of just 5 people, 3 who will work in the Q-ROC, and 2 others to support and train them during a transition period.

In creating the implementation team, remember to use the QRM principle of asking for volunteers, along with other related principles laid down in the previous chapter. (It is to be hoped that some of the members of the planning team are so enthused with the ideas they have created that they will volunteer, and thus help accomplish the preceding goal.) Also, the implementation team must be collocated into a space created for it, and its members must be full-time. On these two issues there can be no compromises.

Step Twelve: Team-Building and Training for the Implementation Team

This step is more critical than for the planning team. Most members of the implementation team may end up working together for several years, so it is essential to put effort into making them a cohesive team. Not only is it necessary to go through team-building exercises, but now it is important to study all the different categories of training listed in the previous chapter and to create a mechanism and plan for the team members to obtain training in those areas.

Step Thirteen: Implement the Recommendations

This is the step you have been waiting for, where the rubber finally hits the road and you start to see some results. All preceding steps have involved theory, planning, training, but now you get real action. Perhaps you have to move or purchase machines to create a manufacturing cell. Or a new office setting is created for a Q-ROC. Whatever the detailed idea, whole new ways of delivering the product are put in place.

Of course this stage is not without difficulties. Unanticipated problems will occur and the team will need to resolve them. This is where the foundation of good team training, both in team dynamics and in problem-solving skills, shows its worth. In addition, organizational or political barriers may surface, and the team, along with the help of the champion or steering committee, will have to tackle these as well.

Be sure to put in place mechanisms, such as those in Chapter 16, to track lead time accurately, so that the team can measure the progress toward its goal, as well as have data for problem-solving and improvement.

Step Fourteen: Progress Review, Presentation, and Recognition

If you did your homework properly in choosing the FTMS, and if the team makes substantial progress toward its goals, then you should get the "splash" that you targeted. Once you have achieved some initial successes, it is time to do a formal review of the team's progress. The team should have a meeting, anywhere from two hours to a day, to assess where it is with respect to its goals and what still needs to be done. In addition, they should brainstorm over what they did well, what were the key contributors to success, what mistakes were made, and what they might have done differently if they were to do it all over again.

The team should document the results of this session and turn them into a presentation. They should deliver this presentation to several audiences, including upper management, but also all other managers and supervisors, and employees in both the office and the shop floor. Senior management should use at least one such occasion to recognize the team; it is not necessary that there be a monetary reward, but public recognition, presentation of plaques or certificates, or even just handshakes with the company president go a long way toward making the team feel good. These presentations and the recognition awarded to the team help stimulate people to volunteer for the next team—which brings us to the last step.

Step Fifteen: Repeat the Process with Additional QRM Projects

After you have obtained concrete results from the first effort, and hopefully the splash too, management should leverage off these results to initiate additional QRM projects. Now the company may be able to engage in several projects, or ones with a broader scope or greater difficulty, because the lessons obtained from the first project should provide insight and help the company be more effective in these additional projects.

In doing the next round of projects you should follow the same set of steps. The company should not become blasé about the need for team-building, training, or project reviews. Each project will have its own needs and its own lessons, and it is worth sticking to the full process.

Again, the best advocates for continued change to the QRM way are both the planning and implementation members from completed projects. You should ask them to speak to people throughout the company to help stimulate even more projects. The managing director of a British company that makes tooling found, after publicizing the results of their first cell, "We were constantly asked by shop floor employees, 'when are you going to start on the next cells, because we want to be included.'"[3]

Eventually, as more and more QRM implementations come on-line, you will see the company transition into the new type of organization described in Chapter 16.

QRM: MANAGEMENT HEADACHE OR OPPORTUNITY?

Despite all the literature on lead time reduction and related manufacturing strategies, companies still struggle with how to successfully implement QRM. Implementation calls for substantial change, and management may well feel the changes are too onerous, and that it cannot put the time into following through with the QRM program.

But this precise point allows us to look at QRM another way: If it was easy to accomplish, why wouldn't all the companies in your industry have done it already? Taking this point further, it is clear that implementing speed is not passé—there is still a substantial competitive opportunity for companies embarking on QRM implementation. If these companies can overcome the obstacles and successfully adopt a QRM strategy, they will not only create a profitable enterprise, but also construct a significant barrier to competition. The very fact that implementing QRM is difficult means it will take the competition a considerable amount of effort and time to catch up, if it can at all.

The QRM journey awaits those companies wanting to begin the twenty-first century with whole new ways of operating, novel organizational structures, hitherto unforeseen levels of performance combined with job satisfaction, and a vision of democracy that flows through our society and into the workplace. What is there to prevent you and your company from embarking on this exciting journey?

Main QRM Lessons:
Key Steps for Implementing QRM

- Lead time reduction cannot be done as a tactic; it has to be part of a QRM strategy led by top management. The first step must be getting upper management's commitment to this strategy.

- Form a high-powered steering committee, and have it pick a potential product and set rough lead time reduction goals.

- Create a planning team consisting of people from the major functional areas that affect the delivery of the product. This team will set more precise goals, analyze the current system, brainstorm solutions by applying QRM principles, and then present its recommendations to management.

- Put together the implementation team that includes all the people who will be involved in any proposed cells, and possibly a few others. Engage in team-building and training, then implement the recommendations. Be sure to put QRM-related measurement systems in place to track progress.

- After initial successes have been achieved by the team, have it present its work to the rest of the company, and also ensure that it receives recognition from management. The team should also document the lessons learned from the implementation process.

- Repeat the process with additional QRM projects, leveraging off the lessons learned to tackle more challenging products.

- A distinguishing point in this whole approach is the recognition that there should be two separate teams, the planning team and the implementation team, with different skill sets.

Appendix

Companies Affiliated with the Center for Quick Response Manufacturing

During the period 1994 to 1997, many companies supported the work of the Center for Quick Response Manufacturing, and contributed to the search for knowledge on QRM principles and practices. I would like to acknowledge the support of the following organizations, and thank all their employees who have worked with us over the years.

ABB Flexible Automation Inc., New Berlin, Wisconsin
A-C Compressor Corporation, Appleton, Wisconsin
Accents Unlimited, Milwaukee, Wisconsin
Alcoa Technical Center, Alcoa Center, Pennsylvania
Alcoa Davenport Works, Davenport, Iowa
ALKAR, Lodi, Wisconsin
Beloit Corporation, Beloit, Wisconsin
Boumatic Division of DEC International, Madison, Wisconsin
Centrifugal Industries, Cambridge, Wisconsin
Converter Concepts Inc., Pardeeville, Wisconsin
Crown International Inc., Elkhart, Indiana
Danfoss Electronic Drives, Rockford, Illinois
Deltrol Controls, Milwaukee, Wisconsin
Electronic Theatre Controls Inc., Middleton, Wisconsin
Fiskars Inc., Sauk City, Wisconsin
Formrite Companies, Two Rivers, Wisconsin
Gerber Products Company, Reedsburg, Wisconsin
Graber Products Inc., Madison, Wisconsin
Greenheck Fan Corporation, Schofield, Wisconsin
Grover Piston Ring Inc., Milwaukee, Wisconsin
HATCO Corporation, Milwaukee, Wisconsin
Honeywell Inc.–Micro Switch Division, Freeport, Illinois
HUFCOR Inc., Janesville, Wisconsin
Ingersoll Cutting Tool Company, Rockford, Illinois
Ingersoll Milling Machine Company, Rockford, Illinois

Isthmus Engineering and Manufacturing Co-op, Madison, Wisconsin
John Deere Horicon Works, Horicon, Wisconsin
Johnson Controls Inc., Milwaukee, Wisconsin
Lyco Manufacturing Inc., Columbus, Wisconsin
Madison-Kipp Corporation, Madison, Wisconsin
Marathon Electric, Wausau, Wisconsin
Microelectronic Modules Corporation, New Berlin, Wisconsin
Midland Plastics Inc., New Berlin, Wisconsin
Modine Manufacturing Company, Racine, Wisconsin
MTS Systems Corporation, Eden Prairie, Minnesota
Panoramic Inc., Janesville, Wisconsin
Pensar Corporation, Appleton, Wisconsin
Phillips Plastics Corporation, New Richmond, Wisconsin
Pillar Industries, Menomonee Falls, Wisconsin
Printing Developments Inc., Racine, Wisconsin
Rexnord Corporation, Milwaukee, Wisconsin
Research Products Corporation, Madison, Wisconsin
Robert Bosch Fluid Power Corporation, Racine, Wisconsin
Rowe Pottery Works, Cambridge, Wisconsin
Senior Flexonics Inc., Bartlett, Illinois
Trek Bicycle Corporation, Waterloo, Wisconsin
Vogel Wood Products Corporation, Madison, Wisconsin
Walnut Hollow Farm Inc., Dodgeville, Wisconsin
Wilson-Hurd, Wausau, Wisconsin
Worzalla Publishing, Stevens Point, Wisconsin

During this same period, over 100 University of Wisconsin students worked on QRM-related projects at these companies. I would like to thank all of them for their contributions to our knowledge of QRM. A few of the students and their projects were partly supported by the State of Wisconsin Industry and Economic Development Research Program administered through the University-Industry Research Program at the University of Wisconsin-Madison.

Endnotes

Chapter 1

1. Examples are the articles by G. Stalk, Jr., "Time–The Next Source of Competitive Advantage," *Harvard Business Review*, July–August 1988, pp. 41–51; and R. Schmenner, "The Merit of Making Things Fast," *Sloan Management Review*, Fall 1988, pp 11–17.
2. Examples are C. Charney, *Time to Market: Reducing Product Lead Time* (Dearborn, MI: Society of Manufacturing Engineers, 1991); J. D. Blackburn, Ed., *Time-Based Competition: The Next Battle Ground in American Manufacturing* (Homewood, IL: Business One Irwin, 1991); and G. Stalk, Jr. and T. M. Hout, *Competing Against Time* (New York, NY: The Free Press, 1992).
3. Detailed data on responses to individual questions were available for only 223 out of the 425 respondents whose data were summarized in Figure 2-1.
4. Two other terms, "cycle time" and flow time," are also in use today to denote the time for jobs to go through the organization. However, "cycle time" has traditionally been used in both industrial engineering and assembly analysis to denote the time for a worker or machine to perform a single task. Also, "flow time" is a relatively new term that is not widely recognized. On the other hand, "lead time" is a classic term used both in industry and the research literature to denote the time to complete a set of operations. Hence I will stick to using the term "lead time" throughout this book.

Chapter 2

1. This and the following quote are from *Forbes*, October 12, 1992.
2. G. Stalk, Jr. and T. M. Hout, *Competing Against Time* (New York, NY: The Free Press, 1992).
3. C. Charney, *Time to Market: Reducing Product Lead Time* (Dearborn, MI: Society of Manufacturing Engineers, 1991).
4. This case study is described in detail in Chapter 12.
5. R. Merrills, "How Northern Telecom Competes on Time," *Harvard Business Review*, July–August 1989, pp. 108–114.
6. For example, see M. Brassard, *The Memory Jogger Plus+* (Methuen, MA: GOAL/QPC, 1989).
7. The quality literature abounds with detailed statistics. As one example, see J. M. Juran and Frank M. Gryna, *Quality Planning and Analysis*, 3d ed. (New York, NY: McGraw-Hill, 1993).
8. L. Ferras, "Continuous Improvements in Electronics Manufacturing," *Production and Inventory Management*, Vol. 35, No. 2, 1994, pp. 1–5.
9. J. H. Hammond and M. G. Kelly, "Quick Response in the Apparel Industry," *Harvard Business School* Case 9-690-038.
10. See Chapters 3 and 6.
11. Stalk and Hout, *Competing Against Time*.

12. R. W. Schmenner, "International Factory Productivity Gains," *Journal of Operations Management*, Vol. 10, No. 2, April 1991, pp. 229–254.

13. J. R. Meredith, D. M. McCutcheon, and J. Hartley, "Enhancing Competitiveness Through the New Market Value Equation," International *Journal of Operations and Production Management*, Vol. 14, No. 11, 1994, pp. 7–22.

14. This statistic is from Stalk and Hout, *Competing Against Time*. A number of related statistics can also be found in J. D. Blackburn (ed.), *Time-Based Competition: The Next Battle Ground in American Manufacturing* (Homewood, IL: Business One Irwin, 1991).

15. See the discussion in D. Veeramani and P. Joshi, "Methodologies for Rapid and Effective Response to Requests for Quotations (RFQs)," *IIE Transactions*, Vol. 29, 1997, pp. 825–838.

16. R. Suri, U. Wemmerlöv, F. Rath, R. Gadh, and R. Veeramani, "Practical Issues in Implementing Quick Response Manufacturing: Insights from Fourteen Projects with Industry," *Proceedings of the MSOM Conference*, Dartmouth, NH, June 1996.

17. W. M. Baker and T. D. Fry, "The Rise and Fall of Time-Based Manufacturing," *Management Accounting*, June 1994, pp. 56–59.

18. This case study is described in detail in Chapter 4.

Chapter 3

1. R. Merrills, "How Northern Telecom Competes on Time," *Harvard Business Review*, July–August 1989, pp. 108–114.

2. G. Stalk, Jr., "Time—The Next Source of Competitive Advantage," *Harvard Business Review*, July–August 1988, pp. 41–51.

3. For example, see the statistics in N. G. McNulty, "European Management Education Comes of Age," *The Conference Board Record*, December 1975, pp. 38–43. This article states that of the more than 200 business schools in Western Europe, almost all were founded after World War II: "It was not until the end of World War II that Europeans, amazed at the industrial strength of the United States, sought the secret. . . . America had the know-how and it was quickly imported. American-oriented schools sprang up, staffed by visiting American professors of business and management." The article goes on to document that until recently, European business schools depended almost entirely on American-developed teaching materials, such as cases and textbooks.

4. The basics of these ideas were discussed by J. W. Forrester in "Industrial Dynamics: A Major Breakthrough for Decision Makers," *Harvard Business Review*, July–August 1958. Later, G. Stalk gave a formal treatment of what he called the "planning loop" in the context of time-based competition in "Time—The Next Source of Competitive Advantage," *Harvard Business Review*, July–August 1988, pp. 41–51. However, these authors only considered the case of companies that make to stock based on a forecast. Here I expand upon these previous ideas by elaborating on specific details for manufacturing firms and clarifying how the planning

problems are also experienced by companies that are in a make-to-order or engineer-to-order environment. I term this more general phenomenon the Response Time Spiral. In subsequent chapters we will use the Response Time Spiral to understand the cause of long lead times in other areas of the company such as purchasing, order processing, and other office operations.

5. This and the next excerpt are from J. P. Womack, D. T. Jones, and D. Roos, *The Machine That Changed the World* (New York, NY: HarperPerennial, 1991).

6. The quotes here are from Chapter XX of Alexis de Tocqueville, *Democracy in America. Part the Second, The Social Influence of Democracy*, originally published by J. & H. G. Langley, New York, 1840. (A more recent version is published by A. A. Knopf, New York, 1945.)

Chapter 4

1. These quotes are based on comments made by users of cellular manufacturing, and reported in the paper by U. Wemmerlöv and D. J. Johnson, "Cellular Manufacturing at 46 User Plants: Implementation Experiences and Performance Improvements," *International Journal of Production Research*, Vol. 35, No. 1, 1997, pp. 29–49.

2. R. Suri, U. Wemmerlöv, F. Rath, R. Gadh, and R. Veeramani, "Practical Issues in Implementing Quick Response Manufacturing: Insights from Fourteen Projects With Industry," *Proceedings of the MSOM Conference*, Dartmouth, NH, June 1996.

3. P. Prickett, "Cell-Based Manufacturing Systems: Design and Implementation," *International Journal of Operations and Production Management*, Vol. 14, No. 2, 1994, pp. 4–17.

4. Some publications equate cellular manufacturing with another concept called group technology (GT), and others state that implementing a GT system is a necessary precursor to forming cells. In my experience, neither of these is true. I will describe GT in Chapter 14; however, it suffices to say for now that while a GT system can support and complement the implementation of cells, it is not necessary to use such a system at all.

5. An example of such a situation was given by Charles Montante, president of Converter Concepts, a manufacturer of power supplies located in Pardeeville, WI. At a seminar at the University of Wisconsin-Madison in November 1996, he shared this fact with the audience: "We thought we had a cell because we put a few workers together around some machines, and especially because we put up a *big sign* in front of the area saying 'SMT CELL'! It wasn't until the faculty from the Center for Quick Response Manufacturing visited us that we realized what we had wasn't a true cell."

6. B. El-Jawhari, R. Khowara, and R. Suri, "Lead Time Reduction for General Purpose Relays," in R. Suri, U. Wemmerlöv, F. Rath, R. Gadh, and R. Veeramani, *Unravelling the Sources of Long Lead Times and the Prospects for Reducing Them: Insights from Fourteen Projects with Industry*,

Technical Report, Center for Quick Response Manufacturing, University of Wisconsin-Madison, 1995.

7. From Chapter XX of Alexis de Tocqueville, *Democracy in America. Part the Second, The Social Influence of Democracy*, originally published by J. & H. G. Langley, New York, 1840. (A more recent version is published by A. A. Knopf, New York, 1945.)

8. D. Davis, "Baxter Creates World-Class Manufacturing Environment," *IIE Solutions*, December 1995, pp. 18–22.

9. Davis, "Baxter Creates World-Class Manufacturing Environment."

10. M. Williams, "Back to the Past," *Wall Street Journal*, October 24, 1994.

11. Described by Mike Borden, president of HUFCOR, at a seminar in Janesville, WI, June 1994.

12. From *ManuFax*, Vol. 4, No. 1, Manufacturing Systems Engineering Program, University of Wisconsin-Madison, 1996, presentations by Jodi Servin, Jim Riihl, and Jim Schneider of Beloit Corporation during seminars held at the Center for Quick Response Manufacturing, University of Wisconsin-Madison, in 1996, and interviews with Beloit employees by Lara Brown of Center for Quick Response Manufacturing.

13. See S. Pantaleo and J. Shands, "Concept IV-MH Headbox—The Next Generation Converflo," *Beloit Technology Update*, Beloit Corp., April 1993. Concept IV-MH is a trademark of Beloit Corporation.

14. This is based on a batch of around 300 tubes. With the cell in place, it is not necessary for the operators to work on a batch of 2,000 tubes at a time, as it was in the prior operations. The smaller batch size helps maintain short lead time and better delivery.

15. These 36 firms are a subset of the 46 firms surveyed in the paper "Cellular Manufacturing at 46 User Plants: Implementation Experiences and Performance Improvements," by Wemmerlöv and Johnson.

16. This quote is reported in the paper "Cellular Manufacturing at 46 User Plants," by U. Wemmerlöv and Johnson.

17. See M. Imai, *Kaizen* (New York, NY: McGraw Hill, 1986).

18. This is illustrated vividly in Davis, "Baxter Creates World-Class Manufacturing Environment."

19. As an aside, this is one more way in which QRM differs from reengineering (or BPR). For instance, M. Hammer and J. Champy state in *Reengineering the Corporation* (New York, NY: Harper Business, 1993), p. 49, that reengineering differs fundamentally from kaizen and quality improvement. QRM, on the other hand, builds on both the kaizen and quality improvement movements and capitalizes on their methods for maximum gain.

20. R. A. Inman, "The Impact of Lot-Size Reduction on Quality," *Production and Inventory Management*, Vol. 35, No. 1, 1994, pp. 5–7.

21. These are additional reasons why I disagree with the managers who were quoted earlier as saying, "Relayout is an expensive, disruptive road to establish flow discipline. It's not necessary 90 percent of the time!" And,

"One should try to increase the throughput velocity of the old system before starting to relocate equipment."

22. N. L. Hyer and U. Wemmerlöv, "Group Technology and Productivity," *Harvard Business Review*, July–August 1984, pp. 140–149; U. Wemmerlöv and N. L. Hyer, "Cellular Manufacturing in the U.S. Industry: A Survey of Users," *International Journal of Production Research*, Vol. 27, No. 9, 1989, pp. 1511–1530.

Chapter 5

1. It can be seen from our definition of "Process Set" that one important difference between the QRM approach and that of JIT or "flow" methods is that we do not require the same sequence of process steps for a part family; while a common sequence helps, it is not necessary for the creation of a cell. In addition, readers familiar with other writings on cells will note that there is another significant difference from the standard procedures. To help companies identify candidate cells, most authors advocate creating a "machine-part matrix." This is a table with machines as columns and parts as rows, with an "X" marking whenever a given part uses a given machine. The matrix is then sorted to create "blocks" of X's. The problem with this procedure is that for companies with hundreds or thousands of part numbers and dozens of machines, the table is too large to comprehend and use. Thus this procedure is not really practical in resolving the problem we have posed—it only works *after* we have narrowed the selection down to a limited number of parts and machines, and are looking to refine the selection. Unfortunately, this practical limitation is overlooked in most other writings on creating cells. The procedure described here gets around this limitation.

2. P. Prickett, "Cell-Based Manufacturing Systems: Design and Implementation," *International Journal of Operations and Production Management*, Vol. 14, No. 2, 1994, pp. 4–17.

3. For more on technical details of layout, material handling, and possibilities for automation, see K. Suzaki, *The New Manufacturing Challenge* (New York, NY: The Free Press, 1987).

4. R. Suri, U. Wemmerlöv, F. Rath, R. Gadh, and R. Veeramani, "Practical Issues in Implementing Quick Response Manufacturing: Insights from Fourteen Projects with Industry," *Proceedings of the MSOM Conference*, Dartmouth, NH, June 1996.

5. T. Deeming, "Cell Mates," *Manufacturing Engineer*, June 1993, pp. 111–113.

6. When a labor union is involved, management and the union need to work together to put in place a short-term agreement, on an experimental basis, to allow the cell operation to begin. The final agreement can be crafted based on the experiences from the cell. The ways to approach such a discussion are described later in this chapter, and a broader perspective is also given in Chapter 16.

7. B. El-Jawhari, R. Khowara, and R. Suri, "Lead Time Reduction for General-Purpose Relays," in *Unravelling the Sources of Long Lead Times*

and the Prospects for Reducing Them: Insights from Fourteen Projects with Industry, by R. Suri, U. Wemmerlöv, F. Rath, R. Gadh, and R. Veeramani, Technical Report, Center for Quick Response Manufacturing, University of Wisconsin-Madison, 1995.

8. From a presentation by Norm Wenker, production manager, Deltrol Controls, during a seminar at the Center for Quick Response Manufacturing, University of Wisconsin-Madison, November 1996.

9. This is illustrated not only in the hypothetical MPC situation, but has been experienced by several companies. See P. Prickett, "Cell-Based Manufacturing Systems: Design and Implementation;" and D. J. Johnson and U. Wemmerlöv, "Cellular Manufacturing Feasibility at Ingersoll Cutting Tool Company," and Chapter 14 in N. C. Suresh and J. M. Kay (eds.), *Group Technology and Cellular Manufacturing: State of the Art Synthesis of Research and Practice* (Hingham, MA: Kluwer Academic Publishers, 1997).

10. D. M. Upton, "What Really Makes Factories Flexible?" *Harvard Business Review*, July–August 1995, pp. 74–84.

11. U. Wemmerlöv and N. L. Hyer, "Cellular Manufacturing in U.S. Industry: A Survey of Users," *International Journal of Production Research*, Vol. 27, No. 9, 1989, pp. 1511–1530.

12. D. Davis, "Baxter Creates World-Class Manufacturing Environment," *IIE Solutions*, December 1995, pp. 18–22.

13 An authority on accounting agrees that such estimated improvements should be included in the justification: "Although intangible benefits may be difficult to quantify, there is no reason to value them at zero in a capital expenditure analysis. Zero is, after all, no less arbitrary than any other number. Conservative accountants who assign zero values to many intangible benefits prefer being precisely wrong to being vaguely right. Managers need not follow their example." (R. S. Kaplan, "Must CIM Be Justified by Faith Alone?" *Harvard Business Review*, March–April 1986, pp. 87–95.)

Chapter 6

1. R. Keshav, R. Suri, and F. Rath, "Quick Response for Aftermarket Parts," in *Unravelling the Sources of Long Lead Times and the Prospects for Reducing Them: Insights from Fourteen Projects with Industry*, by R. Suri, U. Wemmerlöv, F. Rath, R. Gadh, and R. Veeramani, Technical Report, Center for Quick Response Manufacturing, University of Wisconsin-Madison, 1995.

2. This and the following anecdotes were described by Mike Canik, manager–manufacturing administration, and Jim Riihl, planning manager, Beloit Corp., during seminars at the Center for Quick Response Manufacturing, University of Wisconsin-Madison, in 1995 and 1996.

3. *ManuFax*, Vol. 4, No. 1, Manufacturing Systems Engineering Program, University of Wisconsin-Madison, 1996.

4. "The Story Behind Prepainted Blanks," *The Fabricator*, September 1996, pp. 60–61.

5. D. R. Gabe, *Principles of Metal Surface Treatment and Protection* (Oxford, England: Pergamon Press, 1972).

6. J. W. A. Off, "[TC]^2–A demonstration of Agile Virtual Manufacturing," SME Autofact Conference, Detroit, MI, November 1994.

7. "DFM, the C-17's Secret Weapon," *Manufacturing Engineering*, Vol. 117, No. 1, July 1996, pp. 90–95.

8. E. Kubel, "Manufacturers Want More Tailored Blanks," *Manufacturing Engineering*, November 1997, pp. 38–45.

9. *General Electric–Thermocouple Manufacturing (B)*. Case 9-685-062, Harvard Business School, Boston, MA, 1985.

10. V. R. Kannan and S. Ghosh, "A Virtual Cellular Manufacturing Approach to Batch Production," *Decision Sciences*, Vol. 27, No. 3, Summer 1996, pp. 519–539.

11. T. Deeming, "Cell Mates," *Manufacturing Engineer*, June 1993, pp. 111–113.

Chapter 7

1. The queue time for the first part is 0, for the second it is 2 hours, and for the third it is 1 hour. Hence the average over three parts is $(0 + 2 + 1)/3 = 1$ hour. Similarly, the average lead time is $(8 + 10 + 9)/3 = 9$ hours. This also illustrates that average lead time (9) is the sum of average processing time (8) and average queue time (1).

2. The queue time for the first two parts is 0 and for the third it is 1.5 hours. Hence the average over three parts is $(0 + 0 + 1.5)/3$ which equals 0.5 hour. Similarly, the average lead time is $(8 + 8 + 9.5)/3 = 8.5$ hours.

3. The queue time for the first part is 0, for the second it is 3 hours, and for the third it is also 3 hours. Hence the average over three parts is $(0 + 3 + 3)/3$ which equals 2 hours. Similarly, the average lead time is $(13 + 13 + 7)/3 = 11$ hours.

4. The situation where U equals exactly 1 is a bit esoteric mathematically—but we will also eliminate it from consideration because it is untenable in the real world. Essentially, it can be shown that when U = 1, any variability at all, in either arrivals or processing time, will make the backlog of work grow forever.

5. More rigorous statements of these and other formulas used in this chapter can be found in R. Suri, J. L Sanders, and M. Kamath, "Performance Evaluation of Production Networks," pp. 199–286 in S. C. Graves, A. H. G. Rinnooy Kan, and P. H. Zipkin (eds.), *Handbooks in Operations Research and Management Science, Vol. 4: Logistics of Production and Inventory* (Amsterdam: Elsevier, 1993). In that article, mathematically inclined readers will also find detailed references to the derivations of such formulas for manufacturing system dynamics.

6. The costs of these dysfunctional interactions were discussed at length in Chapter 3.

7. The analysis here is motivated by U. Karmarkar, "Lot Sizes, Lead Times and In-Process Inventories," *Management Science*, 1987, pp. 409–418. There, interested readers can find details of the mathematical theory and additional discussion of the assumptions in the analysis.

8. We will also implicitly assume that the variability *ratio* does not change with lot size. This can be justified for manufacturing situations. See the discussion in Karmarkar, "Lot Sizes, Lead Times and In-Process Inventories."

9. For examples, see S. de Treville, "Time Is Money," *OR/MS Today*, October 1992, pp. 30–34; A. Rehman and M. B. Diehl, "Rapid Modeling Helps Focus Setup Reduction at Ingersoll," *Industrial Engineering*, November 1993, pp. 52–55.

10. R. Suri, U. Wemmerlöv, F. Rath, R. Gadh, and R. Veeramani, "Practical Issues in Implementing Quick Response Manufacturing: Insights from Fourteen Projects with Industry," *Proceedings of the MSOM Conference*, Dartmouth, NH, June 1996.

11. R. Suri, "RMT Puts Manufacturing at the Helm," Manufacturing Engineering 100 (2), 1988, pp. 41–44. An example of an RMT-based package is MPX® from Network Dynamics Inc. of Burlington, MA.

12. For examples, see de Treville, "Time Is Money"; Rehman and Diehl, "Rapid Modeling Helps Focus Setup Reduction at Ingersoll"; R. Suri, "Lead Time Reduction Through Rapid Modeling," *Manufacturing Systems*, July 1989, pp. 66–68.

13. *ManuFax*, Vol. 3, No. 1, Manufacturing Systems Engineering Program, University of Wisconsin-Madison, 1994.

14. See S. de Treville, "Using Rapid Modeling to Make Kaizen Work More Effectively," *APICS–The Performance Advantage*, October 1994; and Rehman and Diehl, "Rapid Modeling Helps Focus Setup Reduction at Ingersoll."

15. Each machine takes 30 minutes to set up and 180 minutes (10 $\times$ 18) of cycle (or run) time, for a total of 3 hours and 30 minutes at each machine. Since there are four machines, this gives a total of 14 hours.

16. The first piece comes off the first machine after 40 minutes (30 minutes to set up, 10 minutes to run). It takes another 40 minutes to come off the second machine, similarly for the third and the fourth. Thus the first piece is finished in 160 minutes (4 $\times$ 40). Because all the machines have a cycle time of 10 minutes, each of the remaining pieces arrives at the last machine just as the preceding piece has been completed on that machine. Thus we only have to wait another 17 cycle times after the first piece, or 170 minutes, for the whole batch to be completed. This gives a total time of 160 + 170 minutes, or 5 hours and 30 minutes.

17. A working model to analyze this example is available as the "Metal Cutting Example Evaluation Disk" from Network Dynamics, Inc., 128 Wheeler Road, Burlington, MA 01803, USA.

18. The formula for VRD: VRD = square root of $\{U^2 \times VRJ^2 + (1 - U^2) \times VRA^2\}$. See Suri, Sanders, and Kamath, "Performance Evaluation of Production Networks."

19. This example is from W. J. Hopp and M. L. Spearman, *Factory Physics: Foundations of Manufacturing Management* (Burr Ridge, IL: Richard D. Irwin, 1996), p. 307.

20. Precise application of Little's Law requires several conditions to be satisfied. Specifically, the area should be in "steady state operation," and there should be no material disappearing or being generated in the area. However, for rough application of Little's Law, appropriate for management purposes, we can usually ignore these technicalities, provided that we follow the rules on use of consistent units and clearly demarcate the area being studied, as discussed later in this section.

Chapter 8

1. Another commonly used performance measure is called the "earned-to-paid ratio." This is the ratio of "earned hours," as measured by existing standards, to "paid hours," which is the amount of hours that workers in a department were paid. Thus it is no different than the one described above.

2. R. W. Schmenner, "International Factory Productivity Gains," *Journal of Operations Management*, Vol. 10, No. 2, April 1991, pp. 229–254.

3. J. V. Owen, "Time Is the Yardstick," *Manufacturing Engineering*, November 1993, pp. 65–70; and *ManuFax*, Vol. 4, No. 1, Manufacturing Systems Engineering Program, University of Wisconsin-Madison, 1996.

4. My work in the pipe roll shop was supported by the efforts of Dr. Len Berger of Clemson University, who was working with several teams at Beloit Corporation.

5. T. E. Vollman, W. L. Berry, and D. C. Whybark, *Manufacturing Planning and Control Systems* (Burr Ridge, IL: Richard D. Irwin, 1992).

6. See R. Bakerjian, *Tool and Manufacturing Engineers Handbook, Volume 6, Design for Manufacturing,* Society of Manufacturing Engineers (Dearborn, MI, 1989); G. Boothroyd and P. Dewhurst, *Product Design for Assembly* (Wakefield, RI: Boothroyd Dewhurst Inc., 1989).

7. An interactive computer demonstration that helps to drive home these points for cell teams, via examples of a metal cutting cell and an electronics assembly cell, is available from Network Dynamics, Inc., 128 Wheeler Road, Burlington, MA 01803, USA.

8. From R. Suri, "MRP + Q = QRM," *The Performance Advantage*, August 1996, pp. 68–71.

9. For examples, see: S. de Treville, "Using Rapid Modeling to Make Kaizen Work More Effectively," *APICS—The Performance Advantage*, October 1994; S. de Treville, "Time is Money," *OR/MS Today*, October 1992, pp. 30–34; A. Rehman and M. B. Diehl, "Rapid Modeling Helps Focus Setup Reduction at Ingersoll" *Industrial Engineering*, November 1993, pp. 52–55; and R. Suri, "Lead Time Reduction Through Rapid Modeling," *Manufacturing Systems*, July 1989, pp. 66–68.

10. One reference that covers quality, setup reduction, and productive maintenance is: H. J. Steudel and P. Desruelle, *Manufacturing in the Nineties:*

How to Become a Mean, Lean, World-Class Competitor (New York, NY: Van Nostrand Reinhold, 1992). For lessons on implementing company-wide quality practices see Shoji Shiba, Alan Graham, and David Walden, *A New American TQM: Four Practical Revolutions in Management* (Portland, OR: Productivity Press, 1993). An excellent book focusing on setup reduction is: Shigeo Shingo, *A Revolution in Manufacturing: The SMED System* (Portland, OR: Productivity Press, 1985). For productive maintenance, Seiichi Nakajima, *Introduction to TPM: Total Productive Maintenance* (Portland, OR: Productivity Press, 1988). For a discussion on the use of quality, Kazvo Ozeki and Tetsuichi Asaka, *The Handbook of Quality Tools: The Japanese Approach* (Portland, OR: Productivity Press, 1990). For kaizen, see M. Imai, *Kaizen* (New York, NY: McGraw Hill, 1986).

Chapter 9

1. This is not intended to be a detailed tutorial on push and pull systems, but only an overview sufficient for our discussion. An in-depth description of both types of systems can be found in W. J. Hopp and M. L. Spearman, *Factory Physics: Foundations of Manufacturing Management* (Richard D. Irwin, 1996).
2. For details on this and the preceding points see Hopp and Spearman, *Factory Physics: Foundations of Manufacturing Management.* The authors discuss two theoretical properties called *efficiency* and *robustness.*
3. J. P. Womack and D. T. Jones, *Lean Thinking* (New York, NY: Simon and Schuster, 1996).
4. As stated by one expert, "Line design is a critical part of flow manufacturing. In vertically integrated plants with large product mixes, line design can be a very complex task." J. Bermudez, "Synchronized and Flow: Manufacturing Techniques to Support Supply Chain Management," *The Report on Manufacturing*, March 1996.
5. Statements in two recent articles support this discussion. From Bermudez, "Synchronized and Flow": "JIT ... has actually made the consumer end of the supply chain less demand-driven. A customer who wants to order a car to exact specifications can expect to wait six to eight weeks for delivery even though cars are rolling off the assembly line every couple of minutes." And from R. DeVor, R. Graves, and J. J. Mills, "Agile Manufacturing Research: Accomplishments and Opportunities," *IIE Transactions*, Vol. 29, pp. 813–823: "... mass production, despite improvements brought to the system by JIT and lean production strategies, was essentially a system favoring large-scale and comprehensive cooperate structures ... [on the other hand] future competitive advantage lay in strategies supporting speed to market ... and the ability to satisfy individual customer preferences. ..."
6. Hopp and Spearman, *Factory Physics.*
7. R. J. Schonberger, *Japanese Manufacturing Techniques: Nine Hidden Lessons in Simplicity* (New York, NY: The Free Press, 1982); Hopp and Spearman, *Factory Physics.*

8. The rabbit chase is described in: K. Suzaki, *The New Manufacturing Challenge*, (New York, NY: The Free Press, 1987).

Chapter 10

1. For example, see H. J. Steudel and P. Desruelle, *Manufacturing in the Nineties: How to Become a Mean, Lean, World-Class Competitor* (New York, NY: Van Nostrand Reinhold, 1992); and G. Merli, *Co-Makership: The New Supply Strategy for Manufacturers*. (Portland, OR: Productivity Press, 1991).

2. This approach was introduced to U.S. industry when Japanese companies such as Honda and Toyota set up their own factories in the United States. For instance, see the discussion in D. Davis, "Partnerships Pay Off," *Supply Chain Strategies*, a supplement to *Manufacturing Systems*, November 1994, pp. 4–14.

3. For more on the Taurus and Bandit examples, see C. Charney, *Time to Market: Reducing Product Lead Time* (Dearborn, MI: Society of Manufacturing Engineers, 1991).

4. Dell Computer Corporation provides another case study on the importance of supplier location when implementing a QRM strategy. In 1996 the company reduced lead time by switching to regional suppliers closer to its plants, instead of ordering in bulk from one overseas supplier. Apparently, what it lost in discounts has been more than made up by the reduction in response time. *Business Week*, April 7, 1997, pp. 132–146.

5. In fact, the strategy of requiring broad work commitments from suppliers, rather than placing large specific orders, has been effective in reducing lead times in the apparel industry. See J. H. Hammond and M. G. Kelly, "Quick Response in the Apparel Industry," *Harvard Business School* Case 9-690-038.

6. This quote and the Roadway Express example that follows are from S. Hill, "Logistics Takes New Road," *Supply Chain Strategies*, a supplement to *Manufacturing Systems*, November 1994, pp. 28–32.

7. For example, see S. Gupta, "Supply Chain Management in Complex Manufacturing," *Industrial Engineering Solutions*, March 1997, pp. 18–23.

8. A good example is the success achieved at Hewlett-Packard. See H. L. Lee and C. Billington, "The Evolution of Supply-Chain-Management Models and Practice at Hewlett-Packard," *INTERFACES*, Vol. 25, No. 5, Sept.–Oct. 1995, pp. 42–63.

9. E. Heard, "Quick Response: Technology or Knowledge?" *Industrial Engineering*, August 1994, pp. 28–30.

10. J. W. Forrester, "Industrial Dynamics: A Major Breakthrough for Decision Makers," *Harvard Business Review*, July–August 1958.

11. R. Metters, "Quantifying the Bullwhip Effect in Supply Chains," *Proceedings of the 1996 MSOM Conference*, pp. 264–269.

12. E. G. Anderson Jr., C. H. Fine, and G. G. Parker, "Upstream Volatility in the Supply Chain: The Machine Tool Industry as a Case Study," Working Paper, MIT Sloan School of Management, Cambridge, MA, December 1997.

13. See Hammond and Kelly, "Quick Response in the Apparel Industry."
14. This quote and the Bose story are from Davis, "Partnerships Pay Off."

Chapter 11

1. R. Suri, U. Wemmerlöv, F. Rath, R. Gadh, and R. Veeramani, "Practical Issues in Implementing Quick Response Manufacturing: Insights from Fourteen Projects with Industry," *Proceedings of the MSOM Conference*, Dartmouth, NH, June 1996.

2. See the detailed discussion in the section "Benefits of Quick Response in Securing Orders" in Chapter 2, and also Table 11-2 in this chapter. Also important is the impression created on the customer when you deliver quotes rapidly. This too is underestimated. In a personal communication to me in 1996, Jim Riihl, planning manager at Beloit Corporation, stated: "Too often we think that for the customer, lead time is measured from the time they issued a purchase order. In fact, in the customer's mind, this whole process started when they first got the idea to get the equipment. Thus, to the customer, the clock started ticking when they asked for the quote."

3. See B. P. Shapiro, V. K. Rangan, and J. J. Sviokla,"Staple Yourself to an Order," *Harvard Business Review*, July–August 1992, pp. 113–122. While this article highlights the lack of attention to office operations, the solutions put forth lack a specific focus such as QRM, and do not go far enough from a QRM viewpoint. In a similar way, another article, "Speeding Up the Price Quote System," by J. A. Chalker and K. Bramer, *Management Accounting*, September 1993, pp. 45–49, makes several excellent points about problems in office processing, but it too, does not go far enough in its recommendations. We will return to this point later when we present some of the QRM solutions.

4. F. X. Frei, P. T. Harker, and L. W. Hunter, "Efficiency in Financial Services: Impact of Human Resource, Technology and Process Design in Retail Banking," *Proceedings of the MSOM Conference*, Dartmouth, NH, June 1996, pp. 99–104.

5. Batching is not confined to manufacturing operations. Two examples of batching in the office environment are waiting for a stack of folders to build up before carrying them to the next department, and waiting for a sufficient number of jobs to accumulate before logging on to a special computer program for order entry.

6. This statistic is from "The Technology Payoff," *Business Week*, June 14, 1993, pp. 57–68.

7. A recent survey of manufacturing firms showed that 80 percent of reengineering (BPR) efforts failed. See J. V. Owen, "Mandates for Managers," *Manufacturing Engineering*, April 1996, pp. 65–73. In fact, lack of appreciation for system dynamics may be one of the contributing causes to such failures; the main books on reengineering do not discuss system dynamics principles at all. For more on this point, see the discussion in J. M. Harrison and C. H. Loch, "Operations Management and

Reengineering," working paper, Graduate School of Business, Stanford University, 1995. In addition, it should be noted that the *Harvard Business Review* article, "Staple Yourself to an Order," mentioned in an earlier note, also neglected to present any systems dynamics principles as part of their solution. These principles are a key component of a QRM solution, however, as we discuss in this chapter as well as Chapter 13.

8. There is an ironic twist to this example. There was actually a book published in the early 1980s with the title *Zero Inventories*, which purported to explain the Japanese methods, but later Taiichi Ohno, the father of the Toyota Production System, admitted that he had deliberately used confusing terms to describe JIT in order to retain competitive advantage. "If the U.S. had understood what Toyota was doing, it would have been no good for us," he once said. F. S. Myers, "Japan's Henry Ford," *Scientific American*, Vol. 262, No. 5, 1990, p. 98. Another recent book makes the observation that "terms like JIT, zero inventories, and stockless production may have served to delude Americans into thinking that JIT is far simpler than it is." W. J. Hopp and M. L. Spearman, *Factory Physics: Foundations of Manufacturing Management*, (Burr Ridge, IL: Richard D. Irwin, 1996).

9. Owen, "Mandates for Managers," quotes a survey of manufacturing firms that shows 80 percent of reengineering (BPR) efforts failed. Also see U. Nwabueze, "Renew BPR," *Proceedings of the 1996 IIE Conference*, pp. 57–65.

10. The section on "Staffing and Training Cell Workers," in Chapter 4, reviewed why cross-training results in many benefits in the shopfloor context. All of those reasons apply here as well, and you may find it useful to review that discussion before proceeding.

11. See the examples and quotes in two sections of Chapter 4: "Cells Foster Continuous Improvement" and "Staffing and Training Cell Workers."

12. To see how pervasive this belief is, note that even the previously mentioned articles, "Staple Yourself to an Order" and "Speeding Up the Price Quote System," failed to recognize the pitfalls of this approach. In fact, those articles suggested interdepartmental projects with no mention of changing the organizational structure. In particular, they did not propose the creation of office cells at all, which is why I said, in an earlier note, that they did not go far enough.

13. This term was introduced in the context of office operations by G. Stalk, Jr., and T. M. Hout in their book *Competing Against Time* (New York, NY: The Free Press, 1992).

14. Companies implementing cells have reported that an ideal range for cell size is between 4 and 9 people; 10–15 is a bit large but manageable; over 15 is unwieldy—the benefits of close communication and teamwork seem to decline. At the other end of the scale, fewer than 4 people can be a problem due to dominant personalities preventing the occurrence of true team-based operation.

15. For more details, see D. Veeramani and P. Joshi, "Methodologies for Rapid and Effective Response to Requests for Quotations (RFQs)," *IIE Transactions*, Vol. 29, 1997, pp. 825–838.

16. Two related technologies that can change the paradigms for quoting are group technology (GT) and product-data-management (PDM) systems. See N. L. Hyer and U. Wemmerlöv, "Assessing the Merits of Group Technology," *Manufacturing Engineering*, August 1988, pp. 107–109; and D. Deitz, "Customer-Driven Engineering," *Mechanical Engineering*, May 1996, pp. 68–71.

17. We are not implying that cell workers should have unlimited authority, but rather, that they should have complete authority within well-defined limits. See the discussion in the preceding section under the topic of "closed-loop."

Chapter 12

1. Remark made by F. Bradley of the University of Wisconsin-Madison, in a personal communication to me.

2. Many of these tools are discussed in M. Brassard, *The Memory Jogger Plus+* (Methuen, MA: GOAL/QPC, 1989). A more comprehensive compendium of 222 tools, especially useful to team facilitators, is W. J. Michalski, *Tool Navigator: The Master Guide for Teams* (Portland, OR: Productivity Press, 1997).

3. This approach is also called value-added mapping. It should not be confused with value engineering, which is a different tool from traditional Industrial engineering methods, and has no connection to our charting tool here. Our view of value-added charting does differ somewhat from existing value-added mapping approaches, a point we will elaborate on later.

4. R. Suri, U. Wemmerlöv, F. Rath, R. Gadh, and R. Veeramani, "Practical Issues in Implementing Quick Response Manufacturing: Insights from Fourteen Projects with Industry," *Proceedings of the MSOM Conference*, Dartmouth, NH, June 1996.

5. The article "Staple Yourself to an Order," mentioned in the notes to Chapter 11, also discussed the need for top management support, but did not elaborate on what kind of support is needed. We will describe in precise terms, in Chapters 16 and 17, what top management needs to do to support the Q-ROC concept.

6. The second project, aimed at reducing manufacturing lead time for inserts, was conducted by University of Wisconsin graduate student Aamer Rehman and supervised jointly by Mike Diehl, team leader for insert production at ICTC, and myself. The results of that project can be found in the article by A. Rehman and M. B. Diehl, "Rapid Modeling Helps Focus Setup Reduction at Ingersoll," *Industrial Engineering*, November 1993, pp. 52–55.

7. From M. J. Wayman, "Order Processing Lead Time Reduction: A Case Study," *Proceedings of the 1995 International Industrial Engineering Conference*, pp. 400–409.

Chapter 13

1. The main books on reengineering do not discuss system dynamics principles at all. For more on this point, see the discussion in J. M. Harrison and C. H. Loch, "Operations Management and Reengineering," working paper, Graduate School of Business, Stanford University, 1995.
2. See R. Suri, "RMT Puts Manufacturing at the Helm," *Manufacturing Engineering* 100 (2), 1988, pp. 41–44. An example of a rapid modeling software package is MPX® from Network Dynamics Inc. of Burlington, MA.
3. Managers who are concerned that eliminating efficiency measures will lead to workers slacking off should read the case study in Chapter 8, "An Experiment with Using Time as the Yardstick at Beloit Corporation."
4. G. Stalk, Jr., and T. M. Hout, *Competing Against Time* (New York, NY: The Free Press, 1992).
5. You may find it useful to review the graphs in Chapter 7, which show the huge impact of setup and batch size reduction on lead time.
6. Simple tools that assist with rough capacity estimation and lead time planning were discussed in Chapter 7, and also mentioned in an earlier note in this chapter.

Chapter 14

1. A. Page, presentation to Product Development and Management Association, Chicago, November 1991. Also see: N. Capon, J. U. Farley, D. R. Lehmann, and J. M. Hulbert, "Profiles of Innovators Among Large U.S. Manufacturers," *Management Science*, Vol. 38, February 1992, pp. 157–169.
2. Statement made at a seminar held in September 1997 at the Center for Quick Response Manufacturing, by the sales manager of a company that supplies sheet metal parts to the computer industry.
3. *Electronic Business*, July 1983, p. 86.
4. K. T. Ulrich and S. D. Eppinger, *Product Design and Development* (New York, NY: McGraw-Hill, 1995).
5. S. C. Wheelwright and K. B. Clark, *Revolutionizing Product Development* (New York, NY: The Free Press, 1992).
6. S. C. Wheelwright and K. B. Clark, *Leading Product Development* (New York, NY: The Free Press, 1995).
7. A similar observation is made in the book by S. R. Rosenthal, *Effective Product Design and Development* (Burr Ridge, IL: Business One Irwin, 1992).
8. W. I. Zangwill, *Lightning Strategies for Innovation* (New York: Lexington Books, 1993).
9. S. L. Goldman, R. N. Nagel, and K. Preiss, *Agile Competitors and Virtual Organizations* (New York, NY: Van Nostrand Reinhold, 1995).
10. S. W. Sanderson and M. Uzumeri, *Managing Product Families* (Burr Ridge, IL: Richard D. Irwin, 1997).
11. Another term used for this approach is *simultaneous engineering*.
12. For more discussion on CE and its benefits, see C. Charney, *Time to Market: Reducing Product Lead Time*, (Dearborn, MI: Society of Manufacturing

Engineers, 1991). All the books on NPI, cited in the preceding section contain discussions on the benefits of CE.

13. This example was related to me by the materials manager of a U.S. equipment manufacturing company, and is based on actual performance measures and resulting behavior in that company.

14. Kaizen is the name of a philosophy of continuous improvement. See M. Imai, *Kaizen* (New York, NY: McGraw Hill, 1986).

15. I am indebted to Dr. Suzanne de Treville of Helsinki University of Technology for this insightful observation during one of our collaborative teaching efforts. She also pointed out the resolution of the discipline versus creativity dilemma, which is discussed later in this section.

16. Modern statistical quality improvement methods, including statistical process control and design of experiments, can support improved understanding as well as the "discipline" discussed here. A good overview of such methods can be found in G. Box and S. Bisgaard, "Statistical Tools for Improving Designs," *Mechanical Engineering*, January 1988, pp. 32–40.

17. S. de Treville, "Improving the Innovation Process," *OR/MS Today*, December 1994, pp. 28–30. Dr. de Treville pointed out, in a conversation with me, that such discipline is even needed to support artists in highly creative fields. Imagine a painter who cannot rely on the quality of his paints from one brush stroke to the next, or a cellist who cannot reliably tune her cello before a concert.

18. Several examples are cited in W. I. Zangwill, *Lightning Strategies for Innovation*, (New York: Lexington Books, 1993).

19. M. K. Starr, *Operations Management*, (Danvers, MA: Boyd & Fraser, 1996).

20. For a description of the DSM method, see Ulrich and Eppinger, *Product Design and Development*.

21. Gantt Charts can serve as a visual aid in this stage. See Starr, *Operations Management*.

22. B. J. Pine II, "The Future of Mass Customized Products," epilogue in D. M. Anderson, *Agile Product Development for Mass Customization* (Burr Ridge, IL: Irwin, 1997).

23. See R. N. Stauffer, "Converting Customers to Partners at Ingersoll," *Manufacturing Engineering*, September 1988, pp. 268–271.

24. See Goldman, Nagel, and Preiss, *Agile Competitors and Virtual Organizations*, Chapter 1.

25. This application and the following application for customers are described in D. Dietz, "Customer-Driven Engineering," *Mechanical Engineering*, May 1996, pp. 68–71.

26. This definition, and the strategic benefits, are from Goldman, Nagel, and Preiss, *Agile Competitors and Virtual Organizations*. The virtual organization concept was also discussed in *Business Week*, February 1993. W. H. Davidow and M. S. Malone used a similar term in their book *The Virtual Corporation* (New York, NY: HarperBusiness 1992); however, as the

authors clarify in that book, their notion is different, focusing on a company that produces "virtual products" made from modular elements.

27. Ulrich and Eppinger, *Product Design and Development*.
28. Y. Akao, ed., *Quality Function Deployment: Integrating Customer Requirements into Product Design* (Portland, OR: Productivity Press, 1990). A good overview of QFD can be found in the article, "The House of Quality," *Harvard Business Review*, May–June 1988, pp. 63–73. Also see B. King, *Better Designs in Half the Time: Implementing Quality Function Deployment in America* (Methuen, MA: Goal/QPC, 1989).
29. See Anderson, *Agile Product Development for Mass Customization*, Chapter 11.
30. *Agile Product Development for Mass Customization*.
31. See the different areas of application described in N. L. Hyer and U. Wemmerlöv, "Group Technology and Productivity," *Harvard Business Review*, July–August 1984, pp. 140–149.
32. N. L. Hyer and U. Wemmerlöv, "Assessing the Merits of Group Technology," *Manufacturing Engineering*, August 1988, pp. 107–109.
33. See *Agile Product Development for Mass Customization*, Chapter 11.
34. This example is from H. L. Lee and C. Billington, "The Evolution of Supply-Chain-Management Models and Practice at Hewlett-Packard," *INTERFACES*, Vol. 25, No. 5, Sept.–Oct. 1995, pp. 42–63. As explained there, HP uses the term "postponement" instead of delayed differentiation, and this particular type of postponement strategy is also known as "design for localization."
35. A good overview of the use of such methodology in design can be found in Box and Bisgaard, "Statistical Tools for Improving Designs."
36. See Wheelwright and Clark, *Revolutionizing Product Development*, Chapter 11.
37. J. H. Hammond and M. G. Kelly, "Quick Response in the Apparel Industry," Harvard Business School Case 9-690-038.
38. S. Ashley, "Cutting Costs and Time with DFMA," *Mechanical Engineering*, March 1995, pp. 74–77.
39. R. Bakerjian, *Tool and Manufacturing Engineers Handbook*, Volume 6, *Design for Manufacturability*, (Dearborn, MI: Society of Manufacturing Engineers, 1992); G. Boothroyd and P. Dewhurst, *Product Design for Assembly* (Wakefield, RI: Boothroyd Dewhurst Inc., 1989).
40. D. E. Whitney, "Integrated Design and Manufacturing in Japan," *Manufacturing Review*, Vol. 6, No. 4, December 1993, pp. 329–342.
41. This trade-off between development time and manufacturing cost is described in K. Ulrich, S. Pearson, D. Sartorius, and M. Jakiela, "Including the Value of Time in Design-for-Manufacturing Decision Making," *Management Science*, Vol. 39, No. 4, April 1993.
42. D/A was first proposed in R. Suri, "A New Perspective on Manufacturing Systems Analysis," in *Design and Analysis of Integrated Manufacturing Systems*, ed. W. D. Compton (Washington, D.C.: National Academy Press, 1988), pp. 118–133. This approach was developed further by Masami

Shimizu of Mitsubishi Heavy Industries while he was a graduate student working with me at University of Wisconsin-Madison. See R. Suri and M. Shimizu, "Design for Analysis: A New Strategy to Improve the Design Process," *Research in Engineering Design*, 1989, pp. 105–120.

43. These results are from Suri and Shimizu, "Design for Analysis: A New Strategy to Improve the Design Process."

44. Details of these case studies can be found in "Design for Analysis: A New Strategy to Improve the Design Process."

45. The plating line design was generated by us for an actual situation using the D/A approach, and is described in R. Suri, M. Shimizu, and R. Desiraju, "Design for Analysis Leads to a New Concept in Plating Lines," *Industrial Engineering*, Vol. 25, No. 8, August 1993, pp. 54–60. The other case studies are from the previously cited paper by Suri and Shimizu.

46. This observation is made by G. P. Pisano and S. C. Wheelwright, "High Tech R&D," *Harvard Business Review*, September–October 1995, pp. 93–105.

47. The previous article by Pisano and Wheelwright gives detailed examples of this, as does S. R. Rosenthal in his book *Effective Product Design and Development.*

48. This case study is described in more detail in R. Suri and S. de Treville, "Full Speed Ahead," *OR/MS Today*, June 1991, pp. 34–42.

49. The tool we used is MPX® from Network Dynamics, Inc., Burlington, MA.

50. See the quotes by Dr. Wali Haider in E. Brown, "IBM Combines Rapid Modeling Technique and Simulation to Design PCB Factory-of-the-Future," *Industrial Engineering*, Vol. 20, No. 6, 1988, pp. 23–26.

51. See J. Nymon, "Using Analytical and Simulation Modeling for Early Factory Prototyping," *Proceedings of the Winter Simulation Conference*, pp. 721–724, 1987.

52. See Zangwill, *Lightning Strategies for Innovation*, Chapter 8.

53. This example is from Wheelwright and Clark, *Revolutionizing Product Development*, Chapter 4.

54. J. M. Harrison and C. H. Loch, "Five Principles of Business Process Design," Working Paper, Graduate School of Business, Stanford University, March 1995.

55. This principle and the following example are from Whitney, "Integrated Design and Manufacturing in Japan."

56. See additional discussion of this point in Whitney, "Integrated Design and Manufacturing in Japan."

57. S. W. Sanderson, "Strategies for New Product Design and Product Renewal," *Academy of Management Best Papers Proceedings*, 1989, pp. 296–300.

58. Whitney, "Integrated Design and Manufacturing in Japan."

59. Pine, "The Future of Mass Customized Products."

Chapter 15

1. The ideas in this chapter were developed jointly with Dr. Suzanne de Treville of Helsinki University of Technology, during one of our collaborative teaching efforts.
2. "Advanced Parts Orientation System Has Wide Application," *Assembly Engineering*, August 1988, pp. 147–150.
3. S. W. Sanderson and M. Uzumeri, *Managing Product Families* (Burr Ridge, IL: Richard D. Irwin, 1997).
4. J. F. Krafcik, "Triumph of the Lean Production System," *Sloan Management Review*, Vol. 30, No. 1, Fall 1988, pp. 41–52.
5. G. Stalk, Jr., and A. M. Weber, "Japan's Dark Side of Time," *Harvard Business Review*, July-August 1993, pp. 93–102.
6. These technologies were discussed in more detail in Chapter 14.
7. R. Suri, U. Wemmerlöv, F. Rath, R. Gadh, and R. Veeramani, "Practical Issues in Implementing Quick Response Manufacturing: Insights from Fourteen Projects with Industry," *Proceedings of the MSOM Conference*, Dartmouth, NH, June 1996.

Chapter 16

1. The time horizon can be as short as a week, as seen from this quote: "...the stockholders [are] increasingly represented by institutional traders whose performance is evaluated weekly and who have little interest in the productive work of the organization; they just want the stock price to rise, and their focus is typically short term." From D. Hitchcock and M. Willard, *Why Teams Can Fail and What to Do About It* (Burr Ridge, IL: Irwin, 1995).
2. From B. H. Maskell, *Performance Measurement for World-Class Manufacturing* (Portland, OR: Productivity Press, 1991).
3. W. M. Baker, T. D. Fry, and K. Karwan, "The Rise and Fall of Time-Based Manufacturing," *Management Accounting*, June 1994, pp. 56–59.
4. For detailed examples of these problems, see R. S. Kaplan, "Yesterday's Accounting Undermines Production," *Harvard Business Review*, July–August 1984, pp. 22–28; R. S. Kaplan, "Management Accounting for Advanced Technological Environments," *Science*, 25 August 1989, pp. 819–823; the case study on outsourcing at Briggs and Stratton, reported in J. R. Coleman, "Problem Solving with Software," *Manufacturing Engineering*, March 1991, pp. 35–40; and J. V. Owen, "Justifying Manufacturing Flexibility," *Manufacturing Engineering*, March 1991, pp. 65–71.
5. J. A. Chalker and K. Bramer, "Speeding Up the Price Quote System," *Management Accounting*, September 1993, pp. 45–49; R. Suri, U. Wemmerlöv, F. Rath, R. Gadh, and R. Veeramani, "Practical Issues in Implementing Quick Response Manufacturing: Insights from Fourteen Projects with Industry," *Proceedings of the MSOM Conference*, Dartmouth, NH, June 1996.

6. As quoted in S. L. Goldman, R. N. Nagel, and K. Preiss, A*gile Competitors and Virtual Organizations* (New York, NY: Van Nostrand Reinhold, 1995).
7. Maskell, *Performance Measurement for World Class Manufacturing.*
8. For a more detailed discussion of these issues, see Maskell, *Performance Measurement for World Class Manufacturing.*
9. As an example, see the set of measures in R. S. Kaplan and D. P. Norton, "The Balance Scorecard—Measures That Drive Performance," *Harvard Business Review*, January-February 1992, pp. 71-79.
10. From R. N. Anthony, *Management Accounting* (Burr Ridge, IL: Irwin, 1956).
11. From M. C. Zieke and M. Spann, "Warning: Don't Be Half-Hearted in Your Efforts to Employ Concurrent Engineering," *Industrial Engineering*, February 1991.
12 From "Why Teams Fail," *USA Today*, February 25, 1997, pp. B1–B2.
13. For example: J. V. Owen, "Why Teams Fail," *Manufacturing Engineering*, September 1995, pp. 63–69; Hitchcock and Willard, *Why Teams Can Fail and What to Do About It*; "Why Teams Fail," *USA Today*. This article also mentions that the best-attended session every year at the International Conference on Work Teams is titled "Why Teams Fail."
14. This statistic and the list of fears is from Hitchcock and Willard, *Why Teams Can Fail and What to Do About It.*
15. "Why Teams Fail," *USA Today.*
16. See M. Donovan and L. Letize, "The Changing Role of the Supervisor in a High-Involvement Organization," in W. C. Hauck and V. R. Dingus eds., *Achieving High Commitment Work Systems* (Norcross, GA: Industrial Engineering and Management Press, 1990), pp. 373–379. This book also contains other relevant articles on the subject of teamwork and employee involvement in the new manufacturing organization.
17. H. W. Wain and P. M. O'Connor, "Self-Directed Teams at U.S. Gauge," *Proceedings of the 1996 International Industrial Engineering Conference*, pp. 438-443.
18. "Why Teams Fail," *USA Today.*
19. From "Yup: Help Wanted," *Newsweek*, January 13, 1997, pp. 52–53.
20. See "Business' Black Hole: Spiraling Number of Meetings Consume Time and Productivity," *USA Today*, December 8, 1997, p. 1A. This article confirms that even as recently as December 1997 few people are trained in meeting skills, resulting in many hours lost in wasteful meetings. You can find a number of pointers for conducting meetings and other team-related topics in P. R. Scholtes, *The Team Handbook* (Madison, WI: Joiner Associates Inc., 1995).
21. D. E. Whitney, "Integrated Design and Manufacturing in Japan," *Manufacturing Review*, Vol. 6, No. 4, December 1993, pp. 329-342.
22. D. Hitchcock and M. Willard, *Why Teams Can Fail and What to Do About It.*
23. "How Managers Can Succeed Through Speed," *Fortune*, February 13, 1989, pp. 54–59.

24. From Goldman, Nagel, and Preiss, *Agile Competitors and Virtual Organizations.*

25. Remarks made by G. L. Graf in his presentation, "Teaming and Multiskilling: When Have We Gone Too Far?" Institute of Industrial Engineers Conference, Minneapolis, MN, 1996.

26. From R. Y. Bergstrom, "Take Three People, and Build a Motorcycle," *Production,* November 1995, pp. 60–63.

27. From D. H. Sheldon, "Small Group Improvement Activity–The Path to Empowerment," *Proceedings of the 1996 International Industrial Engineering Conference,* pp. 482–489.

28. See K. Mericle, "Compensation Practices in the New Manufacturing Environment," in F. Frei, ed., *Work Design for the Competent Organization,* (Westport, CT: Quorum Books, 1993), pp. 283–304; and K. Mericle and D.-O. Kim, "Skill-Based Pay and Work Reorganization in High Performance Firms," Final Report of the U.S. Department of Labor National Center for the Workplace, Project No. 4, School for Workers, University of Wisconsin-Madison, March 1995.

29. See R. Cooper and R. S. Kaplan, "Measure Costs Right: Make the Right Decisions," *Harvard Business Review,* Vol. 66, No. 5, 1988, pp. 96–105; and Kaplan, "Management Accounting for Advanced Technological Environments."

30. See the case study in Baker, Fry, and Karwan, "The Rise and Fall of Time-Based Manufacturing."

31. From R. S. Kaplan, "Relevance Regained," *Management Accounting,* September 1988.

32. In fact, it is possible to combine a model of manufacturing system dynamics with an economic model in order to estimate costs under entirely new operating strategies. Such a modeling approach was used in World Bank studies of international manufacturing competitiveness. See A. Mody, R. Suri, and J. L. Sanders, "Keeping Pace with Change: Organizational and Technological Imperatives," *World Development,* Vol. 20, No. 12, 1992, pp. 1797–1816; and R. Suri, J. L. Sanders, P. C. Rao, and A. Mody, "Impact of Manufacturing Practices on the Global Bicycle Industry," *Manufacturing Review,* Vol. 6, No. 1, March 1993, pp. 14–24. This approach can be insightful; however, such an approach is intended primarily for economic justification. It is not a replacement for management accounting systems. The "fix" proposed in the chapter is simpler and it will work along with your existing accounting system.

33. This exact point is eloquently argued by H. T. Johnson in "It's Time to Stop Overselling Activity-Based Concepts," *Management Accounting,* September 1992, pp. 26–35. What is particularly interesting about this article is that Johnson was previously co-author with Kaplan of several articles and a book on ABC: H. T. Johnson and R. S. Kaplan, *Relevance Lost: The Rise and Fall of Management Accounting* (Boston, MA: Harvard Business School Press, 1987).

34. From Hitchcock and Willard, *Why Teams Can Fail and What to Do About It.*

35. From Chapter XX of A. de Tocqueville, *Democracy in America. Part the Second, The Social Influence of Democracy*, originally published by J. & H. G. Langley, New York, 1840. (A more recent version is published by A.A. Knopf, New York, 1945.) Also see the related quotes by de Tocqueville in Chapter 3.
36. These details are from G. F. Will, "United Airlines Rewrites History of Labor, Capital," *Wisconsin State Journal*, December 15, 1996, p. 1B.
37. Groetsch's quote is from "Look Who's Pushing Productivity," *Business Week*, April 7, 1997, pp. 72–75. The next quote is also from this article.
38. Personal communication to the author from Michael Tomsicek of Motorola Lighting Inc., Buffalo Grove, IL.
39. From Hitchcock and Willard, *Why Teams Can Fail and What to Do About It*.
40. R. C. Gulati and B. R. Tucker, "Self-Directed Team Implementation Process," *Proceedings of the 1996 International Industrial Engineering Conference*, pp. 444–448.

Chapter 17

1. Indeed, it is one of the underlying principles of the Center for Quick Response Manufacturing at the University of Wisconsin-Madison that managers and employees at companies implementing QRM can learn a great deal from each other, regardless of their industry. This principle has been proved by the growth and success of the Center.
2. See Chapter 7 for a discussion of tools based on rapid modeling technology (RMT).
3. T. Deeming, "Cell Mates," *Manufacturing Engineer*, June 1993, pp. 111–113.

About the Author

©1997 Bruce Fritz

RAJAN SURI is Professor of Industrial Engineering at the University of Wisconsin-Madison. He received his bachelor's degree from Cambridge University (England) and his M.S. and Ph.D. from Harvard University.

Dr. Suri serves as Director of the Center for Quick Response Manufacturing (QRM), a consortium of more than 40 firms working with the university on understanding and implementing QRM strategies (see QRM Web site at www.engr.wisc.edu/qrm). He is internationally regarded as an expert on the analysis of manufacturing systems, specializing in lead time reduction. Dr. Suri is author of more than 100 technical publications, and has chaired three international conferences on manufacturing systems. He has been instrumental in extending the theories of queuing networks and perturbation analysis for manufacturing applications, and served as Editor-in-Chief of the *Journal of Manufacturing Systems* for five years. He is currently Associate Editor of the *International Journal of Flexible Manufacturing Systems*, and Area Editor of the *Journal of Discrete Event Dynamic Systems*.

Professor Suri is also Director of the Manufacturing Systems Engineering Program. This is an interdisciplinary, practice-oriented M.S. degree program housed within the College of Engineering, with strong ties to the School of Business. The program includes courses in manufacturing processes and control, product design and process planning, industrial engineering and systems, and management of technology. Graduates of the program are highly qualified to assist manufacturing firms in implementing practices that will make them more competitive.

Dr. Suri combines his academic credentials with considerable practical experience. He has consulted for leading firms including 3M, Alcoa, Allen Bradley, ABB, AT&T, Beloit Corporation, Ford, Hewlett-Packard, McDonnell Douglas, IBM, Ingersoll, John Deere, Pratt & Whitney, Siemens, and TREK Bicycle. He is also a principal of Network Dynamics Inc., a firm specializing in software for manufacturing systems.

Consulting assignments in Europe and Asia, along with projects for the World Bank, have given him a substantial international perspective on manufacturing competitiveness.

In 1981 Dr. Suri received the Eckman Award from the American Automatic Control Council for outstanding contributions in his field. He was one of the team of people from his university who received the 1988 LEAD Award (for leadership and excellence in application and development) from the Society of Manufacturing Engineers. In 1988 he also received a Research Award from Ford Motor Company "in recognition of outstanding contributions made to the field of Perturbation Analysis of Discrete Event Systems." He is coauthor of a paper that won the 1990 Outstanding Simulation Publication Award from the Institute of Management Sciences. In 1994 he was co-recipient of the IEEE Control Systems Technology Award "for the creation, development, implementation and management of the manufacturing automation planning software, ManuPlan and its derivative MPX, during the period 1986–1993."

Index

Books from Productivity Press

Productivity Press publishes books that empower individuals and companies to achieve excellence in quality, productivity, and the creative involvement of all employees. Through steadfast efforts to support the vision and strategy of continuous improvement, Productivity Press delivers today's leading-edge tools and techniques gathered directly from industry leaders around the world. Call toll-free (800) 394-6868 for our free catalog.

The Shopfloor Series

Put powerful and proven improvement tools in the hands of your entire workforce!

Progressive shopfloor improvement techniques are imperative for manufacturers who want to stay competitive and to achieve world class excellence. And it's the comprehensive education of all shopfloor workers that ensures full participation and success when implementing new programs. The Shopfloor Series books make practical information accessible to everyone by presenting major concepts and tools in simple, clear language and at a reading level that has been adjusted for operators by skilled instructional designers. One main idea is presented every two to four pages so that the book can be picked up and put down easily. Each chapter begins with an overview and ends with a summary section. Helpful illustrations are used throughout.

Books currently in the Shopfloor Series include:

5S for Operators
5 Pillars of the Visual Workplace
The Productivity Press Development Team
ISBN 1-56327-123-0 /
incl. applic. questions / 133 pages
Order 5SOP-B283 / $25.00

Quick Changeover for Operators
The SMED System
The Productivity Press Development Team
ISBN 1-56327-125-7 /
incl. applic. questions / 93 pages
Order QCOOP-B283 / $25.00

Mistake-Proofing for Operators
The Productivity Press Development Team
ISBN 1-56327-127-3 / 93 pages
Order ZQCOP-B283 / $25.00

TPM for Supervisors
The Productivity Press Development Team
ISBN 1-56327-161-3 / 96 pages
Order TPMSUP-B283 / $25.00

Just-In-Time for Operators
The Productivity Press Development Team
ISBN 1-56327-133-8 / 84 pages
Order JITOP-B283 / $25.00

TPM Team Guide
Kunio Shirose
ISBN 1-56327-079-X / 175 pages
Order TGUIDE-B283 / $25.00

TPM for Every Operator
Japan Institute of Plant Maintenance
ISBN 1-56327-080-3 / 136 pages
Order TPMEO-B283 / $25.00

Autonomous Maintenance
Japan Institute of Plant Maintenance
ISBN 1-56327-082-X / 138 pages
Order AUTMOP-B283 / $25.00

Focused Equipment Improvement for TPM Teams
Japan Institute of Plant Maintenance
ISBN 1-56327-081-1 / 138 pages
Order FEIOP-B283 / $25.00

TO ORDER: Write, phone, or fax Productivity Press, Dept. BK, P.O. Box 13390, Portland, OR 97213-0390, phone 1-800-394-6868, fax 1-800-394-6286.

Outside the U.S. phone (503) 235-0600; fax (503) 235-0909

Send check or charge to your credit card (American Express, Visa, MasterCard accepted).

U.S. ORDERS: Add $5 shipping for first book, $2 each additional for UPS surface delivery. Add $5 for each AV program containing 1 or 2 tapes; add $12 for each AV program containing 3 or more tapes. We offer attractive quantity discounts for bulk purchases of individual titles; call for more information.

ORDER BY E-MAIL: Order 24 hours a day from anywhere in the world. Use either address:

To order: **service@ppress.com**
To view the online catalog and/or order: **http://www.ppress.com/**

QUANTITY DISCOUNTS: For information on quantity discounts, please contact our sales department.

INTERNATIONAL ORDERS: Write, phone, or fax for quote and indicate shipping method desired. For international callers, telephone number is 503-235-0600 and fax number is 503-235-0909. Prepayment in U.S. dollars must accompany your order (checks must be drawn on U.S. banks). When quote is returned with payment, your order will be shipped promptly by the method requested.

NOTE: Prices are in U.S. dollars and are subject to change without notice.